A TREATISE ON
ELECTRICITY AND MAGNETISM

VOL. 2 – FIRST EDITION

James Clerk Maxwell
(June 13, 1831 – November 5, 1879)

JAMES CLERK MAXWELL

A Treatise on Electricity and Magnetism
- James Clerk Maxwell

IN TWO VOLUMES AND IN THREE EDITIONS

VOLUME 1
- First Edition (1873)
- Second Edition (1881)
- Third Edition (1892)

VOLUME 2
- First Edition (1873)
- Second Edition (1881)
- Third Edition (1892)

-------------- AVAILABLE IN PRINT NOW --------------

VOL. 1 – 1ST EDITION - ISBN : 978-1773751443
VOL. 2 – 1ST EDITION - ISBN : 978-1773751467

VOL. 1 – 2ND EDITION - ISBN : 978-1773751436
VOL. 2 – 2ND EDITION - ISBN : 978-1773751474

VOL. 1 – 3RD EDITION - ISBN : 978-1773751450
VOL. 2 – 3RD EDITION - ISBN : 978-1773751481

ALL EDITIONS ARE ORIGINAL, AUTHENTIC, AND UNALTERED.
FIRST PUBLISHED IN 1873, 1881, AND 1892 RESPECTIVELY.

A TREATISE ON
ELECTRICITY AND MAGNETISM

VOL. 2 - FIRST EDITION

BY

JAMES CLERK MAXWELL

(June 13, 1831 – November 5, 1879)

James Clerk Maxwell FRSE FRS) was a Scottish scientist in the field of mathematical physics. His most notable achievement was to formulate the classical theory of electromagnetic radiation, bringing together for the first time electricity, magnetism, and light as different manifestations of the same phenomenon.

DIAMOND™
BOOKS

www.diamondbooks .ca

TORONTO, CANADA – 2020

DIAMOND BOOKS - CANADA

DIAMOND™
BOOKS

Toronto, ON, CANADA
http://www.**diamondbooks** .ca

BIBLIOGRAPHIC INFORMATION

A TREATISE ON ELECTRICITY AND MAGNETISM

VOL. 2 - FIRST EDITION

FIRST PUBLISHED IN 1873
OXFORD
AT THE CLARENDON PRESS
LONDON

PUBLISHED IN CANADA

Published in Canada by DIAMOND BOOKS ™, an imprint
of
DIAMOND PUBLISHERS CANADA
http://www.diamondpublishers .com

REPUBLISHED EDITION : AUGUST, 2020.

PAPERBACK EDITION : ISBN: 978-1-77375-146-7

PRINTED IN CANADA

Clarendon Press Series

A TREATISE

ON

ELECTRICITY AND MAGNETISM

BY

JAMES CLERK MAXWELL, M.A.

LL.D. EDIN., D.C.L., F.R.SS. LONDON AND EDINBURG

HONORARY FELLOW OF TRINITY COLLEGE

AND LATE PROFESSOR OF EXPERIMENTAL PHYSICS IN THE UNIVERSITY OF CAMBRIDGE

VOL. II

FIRST EDITION

Oxford

AT THE CLARENDON PRESS

1873

Clarendon Press Series

A TREATISE

ON

ELECTRICITY AND MAGNETISM

BY

JAMES CLERK MAXWELL, M.A.

LLD. EDIN., F.R.SS. LONDON AND EDINBURGH

HONORARY FELLOW OF TRINITY COLLEGE,

AND PROFESSOR OF EXPERIMENTAL PHYSICS

IN THE UNIVERSITY OF CAMBRIDGE

VOL. II

Oxford

AT THE CLARENDON PRESS

1873

CONTENTS.

PART III.

MAGNETISM.

CHAPTER I.

ELEMENTARY THEORY OF MAGNETISM.

Art.		Page
371.	Properties of a magnet when acted on by the earth	1
372.	Definition of the axis of the magnet and of the direction of magnetic force	1
373.	Action of magnets on one another. Law of magnetic force ..	2
374.	Definition of magnetic units and their dimensions	3
375.	Nature of the evidence for the law of magnetic force	4
376.	Magnetism as a mathematical quantity	4
377.	The quantities of the opposite kinds of magnetism in a magnet are always exactly equal	4
378.	Effects of breaking a magnet	5
379.	A magnet is built up of particles each of which is a magnet ..	5
380.	Theory of magnetic 'matter'	5
381.	Magnetization is of the nature of a vector	7
382.	Meaning of the term 'Magnetic Polarization'	8
383.	Properties of a magnetic particle	8
384.	Definitions of Magnetic Moment, Intensity of Magnetization, and Components of Magnetization	8
385.	Potential of a magnetized element of volume	9
386.	Potential of a magnet of finite size. Two expressions for this potential, corresponding respectively to the theory of polarization, and to that of magnetic 'matter'	9
387.	Investigation of the action of one magnetic particle on another	10
388.	Particular cases	12
389.	Potential energy of a magnet in any field of force	14
390.	On the magnetic moment and axis of a magnet	15

Art. Page

391. Expansion of the potential of a magnet in spherical harmonics .. 16
392. The centre of a magnet and the primary and secondary axes through the centre .. 17
393. The north end of a magnet in this treatise is that which points north, and the south end that which points south. Boreal magnetism is that which is supposed to exist near the north pole of the earth and the south end of a magnet. Austral magnetism is that which belongs to the south pole of the earth and the north end of a magnet. Austral magnetism is considered positive .. 19
394. The direction of magnetic force is that in which austral magnetism tends to move, that is, from south to north, and this is the positive direction of magnetic lines of force. A magnet is said to be magnetized from its south end towards its north end .. 19

CHAPTER II.

MAGNETIC FORCE AND MAGNETIC INDUCTION.

395. Magnetic force defined with reference to the magnetic potential 21
396. Magnetic force in a cylindric cavity in a magnet uniformly magnetized parallel to the axis of the cylinder .. 22
397. Application to any magnet .. 22
398. An elongated cylinder.—Magnetic force .. 23
399. A thin disk.—Magnetic induction .. 23
400. Relation between magnetic force, magnetic induction, and magnetization .. 24
401. Line-integral of magnetic force, or magnetic potential .. 24
402. Surface-integral of magnetic induction .. 25
403. Solenoidal distribution of magnetic induction .. 26
404. Surfaces and tubes of magnetic induction .. 27
405. Vector-potential of magnetic induction .. 27
406. Relations between the scalar and the vector-potential .. 28

CHAPTER III.

PARTICULAR FORMS OF MAGNETS.

407. Definition of a magnetic solenoid .. 31
408. Definition of a complex solenoid and expression for its potential at any point .. 32

Art. Page

409. The potential of a magnetic shell at any point is the product of its strength multiplied by the solid angle its boundary subtends at the point 32
410. Another method of proof 33
411. The potential at a point on the positive side of a shell of strength Φ exceeds that on the nearest point on the negative side by $4\pi\Phi$ 34
412. Lamellar distribution of magnetism 34
413. Complex lamellar distribution 34
414. Potential of a solenoidal magnet 35
415. Potential of a lamellar magnet 35
416. Vector-potential of a lamellar magnet 36
417. On the solid angle subtended at a given point by a closed curve 36
418. The solid angle expressed by the length of a curve on the sphere 37
419. Solid angle found by two line-integrations 38
420. Π expressed as a determinant 39
421. The solid angle is a cyclic function 40
422. Theory of the vector-potential of a closed curve 41
423. Potential energy of a magnetic shell placed in a magnetic field 42

CHAPTER IV.

INDUCED MAGNETIZATION.

424. When a body under the action of magnetic force becomes itself magnetized the phenomenon is called magnetic induction .. 44
425. Magnetic induction in different substances 45
426. Definition of the coefficient of induced magnetization 47
427. Mathematical theory of magnetic induction. Poisson's method 47
428. Faraday's method 49
429. Case of a body surrounded by a magnetic medium 51
430. Poisson's physical theory of the cause of induced magnetism .. 53

CHAPTER V.

MAGNETIC PROBLEMS.

431. Theory of a hollow spherical shell 56
432. Case when κ is large 58
433. When $i = 1$ 58
434. Corresponding case in two dimensions. Fig. XV 59
435. Case of a solid sphere, the coefficients of magnetization being different in different directions 60

Art. Page
436. The nine coefficients reduced to six. Fig. XVI 61
437. Theory of an ellipsoid acted on by a uniform magnetic force .. 62
438. Cases of very flat and of very long ellipsoids 65
439. Statement of problems solved by Neumann, Kirchhoff and Green 67
440. Method of approximation to a solution of the general problem when κ is very small. Magnetic bodies tend towards places of most intense magnetic force, and diamagnetic bodies tend to places of weakest force 69
441. On ship's magnetism 70

CHAPTER VI.

WEBER'S THEORY OF MAGNETIC INDUCTION.

442. Experiments indicating a maximum of magnetization 74
443. Weber's mathematical theory of temporary magnetization .. 75
444. Modification of the theory to account for residual magnetization 79
445. Explanation of phenomena by the modified theory 81
446. Magnetization, demagnetization, and remagnetization 83
447. Effects of magnetization on the dimensions of the magnet .. 85
448. Experiments of Joule 86

CHAPTER VII.

MAGNETIC MEASUREMENTS.

449. Suspension of the magnet 88
450. Methods of observation by mirror and scale. Photographic method 89
451. Principle of collimation employed in the Kew magnetometer .. 93
452. Determination of the axis of a magnet and of the direction of the horizontal component of the magnetic force 94
453. Measurement of the moment of a magnet and of the intensity of the horizontal component of magnetic force 97
454. Observations of deflexion 99
455. Method of tangents and method of sines 101
456. Observation of vibrations 102
457. Elimination of the effects of magnetic induction 105
458. Statical method of measuring the horizontal force 106
459. Bifilar suspension 107
460. System of observations in an observatory 111
461. Observation of the dip-circle 111

Art. Page
462. J. A. Broun's method of correction 115
463. Joule's suspension 115
464. Balance vertical force magnetometer 117

CHAPTER VIII.

TERRESTRIAL MAGNETISM.

465. Elements of the magnetic force 120
466. Combination of the results of the magnetic survey of a country 121
467. Deduction of the expansion of the magnetic potential of the earth in spherical harmonics 123
468. Definition of the earth's magnetic poles. They are not at the extremities of the magnetic axis. False poles. They do not exist on the earth's surface 123
469. Gauss' calculation of the 24 coefficients of the first four harmonics 124
470. Separation of external from internal causes of magnetic force .. 124
471. The solar and lunar variations 125
472. The periodic variations 125
473. The disturbances and their period of 11 years 126
474. Reflexions on magnetic investigations 126

PART IV.

ELECTROMAGNETISM.

CHAPTER I.

ELECTROMAGNETIC FORCE.

475. Örsted's discovery of the action of an electric current on a magnet 128
476. The space near an electric current is a magnetic field 128
477. Action of a vertical current on a magnet 129
478. Proof that the force due to a straight current of indefinitely great length varies inversely as the distance 129
479. Electromagnetic measure of the current 130

Art. Page

480. Potential function due to a straight current. It is a function of many values 130
481. The action of this current compared with that of a magnetic shell having an infinite straight edge and extending on one side of this edge to infinity 131
482. A small circuit acts at a great distance like a magnet 131
483. Deduction from this of the action of a closed circuit of any form and size on any point not in the current itself 131
484. Comparison between the circuit and a magnetic shell 132
485. Magnetic potential of a closed circuit 133
486. Conditions of continuous rotation of a magnet about a current 133
487. Form of the magnetic equipotential surfaces due to a closed circuit. Fig. XVIII 134
488. Mutual action between any system of magnets and a closed current 135
489. Reaction on the circuit 135
490. Force acting on a wire carrying a current and placed in the magnetic field 136
491. Theory of electromagnetic rotations 138
492. Action of one electric circuit on the whole or any portion of another 139
493. Our method of investigation is that of Faraday 140
494. Illustration of the method applied to parallel currents 140
495. Dimensions of the unit of current 141
496. The wire is urged from the side on which its magnetic action strengthens the magnetic force and towards the side on which it opposes it 141
497. Action of an infinite straight current on any current in its plane 142
498. Statement of the laws of electromagnetic force. Magnetic force due to a current 142
499. Generality of these laws 143
500. Force acting on a circuit placed in the magnetic field 144
501. Electromagnetic force is a mechanical force acting on the conductor, not on the electric current itself 144

CHAPTER II.

MUTUAL ACTION OF ELECTRIC CURRENTS.

502. Ampère's investigation of the law of force between the elements of electric currents 146

Art. Page

503. His method of experimenting 146
504. Ampère's balance 147
505. Ampère's first experiment. Equal and opposite currents neutralize each other 147
506. Second experiment. A crooked conductor is equivalent to a straight one carrying the same current 148
507. Third experiment. The action of a closed current as an element of another current is perpendicular to that element .. 148
508. Fourth experiment. Equal currents in systems geometrically similar produce equal forces 149
509. In all of these experiments the acting current is a closed one .. 151
510. Both circuits may, however, for mathematical purposes be conceived as consisting of elementary portions, and the action of the circuits as the resultant of the action of these elements 151
511. Necessary form of the relations between two elementary portions of lines 151
512. The geometrical quantities which determine their relative position 152
513. Form of the components of their mutual action 153
514. Resolution of these in three directions, parallel, respectively, to the line joining them and to the elements themselves 154
515. General expression for the action of a finite current on the element of another 154
516. Condition furnished by Ampère's third case of equilibrium .. 155
517. Theory of the directrix and the determinants of electrodynamic action 156
518. Expression of the determinants in terms of the components of the vector-potential of the current 157
519. The part of the force which is indeterminate can be expressed as the space-variation of a potential 157
520. Complete expression for the action between two finite currents 158
521. Mutual potential of two closed currents 158
522. Appropriateness of quaternions in this investigation 158
523. Determination of the form of the functions by Ampère's fourth case of equilibrium 159
524. The electrodynamic and electromagnetic units of currents .. 159
525. Final expressions for electromagnetic force between two elements 160
526. Four different admissible forms of the theory 160
527. Of these Ampère's is to be preferred 161

CHAPTER III.

INDUCTION OF ELECTRIC CURRENTS.

Art. Page

528. Faraday's discovery. Nature of his methods 162
529. The method of this treatise founded on that of Faraday 163
530. Phenomena of magneto-electric induction 164
531. General law of induction of currents 166
532. Illustrations of the direction of induced currents 166
533. Induction by the motion of the earth 167
534. The electromotive force due to induction does not depend on the material of the conductor 168
535. It has no tendency to move the conductor 168
536. Felici's experiments on the laws of induction 168
537. Use of the galvanometer to determine the time-integral of the electromotive force 170
538. Conjugate positions of two coils 171
539. Mathematical expression for the total current of induction .. 172
540. Faraday's conception of an electrotonic state 173
541. His method of stating the laws of induction with reference to the lines of magnetic force 174
542. The law of Lenz, and Neumann's theory of induction 176
543. Helmholtz's deduction of induction from the mechanical action of currents by the principle of conservation of energy 176
544. Thomson's application of the same principle 178
545. Weber's contributions to electrical science 178

CHAPTER IV.

INDUCTION OF A CURRENT ON ITSELF.

546. Shock given by an electromagnet 180
547. Apparent momentum of electricity 180
548. Difference between this case and that of a tube containing a current of water 181
549. If there is momentum it is not that of the moving electricity .. 181
550. Nevertheless the phenomena are exactly analogous to those of momentum 181
551. An electric current has energy, which may be called electrokinetic energy 182
552. This leads us to form a dynamical theory of electric currents .. 182

CHAPTER V.

GENERAL EQUATIONS OF DYNAMICS.

Art.		Page
553.	Lagrange's method furnishes appropriate ideas for the study of the higher dynamical sciences	184
554.	These ideas must be translated from mathematical into dynamical language	184
555.	Degrees of freedom of a connected system	185
556.	Generalized meaning of velocity	186
557.	Generalized meaning of force	186
558.	Generalized meaning of momentum and impulse	186
559.	Work done by a small impulse	187
560.	Kinetic energy in terms of momenta, (T_p)	188
561.	Hamilton's equations of motion	189
562.	Kinetic energy in terms of the velocities and momenta, $(T_{p\dot{q}})$ ..	190
563.	Kinetic energy in terms of velocities, $(T_{\dot{q}})$	191
564.	Relations between T_p and $T_{\dot{q}}$, p and $\dot{q}$	191
565.	Moments and products of inertia and mobility	192
566.	Necessary conditions which these coefficients must satisfy ..	193
567.	Relation between mathematical, dynamical, and electrical ideas	193

CHAPTER VI.

APPLICATION OF DYNAMICS TO ELECTROMAGNETISM.

568.	The electric current possesses energy	195
569.	The current is a kinetic phenomenon	195
570.	Work done by electromotive force	196
571.	The most general expression for the kinetic energy of a system including electric currents	197
572.	The electrical variables do not appear in this expression	198
573.	Mechanical force acting on a conductor	198
574.	The part depending on products of ordinary velocities and strengths of currents does not exist	200
575.	Another experimental test	202
576.	Discussion of the electromotive force	204
577.	If terms involving products of velocities and currents existed they would introduce electromotive forces, which are not observed	204

CHAPTER VII.

ELECTROKINETICS.

578.	The electrokinetic energy of a system of linear circuits	206
579.	Electromotive force in each circuit	207

Art. Page

580. Electromagnetic force 208
581. Case of two circuits 208
582. Theory of induced currents 209
583. Mechanical action between the circuits 210
584. All the phenomena of the mutual action of two circuits depend on a single quantity, the potential of the two circuits 210

CHAPTER VIII.

EXPLORATION OF THE FIELD BY MEANS OF THE SECONDARY CIRCUIT.

585. The electrokinetic momentum of the secondary circuit 211
586. Expressed as a line-integral 211
587. Any system of contiguous circuits is equivalent to the circuit formed by their exterior boundary 212
588. Electrokinetic momentum expressed as a surface-integral .. 212
589. A crooked portion of a circuit equivalent to a straight portion 213
590. Electrokinetic momentum *at a point* expressed as a vector, $\mathfrak{A}$ .. 214
591. Its relation to the magnetic induction, $\mathfrak{B}$. Equations (A) .. 214
592. Justification of these names 215
593. Conventions with respect to the signs of translations and rotations 216
594. Theory of a sliding piece 217
595. Electromotive force due to the motion of a conductor 218
596. Electromagnetic force on the sliding piece 218
597. Four definitions of a line of magnetic induction 219
598. General equations of electromotive force, (B) 219
599. Analysis of the electromotive force 222
600. The general equations referred to moving axes 223
601. The motion of the axes changes nothing but the apparent value of the electric potential 224
602. Electromagnetic force on a conductor 224
603. Electromagnetic force on an element of a conducting body. Equations (C) 226

CHAPTER IX.

GENERAL EQUATIONS.

604. Recapitulation 227
605. Equations of magnetization, (D) 228
606. Relation between magnetic force and electric currents 229
607. Equations of electric currents, (E) 230
608. Equations of electric displacement, (F) 232

Art. Page
609. Equations of electric conductivity, (G) 232
610. Equations of total currents, (H) 232
611. Currents in terms of electromotive force, (I) 233
612. Volume-density of free electricity, (J) 233
613. Surface-density of free electricity, (K) 233
614. Equations of magnetic permeability, (L) 233
615. Ampère's theory of magnets 234
616. Electric currents in terms of electrokinetic momentum 234
617. Vector-potential of electric currents 236
618. Quaternion expressions for electromagnetic quantities 236
619. Quaternion equations of the electromagnetic field 237

CHAPTER X.

DIMENSIONS OF ELECTRIC UNITS.

620. Two systems of units 239
621. The twelve primary quantities 239
622. Fifteen relations among these quantities 240
623. Dimensions in terms of [e] and [m] 241
624. Reciprocal properties of the two systems 241
625. The electrostatic and the electromagnetic systems 241
626. Dimensions of the 12 quantities in the two systems 242
627. The six derived units 243
628. The ratio of the corresponding units in the two systems .. 243
629. Practical system of electric units. Table of practical units .. 244

CHAPTER XI.

ENERGY AND STRESS.

630. The electrostatic energy expressed in terms of the free electricity and the potential 246
631. The electrostatic energy expressed in terms of the electromotive force and the electric displacement 246
632. Magnetic energy in terms of magnetization and magnetic force 247
633. Magnetic energy in terms of the square of the magnetic force .. 247
634. Electrokinetic energy in terms of electric momentum and electric current 248
635. Electrokinetic energy in terms of magnetic induction and magnetic force 248
636. Method of this treatise 249
637. Magnetic energy and electrokinetic energy compared 249
638. Magnetic energy reduced to electrokinetic energy 250

Art. Page

639. The force acting on a particle of a substance due to its magnetization 251
640. Electromagnetic force due to an electric current passing through it 252
641. Explanation of these forces by the hypothesis of stress in a medium 253
642. General character of the stress required to produce the phenomena 255
643. When there is no magnetization the stress is a tension in the direction of the lines of magnetic force, combined with a pressure in all directions at right angles to these lines, the magnitude of the tension and pressure being $\frac{1}{8\pi}\mathfrak{H}^2$, where $\mathfrak{H}$ is the magnetic force 256
644. Force acting on a conductor carrying a current 257
645. Theory of stress in a medium as stated by Faraday 257
646. Numerical value of magnetic tension 258

CHAPTER XII.

CURRENT-SHEETS.

647. Definition of a current-sheet 259
648. Current-function 259
649. Electric potential 260
650. Theory of steady currents 260
651. Case of uniform conductivity 260
652. Magnetic action of a current-sheet with closed currents 261
653. Magnetic potential due to a current-sheet 262
654. Induction of currents in a sheet of infinite conductivity 262
655. Such a sheet is impervious to magnetic action 263
656. Theory of a plane current-sheet 263
657. The magnetic functions expressed as derivatives of a single function 264
658. Action of a variable magnetic system on the sheet 266
659. When there is no external action the currents decay, and their magnetic action diminishes as if the sheet had moved off with constant velocity R 267
660. The currents, excited by the instantaneous introduction of a magnetic system, produce an effect equivalent to an image of that system 267
661. This image moves away from its original position with velocity R 268
662. Trail of images formed by a magnetic system in continuous motion 268

Art. Page
663. Mathematical expression for the effect of the induced currents 269
664. Case of the uniform motion of a magnetic pole 269
665. Value of the force acting on the magnetic pole 270
666. Case of curvilinear motion 271
667. Case of motion near the edge of the sheet 271
668. Theory of Arago's rotating disk 271
669. Trail of images in the form of a helix 274
670. Spherical current-sheets 275
671. The vector-potential 276
672. To produce a field of constant magnetic force within a spherical shell 277
673. To produce a constant force on a suspended coil 278
674. Currents parallel to a plane 278
675. A plane electric circuit. A spherical shell. An ellipsoidal shell 279
676. A solenoid 280
677. A long solenoid 281
678. Force near the ends 282
679. A pair of induction coils 282
680. Proper thickness of wire 283
681. An endless solenoid 284

CHAPTER XIII.

PARALLEL CURRENTS.

682. Cylindrical conductors 286
683. The external magnetic action of a cylindric wire depends only on the whole current through it 287
684. The vector-potential 288
685. Kinetic energy of the current 288
686. Repulsion between the direct and the return current 289
687. Tension of the wires. Ampère's experiment 289
688. Self-induction of a wire doubled on itself 290
689. Currents of varying intensity in a cylindric wire 291
690. Relation between the electromotive force and the total current 292
691. Geometrical mean distance of two figures in a plane 294
692. Particular cases 294
693. Application of the method to a coil of insulated wires 296

CHAPTER XIV.

CIRCULAR CURRENTS.

694. Potential due to a spherical bowl 299
695. Solid angle subtended by a circle at any point 301

Art. Page
696. Potential energy of two circular currents 302
697. Moment of the couple acting between two coils 303
698. Values of Q_i' 303
699. Attraction between two parallel circular currents 304
700. Calculation of the coefficients for a coil of finite section 304
701. Potential of two parallel circles expressed by elliptic integrals 305
702. Lines of force round a circular current. Fig. XVIII 307
703. Differential equation of the potential of two circles 307
704. Approximation when the circles are very near one another .. 309
705. Further approximation 310
706. Coil of maximum self-induction 311

CHAPTER XV.

ELECTROMAGNETIC INSTRUMENTS.

707. Standard galvanometers and sensitive galvanometers 313
708. Construction of a standard coil 314
709. Mathematical theory of the galvanometer 315
710. Principle of the tangent galvanometer and the sine galvanometer 316
711. Galvanometer with a single coil 316
712. Gaugain's eccentric suspension 317
713. Helmholtz's double coil. Fig. XIX 318
714. Galvanometer with four coils 319
715. Galvanometer with three coils 319
716. Proper thickness of the wire of a galvanometer 321
717. Sensitive galvanometers 322
718. Theory of the galvanometer of greatest sensibility 322
719. Law of thickness of the wire 323
720. Galvanometer with wire of uniform thickness 325
721. Suspended coils. Mode of suspension 326
722. Thomson's sensitive coil 326
723. Determination of magnetic force by means of suspended coil and tangent galvanometer 327
724. Thomson's suspended coil and galvanometer combined 328
725. Weber's electrodynamometer 328
726. Joule's current-weigher 332
727. Suction of solenoids 333
728. Uniform force normal to suspended coil 333
729. Electrodynamometer with torsion-arm 334

CHAPTER XVI.

ELECTROMAGNETIC OBSERVATIONS.

Art.		Page
730.	Observation of vibrations	335
731.	Motion in a logarithmic spiral	336
732.	Rectilinear oscillations in a resisting medium	337
733.	Values of successive elongations	338
734.	Data and quæsita	338
735.	Position of equilibrium determined from three successive elongations	338
736.	Determination of the logarithmic decrement	339
737.	When to stop the experiment	339
738.	Determination of the time of vibration from three transits	339
739.	Two series of observations	340
740.	Correction for amplitude and for damping	341
741.	Dead beat galvanometer	341
742.	To measure a constant current with the galvanometer	342
743.	Best angle of deflexion of a tangent galvanometer	343
744.	Best method of introducing the current	343
745.	Measurement of a current by the first elongation	344
746.	To make a series of observations on a constant current	345
747.	Method of multiplication for feeble currents	345
748.	Measurement of a transient current by first elongation	346
749.	Correction for damping	347
750.	Series of observations. *Zurückwerfungs methode*	348
751.	Method of multiplication	350

CHAPTER XVII.

ELECTRICAL MEASUREMENT OF COEFFICIENTS OF INDUCTION.

752.	Electrical measurement sometimes more accurate than direct measurement	352
753.	Determination of G_1	353
754.	Determination of g_1	354
755.	Determination of the mutual induction of two coils	354
756.	Determination of the self-induction of a coil	356
757.	Comparison of the self-induction of two coils	357

CHAPTER XVIII.

DETERMINATION OF RESISTANCE IN ELECTROMAGNETIC MEASURE.

758.	Definition of resistance	358
759.	Kirchhoff's method	358

Art. Page

760. Weber's method by transient currents 360
761. His method of observation 361
762. Weber's method by damping 361
763. Thomson's method by a revolving coil 364
764. Mathematical theory of the revolving coil 364
765. Calculation of the resistance 365
766. Corrections 366
767. Joule's calorimetric method 367

CHAPTER XIX.

COMPARISON OF ELECTROSTATIC WITH ELECTROMAGNETIC UNITS.

768. Nature and importance of the investigation 368
769. The ratio of the units is a velocity 369
770. Current by convection 370
771. Weber and Kohlrausch's method 370
772. Thomson's method by separate electrometer and electrodynamometer 372
773. Maxwell's method by combined electrometer and electrodynamometer 372
774. Electromagnetic measurement of the capacity of a condenser. Jenkin's method 373
775. Method by an intermittent current 374
776. Condenser and Wippe as an arm of Wheatstone's bridge .. 375
777. Correction when the action is too rapid 376
778. Capacity of a condenser compared with the self-induction of a coil 377
779. Coil and condenser combined 379
780. Electrostatic measure of resistance compared with its electromagnetic measure 382

CHAPTER XX.

ELECTROMAGNETIC THEORY OF LIGHT.

781. Comparison of the properties of the electromagnetic medium with those of the medium in the undulatory theory of light 383
782. Energy of light during its propagation 384
783. Equation of propagation of an electromagnetic disturbance .. 384
784. Solution when the medium is a non-conductor 386
785. Characteristics of wave-propagation 386
786. Velocity of propagation of electromagnetic disturbances 387
787. Comparison of this velocity with that of light 387

Art. Page

788. The specific inductive capacity of a dielectric is the square of its index of refraction 388
789. Comparison of these quantities in the case of paraffin 388
790. Theory of plane waves 389
791. The electric displacement and the magnetic disturbance are in the plane of the wave-front, and perpendicular to each other 390
792. Energy and stress during radiation 391
793. Pressure exerted by light 391
794. Equations of motion in a crystallized medium 392
795. Propagation of plane waves 393
796. Only two waves are propagated 393
797. The theory agrees with that of Fresnel 394
798. Relation between electric conductivity and opacity 394
799. Comparison with facts 395
800. Transparent metals 395
801. Solution of the equations when the medium is a conductor .. 395
802. Case of an infinite medium, the initial state being given .. 396
803. Characteristics of diffusion 397
804. Disturbance of the electromagnetic field when a current begins to flow 397
805. Rapid approximation to an ultimate state 398

CHAPTER XXI.

MAGNETIC ACTION ON LIGHT.

806. Possible forms of the relation between magnetism and light .. 399
807. The rotation of the plane of polarization by magnetic action .. 400
808. The laws of the phenomena 400
809. Verdet's discovery of negative rotation in ferromagnetic media 400
810. Rotation produced by quartz, turpentine, &c., independently of magnetism 401
811. Kinematical analysis of the phenomena 402
812. The velocity of a circularly-polarized ray is different according to its direction of rotation 402
813. Right and left-handed rays 403
814. In media which of themselves have the rotatory property the velocity is different for right and left-handed configurations 403
815. In media acted on by magnetism the velocity is different for opposite directions of rotation 404
816. The luminiferous disturbance, mathematically considered, is a vector 404
817. Kinematic equations of circularly-polarized light 405

Art. Page
818. Kinetic and potential energy of the medium 406
819. Condition of wave-propagation 406
820. The action of magnetism must depend on a real rotation about the direction of the magnetic force as an axis 407
821. Statement of the results of the analysis of the phenomenon .. 407
822. Hypothesis of molecular vortices 408
823. Variation of the vortices according to Helmholtz's law 409
824. Variation of the kinetic energy in the disturbed medium .. 409
825. Expression in terms of the current and the velocity 410
826. The kinetic energy in the case of plane waves 410
827. The equations of motion 411
828. Velocity of a circularly-polarized ray 411
829. The magnetic rotation 412
830. Researches of Verdet 413
831. Note on a mechanical theory of molecular vortices 415

CHAPTER XXII.

ELECTRIC THEORY OF MAGNETISM.

832. Magnetism is a phenomenon of molecules 418
833. The phenomena of magnetic molecules may be imitated by electric currents 419
834. Difference between the elementary theory of continuous magnets and the theory of molecular currents 419
835. Simplicity of the electric theory 420
836. Theory of a current in a perfectly conducting circuit 420
837. Case in which the current is entirely due to induction 421
838. Weber's theory of diamagnetism 421
839. Magnecrystallic induction 422
840. Theory of a perfect conductor 422
841. A medium containing perfectly conducting spherical molecules 423
842. Mechanical action of magnetic force on the current which it excites 423
843. Theory of a molecule with a primitive current 424
844. Modifications of Weber's theory 425
845. Consequences of the theory 425

CHAPTER XXIII.

THEORIES OF ACTION AT A DISTANCE.

846. Quantities which enter into Ampère's formula 426
847. Relative motion of two electric particles 426

Art. Page
848. Relative motion of four electric particles. Fechner's theory .. 427
849. Two new forms of Ampère's formula 428
850. Two different expressions for the force between two electric particles in motion 428
851. These are due to Gauss and to Weber respectively 429
852. All forces must be consistent with the principle of the conservation of energy 429
853. Weber's formula is consistent with this principle but that of Gauss is not 429
854. Helmholtz's deductions from Weber's formula 430
855. Potential of two currents 431
856. Weber's theory of the induction of electric currents 431
857. Segregating force in a conductor 432
858. Case of moving conductors 433
859. The formula of Gauss leads to an erroneous result 434
860. That of Weber agrees with the pheuomena 434
861. Letter of Gauss to Weber 435
862. Theory of Riemann 435
863. Theory of C. Neumann 435
864. Theory of Betti 436
865. Repugnance to the idea of a medium 437
866. The idea of a medium cannot be got rid of 437

PART III.

MAGNETISM.

CHAPTER I.

ELEMENTARY THEORY OF MAGNETISM.

371.] CERTAIN bodies, as, for instance, the iron ore called loadstone, the earth itself, and pieces of steel which have been subjected to certain treatment, are found to possess the following properties, and are called Magnets.

If, near any part of the earth's surface except the Magnetic Poles, a magnet be suspended so as to turn freely about a vertical axis, it will in general tend to set itself in a certain azimuth, and if disturbed from this position it will oscillate about it. An unmagnetized body has no such tendency, but is in equilibrium in all azimuths alike.

372.] It is found that the force which acts on the body tends to cause a certain line in the body, called the Axis of the Magnet, to become parallel to a certain line in space, called the Direction of the Magnetic Force.

Let us suppose the magnet suspended so as to be free to turn in all directions about a fixed point. To eliminate the action of its weight we may suppose this point to be its centre of gravity. Let it come to a position of equilibrium. Mark two points on the magnet, and note their positions in space. Then let the magnet be placed in a new position of equilibrium, and note the positions in space of the two marked points on the magnet.

Since the axis of the magnet coincides with the direction of magnetic force in both positions, we have to find that line in the magnet which occupies the same position in space before and after the motion. It appears, from the theory of the motion of bodies of invariable form, that such a line always exists, and that a motion equivalent to the actual motion might have taken place by simple rotation round this line.

To find the line, join the first and last positions of each of the marked points, and draw planes bisecting these lines at right angles. The intersection of these planes will be the line required, which indicates the direction of the axis of the magnet and the direction of the magnetic force in space.

The method just described is not convenient for the practical determination of these directions. We shall return to this subject when we treat of Magnetic Measurements.

The direction of the magnetic force is found to be different at different parts of the earth's surface. If the end of the axis of the magnet which points in a northerly direction be marked, it has been found that the direction in which it sets itself in general deviates from the true meridian to a considerable extent, and that the marked end points on the whole downwards in the northern hemisphere and upwards in the southern.

The azimuth of the direction of the magnetic force, measured from the true north in a westerly direction, is called the Variation, or the Magnetic Declination. The angle between the direction of the magnetic force and the horizontal plane is called the Magnetic Dip. These two angles determine the direction of the magnetic force, and, when the magnetic intensity is also known, the magnetic force is completely determined. The determination of the values of these three elements at different parts of the earth's surface, the discussion of the manner in which they vary according to the place and time of observation, and the investigation of the causes of the magnetic force and its variations, constitute the science of Terrestrial Magnetism.

373.] Let us now suppose that the axes of several magnets have been determined, and the end of each which points north marked. Then, if one of these be freely suspended and another brought near it, it is found that two marked ends repel each other, that a marked and an unmarked end attract each other, and that two unmarked ends repel each other.

If the magnets are in the form of long rods or wires, uniformly and longitudinally magnetized, see below, Art. 384, it is found that the greatest manifestation of force occurs when the end of one magnet is held near the end of the other, and that the phenomena can be accounted for by supposing that like ends of the magnets repel each other, that unlike ends attract each other, and that the intermediate parts of the magnets have no sensible mutual action.

The ends of a long thin magnet are commonly called its Poles. In the case of an indefinitely thin magnet, uniformly magnetized throughout its length, the extremities act as centres of force, and the rest of the magnet appears devoid of magnetic action. In all actual magnets the magnetization deviates from uniformity, so that no single points can be taken as the poles. Coulomb, however, by using long thin rods magnetized with care, succeeded in establishing the law of force between two magnetic poles*.

The repulsion between two magnetic poles is in the straight line joining them, and is numerically equal to the product of the strengths of the poles divided by the square of the distance between them.

374.] This law, of course, assumes that the strength of each pole is measured in terms of a certain unit, the magnitude of which may be deduced from the terms of the law.

The unit-pole is a pole which points north, and is such that, when placed at unit distance from another unit-pole, it repels it with unit of force, the unit of force being defined as in Art. 6. A pole which points south is reckoned negative.

If m_1 and m_2 are the strengths of two magnetic poles, l the distance between them, and f the force of repulsion, all expressed numerically, then

$$f = \frac{m_1 m_2}{l^2}.$$

But if $[m]$, $[L]$ and $[F]$ be the concrete units of magnetic pole, length and force, then

$$f[F] = \left[\frac{m}{L}\right]^2 \frac{m_1 m_2}{l^2},$$

whence it follows that

$$[m^2] = [L^2 F] = \left[L^2 \frac{ML}{T^2}\right],$$

$$\text{or} \qquad [m] = [L^{\frac{3}{2}} T^{-1} M^{\frac{1}{2}}].$$

The dimensions of the unit pole are therefore $\frac{3}{2}$ as regards length, (-1) as regards time, and $\frac{1}{2}$ as regards mass. These dimensions are the same as those of the electrostatic unit of electricity, which is specified in exactly the same way in Arts. 41, 42.

* His experiments on magnetism with the Torsion Balance are contained in the *Memoirs of the Academy of Paris*, 1780–9, and in Biot's *Traité de Physique*, tom. iii.

375.] The accuracy of this law may be considered to have been established by the experiments of Coulomb with the Torsion Balance, and confirmed by the experiments of Gauss and Weber, and of all observers in magnetic observatories, who are every day making measurements of magnetic quantities, and who obtain results which would be inconsistent with each other if the law of force had been erroneously assumed. It derives additional support from its consistency with the laws of electromagnetic phenomena.

376.] The quantity which we have hitherto called the strength of a pole may also be called a quantity of 'Magnetism,' provided we attribute no properties to 'Magnetism' except those observed in the poles of magnets.

Since the expression of the law of force between given quantities of 'Magnetism' has exactly the same mathematical form as the law of force between quantities of 'Electricity' of equal numerical value, much of the mathematical treatment of magnetism must be similar to that of electricity. There are, however, other properties of magnets which must be borne in mind, and which may throw some light on the electrical properties of bodies.

Relation between the Poles of a Magnet.

377.] The quantity of magnetism at one pole of a magnet is always equal and opposite to that at the other, or more generally thus:—

In every Magnet the total quantity of Magnetism (reckoned algebraically) *is zero.*

Hence in a field of force which is uniform and parallel throughout the space occupied by the magnet, the force acting on the marked end of the magnet is exactly equal, opposite and parallel to that on the unmarked end, so that the resultant of the forces is a statical couple, tending to place the axis of the magnet in a determinate direction, but not to move the magnet as a whole in any direction.

This may be easily proved by putting the magnet into a small vessel and floating it in water. The vessel will turn in a certain direction, so as to bring the axis of the magnet as near as possible to the direction of the earth's magnetic force, but there will be no motion of the vessel as a whole in any direction; so that there can be no excess of the force towards the north over that towards the south, or the reverse. It may also be shewn from the fact that magnetizing a piece of steel does not alter its weight. It does alter the apparent position of its centre of gravity, causing it in these latitudes to shift along the axis towards the north. The centre of inertia, as determined by the phenomena of rotation, remains unaltered.

378.] If the middle of a long thin magnet be examined, it is found to possess no magnetic properties, but if the magnet be broken at that point, each of the pieces is found to have a magnetic pole at the place of fracture, and this new pole is exactly equal and opposite to the other pole belonging to that piece. It is impossible, either by magnetization, or by breaking magnets, or by any other means, to procure a magnet whose poles are unequal.

If we break the long thin magnet into a number of short pieces we shall obtain a series of short magnets, each of which has poles of nearly the same strength as those of the original long magnet. This multiplication of poles is not necessarily a creation of energy, for we must remember that after breaking the magnet we have to do work to separate the parts, in consequence of their attraction for one another.

379.] Let us now put all the pieces of the magnet together as at first. At each point of junction there will be two poles exactly equal and of opposite kinds, placed in contact, so that their united action on any other pole will be null. The magnet, thus rebuilt, has therefore the same properties as at first, namely two poles, one at each end, equal and opposite to each other, and the part between these poles exhibits no magnetic action.

Since, in this case, we know the long magnet to be made up of little short magnets, and since the phenomena are the same as in the case of the unbroken magnet, we may regard the magnet, even before being broken, as made up of small particles, each of which has two equal and opposite poles. If we suppose all magnets to be made up of such particles, it is evident that since the algebraical quantity of magnetism in each particle is zero, the quantity in the whole magnet will also be zero, or in other words, its poles will be of equal strength but of opposite kind.

Theory of Magnetic 'Matter.'

380.] Since the form of the law of magnetic action is identical with that of electric action, the same reasons which can be given for attributing electric phenomena to the action of one 'fluid' or two 'fluids' can also be used in favour of the existence of a magnetic matter, or of two kinds of magnetic matter, fluid or otherwise. In fact, a theory of magnetic matter, if used in a purely mathematical sense, cannot fail to explain the phenomena, provided new laws are freely introduced to account for the actual facts.

One of these new laws must be that the magnetic fluids cannot pass from one molecule or particle of the magnet to another, but that the process of magnetization consists in separating to a certain extent the two fluids within each particle, and causing the one fluid to be more concentrated at one end, and the other fluid to be more concentrated at the other end of the particle. This is the theory of Poisson.

A particle of a magnetizable body is, on this theory, analogous to a small insulated conductor without charge, which on the two-fluid theory contains indefinitely large but exactly equal quantities of the two electricities. When an electromotive force acts on the conductor, it separates the electricities, causing them to become manifest at opposite sides of the conductor. In a similar manner, according to this theory, the magnetizing force causes the two kinds of magnetism, which were originally in a neutralized state, to be separated, and to appear at opposite sides of the magnetized particle.

In certain substances, such as soft iron and those magnetic substances which cannot be permanently magnetized, this magnetic condition, like the electrification of the conductor, disappears when the inducing force is removed. In other substances, such as hard steel, the magnetic condition is produced with difficulty, and, when produced, remains after the removal of the inducing force.

This is expressed by saying that in the latter case there is a Coercive Force, tending to prevent alteration in the magnetization, which must be overcome before the power of a magnet can be either increased or diminished. In the case of the electrified body this would correspond to a kind of electric resistance, which, unlike the resistance observed in metals, would be equivalent to complete insulation for electromotive forces below a certain value.

This theory of magnetism, like the corresponding theory of electricity, is evidently too large for the facts, and requires to be restricted by artificial conditions. For it not only gives no reason why one body may not differ from another on account of having more of both fluids, but it enables us to say what would be the properties of a body containing an excess of one magnetic fluid. It is true that a reason is given why such a body cannot exist, but this reason is only introduced as an after-thought to explain this particular fact. It does not grow out of the theory.

381.] We must therefore seek for a mode of expression which shall not be capable of expressing too much, and which shall leave room for the introduction of new ideas as these are developed from new facts. This, I think, we shall obtain if we begin by saying that the particles of a magnet are Polarized.

Meaning of the term 'Polarization.'

When a particle of a body possesses properties related to a certain line or direction in the body, and when the body, retaining these properties, is turned so that this direction is reversed, then if as regards other bodies these properties of the particle are reversed, the particle, in reference to these properties, is said to be polarized, and the properties are said to constitute a particular kind of polarization.

Thus we may say that the rotation of a body about an axis constitutes a kind of polarization, because if, while the rotation continues, the direction of the axis is turned end for end, the body will be rotating in the opposite direction as regards space.

A conducting particle through which there is a current of electricity may be said to be polarized, because if it were turned round, and if the current continued to flow in the same direction as regards the particle, its direction in space would be reversed.

In short, if any mathematical or physical quantity is of the nature of a vector, as defined in Art. 11, then any body or particle to which this directed quantity or vector belongs may be said to be Polarized*, because it has opposite properties in the two opposite directions or poles of the directed quantity.

The poles of the earth, for example, have reference to its rotation, and have accordingly different names.

* The word Polarization has been used in a sense not consistent with this in Optics, where a ray of light is said to be polarized when it has properties relating to its sides, which are identical on opposite sides of the ray. This kind of polarization refers to another kind of Directed Quantity, which may be called a Dipolar Quantity, in opposition to the former kind, which may be called Unipolar.

When a dipolar quantity is turned end for end it remains the same as before. Tensions and Pressures in solid bodies, Extensions, Compressions and Distortions and most of the optical, electrical, and magnetic properties of crystallized bodies are dipolar quantities.

The property produced by magnetism in transparent bodies of twisting the plane of polarization of the incident light, is, like magnetism itself, a unipolar property. The rotatory property referred to in Art. 303 is also unipolar.

Meaning of the term 'Magnetic Polarization.'

382.] In speaking of the state of the particles of a magnet as magnetic polarization, we imply that each of the smallest parts into which a magnet may be divided has certain properties related to a definite direction through the particle, called its Axis of Magnetization, and that the properties related to one end of this axis are opposite to the properties related to the other end.

The properties which we attribute to the particle are of the same kind as those which we observe in the complete magnet, and in assuming that the particles possess these properties, we only assert what we can prove by breaking the magnet up into small pieces, for each of these is found to be a magnet.

Properties of a Magnetized Particle.

383.] Let the element $dx\,dy\,dz$ be a particle of a magnet, and let us assume that its magnetic properties are those of a magnet the strength of whose positive pole is m, and whose length is ds. Then if P is any point in space distant r from the positive pole and r' from the negative pole, the magnetic potential at P will be $\frac{m}{r}$ due to the positive pole, and $-\frac{m}{r'}$ due to the negative pole, or

$$V = \frac{m}{rr'}(r'-r). \tag{1}$$

If ds, the distance between the poles, is very small, we may put

$$r'-r = ds\cos\epsilon, \tag{2}$$

where ϵ is the angle between the vector drawn from the magnet to P and the axis of the magnet, or

$$V = \frac{m\,ds}{r^2}\cos\epsilon. \tag{3}$$

Magnetic Moment.

384.] The product of the length of a uniformly and longitudinally magnetized bar magnet into the strength of its positive pole is called its Magnetic Moment.

Intensity of Magnetization.

The intensity of magnetization of a magnetic particle is the ratio of its magnetic moment to its volume. We shall denote it by I.

The magnetization at any point of a magnet may be defined by its intensity and its direction. Its direction may be defined by its direction-cosines λ, μ, ν.

Components of Magnetization.

The magnetization at a point of a magnet (being a vector or directed quantity) may be expressed in terms of its three components referred to the axes of coordinates. Calling these A, B, C,

$$A = I\lambda, \qquad B = I\mu, \qquad C = I\nu,$$

and the numerical value of I is given by the equation (4)

$$I^2 = A^2 + B^2 + C^2. \tag{5}$$

385.] If the portion of the magnet which we consider is the differential element of volume $dx\,dy\,dz$, and if I denotes the intensity of magnetization of this element, its magnetic moment is $I\,dx\,dy\,dz$. Substituting this for $m\,ds$ in equation (3), and remembering that

$$r\cos\epsilon = \lambda(\xi-x)+\mu(\eta-y)+\nu(\zeta-z), \tag{6}$$

where ξ, η, ζ are the coordinates of the extremity of the vector r drawn from the point (x, y, z), we find for the potential at the point (ξ, η, ζ) due to the magnetized element at (x, y, z),

$$\delta V = \{A(\xi-x)+B(\eta-y)+C(\zeta-z)\}\frac{1}{r^3}\,dx\,dy\,dz. \tag{7}$$

To obtain the potential at the point (ξ, η, ζ) due to a magnet of finite dimensions, we must find the integral of this expression for every element of volume included within the space occupied by the magnet, or

$$V = \iiint \{A(\xi-x)+B(\eta-y)+C(\zeta-z)\}\frac{1}{r^3}\,dx\,dy\,dz. \tag{8}$$

Integrating by parts, this becomes

$$V = \iint A\frac{1}{r}\,dy\,dz + \iint B\frac{1}{r}\,dz\,dx + \iint C\frac{1}{r}\,dx\,dy$$
$$-\iiint\frac{1}{r}\left(\frac{dA}{dx}+\frac{dB}{dy}+\frac{dC}{dz}\right)dx\,dy\,dz,$$

where the double integration in the first three terms refers to the surface of the magnet, and the triple integration in the fourth to the space within it.

If l, m, n denote the direction-cosines of the normal drawn outwards from the element of surface dS, we may write, as in Art. 21, the sum of the first three terms,

$$\iint(lA+mB+nC)\frac{1}{r}\,dS,$$

where the integration is to be extended over the whole surface of the magnet.

If we now introduce two new symbols σ and ρ, defined by the equations

$$\sigma = lA + mB + nC,$$

$$\rho = -\left(\frac{dA}{dx} + \frac{dB}{dy} + \frac{dC}{dz}\right),$$

the expression for the potential may be written

$$V = \iint \frac{\sigma}{r} dS + \iiint \frac{\rho}{r} dx\,dy\,dz.$$

386.] This expression is identical with that for the electric potential due to a body on the surface of which there is an electrification whose surface-density is σ, while throughout its substance there is a bodily electrification whose volume-density is ρ. Hence, if we assume σ and ρ to be the surface- and volume-densities of the distribution of an imaginary substance, which we have called 'magnetic matter,' the potential due to this imaginary distribution will be identical with that due to the actual magnetization of every element of the magnet.

The surface-density σ is the resolved part of the intensity of magnetization I in the direction of the normal to the surface drawn outwards, and the volume-density ρ is the 'convergence' (see Art. 25) of the magnetization at a given point in the magnet.

This method of representing the action of a magnet as due to a distribution of 'magnetic matter' is very convenient, but we must always remember that it is only an artificial method of representing the action of a system of polarized particles.

On the Action of one Magnetic Molecule on another.

387.] If, as in the chapter on Spherical Harmonics, Art. 129, we make

$$\frac{d}{dh} = l\frac{d}{dx} + m\frac{d}{dy} + n\frac{d}{dz}, \tag{1}$$

where l, m, n are the direction-cosines of the axis h, then the potential due to a magnetic molecule at the origin, whose axis is parallel to h_1, and whose magnetic moment is m_1, is

$$V_1 = -\frac{d}{dh_1}\frac{m_1}{r} = \frac{m_1}{r^2}\lambda_1, \tag{2}$$

where λ_1 is the cosine of the angle between h_1 and r.

Again, if a second magnetic molecule whose moment is m_2, and whose axis is parallel to h_2, is placed at the extremity of the radius vector r, the potential energy due to the action of the one magnet on the other is

$$W = -m_2 \frac{dV_1}{dh_2} = m_1 m_2 \frac{d^2}{dh_1 \, dh_2}\left(\frac{1}{r}\right), \tag{3}$$

$$= \frac{m_1 m_2}{r^3}(\mu_{12} - 3\lambda_1 \lambda_2), \tag{4}$$

where μ_{12} is the cosine of the angle which the axes make with each other, and λ_1, λ_2 are the cosines of the angles which they make with r.

Let us next determine the moment of the couple with which the first magnet tends to turn the second round its centre.

Let us suppose the second magnet turned through an angle $d\phi$ in a plane perpendicular to a third axis h_3, then the work done against the magnetic forces will be $\frac{dW}{d\phi} d\phi$, and the moment of the forces on the magnet in this plane will be

$$-\frac{dW}{d\phi} = -\frac{m_1 m_2}{r^3}\left(\frac{d\mu_{12}}{d\phi} - 3\lambda_1 \frac{d\lambda_2}{d\phi}\right). \tag{5}$$

The actual moment acting on the second magnet may therefore be considered as the resultant of two couples, of which the first acts in a plane parallel to the axes of both magnets, and tends to *increase* the angle between them with a force whose moment is

$$\frac{m_1 m_2}{r^3} \sin(h_1 h_2), \tag{6}$$

while the second couple acts in the plane passing through r and the axis of the second magnet, and tends to *diminish* the angle between these directions with a force

$$\frac{3 m_1 m_2}{r^3} \cos(r h_1) \sin(r h_2), \tag{7}$$

where $(r h_1)$, $(r h_2)$, $(h_1 h_2)$ denote the angles between the lines r, h_1, h_2.

To determine the force acting on the second magnet in a direction parallel to a line h_3, we have to calculate

$$\frac{dW}{dh_3} = m_1 m_2 \frac{d^3}{dh_1 \, dh_2 \, dh_3}\left(\frac{1}{r}\right), \tag{8}$$

$$= 3 \frac{m_1 m_2}{r^4} \{\lambda_1 \mu_{23} + \lambda_2 \mu_{31} + \lambda_3 \mu_{12} - 5\lambda_1 \lambda_2 \lambda_3\}, \tag{9}$$

$$= 3\lambda_3 \frac{m_1 m_2}{r^4}(\mu_{12} - 5\lambda_1 \lambda_2) + 3\mu_{13} \frac{m_1 m_2}{r^4} \lambda_2 + 3\mu_{23} \frac{m_1 m_2}{r^4} \lambda_1. \tag{10}$$

If we suppose the actual force compounded of three forces, R, H_1 and H_2, in the directions of r, h_1 and h_2 respectively, then the force in the direction of h_3 is

$$\lambda_3 R + \mu_{13} H_1 + \mu_{23} H_2. \tag{11}$$

Since the direction of h_3 is arbitrary, we must have

$$\left.\begin{aligned} R &= \frac{3\,m_1\,m_2}{r^4}(\mu_{12}-5\,\lambda_1\,\lambda_2), \\ H_1 = \frac{3\,m_1\,m_2}{r^4}\lambda_2, &\qquad H_2 = \frac{3\,m_1\,m_2}{r^4}\lambda_1. \end{aligned}\right\} \tag{12}$$

The force R is a repulsion, tending to increase r; H_1 and H_2 act on the second magnet in the directions of the axes of the first and second magnet respectively.

This analysis of the forces acting between two small magnets was first given in terms of the Quaternion Analysis by Professor Tait in the *Quarterly Math. Journ.* for Jan. 1860. See also his work on *Quaternions*, Art. 414.

Particular Positions.

388.] (1) If λ_1 and λ_2 are each equal to 1, that is, if the axes of the magnets are in one straight line and in the same direction, $\mu_{12} = 1$, and the force between the magnets is a repulsion

$$R+H_1+H_2 = -\frac{6\,m_1\,m_2}{r^4}. \tag{13}$$

The negative sign indicates that the force is an attraction.

(2) If λ_1 and λ_2 are zero, and μ_{12} unity, the axes of the magnets are parallel to each other and perpendicular to r, and the force is a repulsion

$$\frac{3\,m_1\,m_2}{r^4}. \tag{14}$$

In neither of these cases is there any couple.

(3) If $\lambda_1 = 1$ and $\lambda_2 = 0$, then $\mu_{12} = 0$. (15)

The force on the second magnet will be $\frac{3\,m_1\,m_2}{r^4}$ in the direction of its axis, and the couple will be $\frac{2\,m_1\,m_2}{r^3}$, tending to turn it parallel to the first magnet. This is equivalent to a single force $\frac{3\,m_1\,m_2}{r^4}$ acting parallel to the direction of the axis of the second magnet, and cutting r at a point two-thirds of its length from m_2.

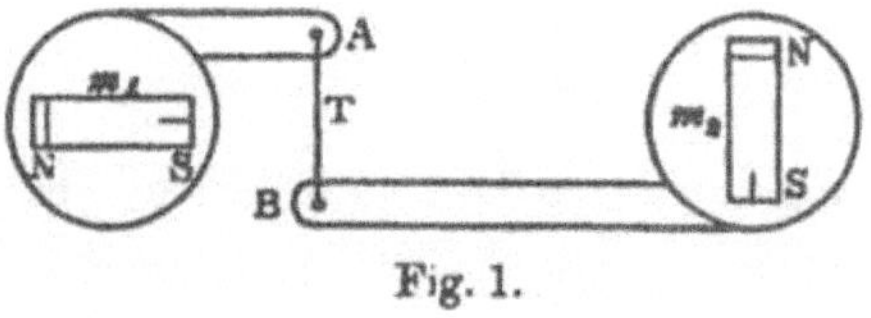

Fig. 1.

Thus in the figure (1) two magnets are made to float on water, m_2

being in the direction of the axis of m_1, but having its own axis at right angles to that of m_1. If two points, A, B, rigidly connected with m_1 and m_2 respectively, are connected by means of a string T, the system will be in equilibrium, provided T cuts the line $m_1 m_2$ at right angles at a point one-third of the distance from m_1 to m_2.

(4) If we allow the second magnet to turn freely about its centre till it comes to a position of stable equilibrium, W will then be a minimum as regards h_2, and therefore the resolved part of the force due to m_2, taken in the direction of h_1, will be a maximum. Hence, if we wish to produce the greatest possible magnetic force at a given point in a given direction by means of magnets, the positions of whose centres are given, then, in order to determine the proper directions of the axes of these magnets to produce this effect, we have only to place a magnet in the given direction at the given point, and to observe the direction of stable equilibrium of the axis of a second magnet when its centre is placed at each of the other given points. The magnets must then be placed with their axes in the directions indicated by that of the second magnet.

Of course, in performing this experiment we must take account of terrestrial magnetism, if it exists.

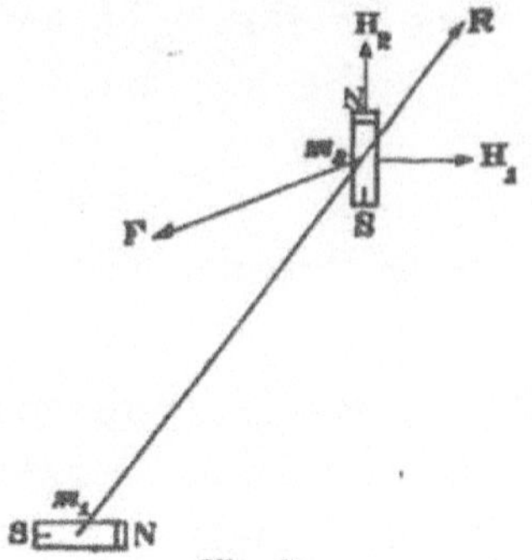

Fig. 2.

Let the second magnet be in a position of stable equilibrium as regards its direction, then since the couple acting on it vanishes, the axis of the second magnet must be in the same plane with that of the first. Hence

$$(h_1 h_2) = (h_1 r) + (r h_2), \qquad (16)$$

and the couple being

$$\frac{m_1 m_2}{r^3}\left(\sin (h_1 h_2) - 3 \cos (h_1 r) \sin (r h_2)\right), \qquad (17)$$

we find when this is zero

$$\tan (h_1 r) = 2 \tan (r h_2), \qquad (18)$$

or

$$\tan H_1 m_2 R = 2 \tan R\, m_2 H_2. \qquad (19)$$

When this position has been taken up by the second magnet the value of W becomes

$$-m_2 \frac{dV}{dh_2},$$

where h_2 is in the direction of the line of force due to m_1 at m_2.

Hence $$W = -m_2 \sqrt{\overline{\frac{dV}{dx}}\bigg|^2 + \overline{\frac{dV}{dy}}\bigg|^2 + \overline{\frac{dV}{dz}}\bigg|^2}. \qquad (20)$$

Hence the second magnet will tend to move towards places of greater resultant force.

The force on the second magnet may be decomposed into a force R, which in this case is always attractive towards the first magnet, and a force H_1 parallel to the axis of the first magnet, where

$$R = -3\frac{m_1 m_2}{r^4}\frac{4\lambda_1^2+1}{\sqrt{3\lambda_1^2+1}}, \qquad H_1 = 3\frac{m_1 m_2}{r^4}\frac{\lambda_1}{\sqrt{3\lambda_1^2+1}}. \qquad (21)$$

In Fig. XVII, at the end of this volume, the lines of force and equipotential surfaces in two dimensions are drawn. The magnets which produce them are supposed to be two long cylindrical rods the sections of which are represented by the circular blank spaces, and these rods are magnetized transversely in the direction of the arrows.

If we remember that there is a tension along the lines of force, it is easy to see that each magnet will tend to turn in the direction of the motion of the hands of a watch.

That on the right hand will also, as a whole, tend to move towards the top, and that on the left hand towards the bottom of the page.

On the Potential Energy of a Magnet placed in a Magnetic Field.

389.] Let V be the magnetic potential due to any system of magnets acting on the magnet under consideration. We shall call V the potential of the external magnetic force.

If a small magnet whose strength is m, and whose length is ds, be placed so that its positive pole is at a point where the potential is V, and its negative pole at a point where the potential is V', the potential energy of this magnet will be $m(V-V')$, or, if ds is measured from the negative pole to the positive,

$$m\frac{dV}{ds}ds. \qquad (1)$$

If I is the intensity of the magnetization, and λ, μ, ν its direction-cosines, we may write,

$$m\,ds = I\,dx\,dy\,dz,$$

and $$\frac{dV}{ds} = \lambda\frac{dV}{dx} + \mu\frac{dV}{dy} + \nu\frac{dV}{dz},$$

and, finally, if A, B, C are the components of magnetization,

$$A = \lambda I, \qquad B = \mu I, \qquad C = \nu I,$$

so that the expression (1) for the potential energy of the element of the magnet becomes

$$\left(A\frac{dV}{dx}+B\frac{dV}{dy}+C\frac{dV}{dz}\right)dx\,dy\,dz. \qquad (2)$$

To obtain the potential energy of a magnet of finite size, we must integrate this expression for every element of the magnet. We thus obtain

$$W=\iiint\left(A\frac{dV}{dx}+B\frac{dV}{dy}+C\frac{dV}{dz}\right)dx\,dy\,dz \qquad (3)$$

as the value of the potential energy of the magnet with respect to the magnetic field in which it is placed.

The potential energy is here expressed in terms of the components of magnetization and of those of the magnetic force arising from external causes.

By integration by parts we may express it in terms of the distribution of magnetic matter and of magnetic potential

$$W=\iint(Al+Bm+Cn)V\,dS-\iiint V\left(\frac{dA}{dx}+\frac{dB}{dy}+\frac{dC}{dz}\right)dx\,dy\,dz, \qquad (4)$$

where l, m, n are the direction-cosines of the normal at the element of surface dS. If we substitute in this equation the expressions for the surface- and volume-density of magnetic matter as given in Art. 386, the expression becomes

$$W=\iint V\sigma\,dS+\iiint V\rho\,dS. \qquad (5)$$

We may write equation (3) in the form

$$W=-\iiint(A\alpha+B\beta+C\gamma)\,dx\,dy\,dz, \qquad (6)$$

where α, β and γ are the components of the external magnetic force.

On the Magnetic Moment and Axis of a Magnet.

390.] If throughout the whole space occupied by the magnet the external magnetic force is uniform in direction and magnitude, the components α, β, γ will be constant quantities, and if we write

$$\iiint A\,dx\,dy\,dz=lK,\quad \iiint B\,dx\,dy\,dz=mK,\quad \iiint C\,dx\,dy\,dz=nK, \qquad (7)$$

the integrations being extended over the whole substance of the magnet, the value of W may be written

$$W=-K(l\alpha+m\beta+n\gamma). \qquad (8)$$

In this expression l, m, n are the direction-cosines of the axis of the magnet, and K is the magnetic moment of the magnet. If ϵ is the angle which the axis of the magnet makes with the direction of the magnetic force $\mathfrak{H}$, the value of W may be written

$$W = -K\,\mathfrak{H} \cos\epsilon. \tag{9}$$

If the magnet is suspended so as to be free to turn about a vertical axis, as in the case of an ordinary compass needle, let the azimuth of the axis of the magnet be ϕ, and let it be inclined θ to the horizontal plane. Let the force of terrestrial magnetism be in a direction whose azimuth is δ and dip ζ, then

$$\alpha = \mathfrak{H} \cos\zeta \cos\delta, \quad \beta = \mathfrak{H} \cos\zeta \sin\delta, \quad \gamma = \mathfrak{H} \sin\zeta; \tag{10}$$

$$l = \cos\theta \cos\phi, \qquad m = \cos\theta \sin\phi, \qquad n = \sin\theta; \tag{11}$$

whence
$$W = -K\,\mathfrak{H}\,(\cos\zeta \cos\theta \cos(\phi-\delta) + \sin\zeta \sin\theta). \tag{12}$$

The moment of the force tending to increase ϕ by turning the magnet round a vertical axis is

$$-\frac{dW}{d\phi} = -K\,\mathfrak{H} \cos\zeta \cos\theta \sin(\phi-\delta). \tag{13}$$

On the Expansion of the Potential of a Magnet in Solid Harmonics.

391.] Let V be the potential due to a unit pole placed at the point (ξ, η, ζ). The value of V at the point x, y, z is

$$V = \{(\xi-x)^2 + (\eta-y)^2 + (\zeta-z)^2\}^{-\frac{1}{2}}. \tag{1}$$

This expression may be expanded in terms of spherical harmonics, with their centre at the origin. We have then

$$V = V_0 + V_1 + V_2 + \&c., \tag{2}$$

when $V_0 = \dfrac{1}{r}$, r being the distance of (ξ, η, ζ) from the origin, (3)

$$V_1 = \frac{\xi x + \eta y + \zeta z}{r^3}, \tag{4}$$

$$V_2 = \frac{3(\xi x + \eta y + \zeta z)^2 - (x^2+y^2+z^2)(\xi^2+\eta^2+\zeta^2)}{2r^5}, \tag{5}$$

&c.

To determine the value of the potential energy when the magnet is placed in the field of force expressed by this potential, we have to integrate the expression for W in equation (3) with respect to x, y and z, considering ξ, η, ζ and r as constants.

If we consider only the terms introduced by V_0, V_1 and V_2 the result will depend on the following volume-integrals,

$$lK = \iiint A\,dx\,dy\,dz, \quad mK = \iiint B\,dx\,dy\,dz, \quad nK = \iiint C\,dx\,dy\,dz; \quad (6)$$

$$L = \iiint Ax\,dx\,dy\,dz, \quad M = \iiint By\,dx\,dy\,dz, \quad N = \iiint Cz\,dx\,dy\,dz; \quad (7)$$

$$P = \iiint (Bz + Cy)\,dx\,dy\,dz, \quad Q = \iiint (Cx + Az)\,dx\,dy\,dz,$$

$$R = \iiint (Ay + Bx)\,dx\,dy\,dz. \quad (8)$$

We thus find for the value of the potential energy of the magnet placed in presence of the unit pole at the point (ξ, η, ζ),

$$W = K\frac{l\xi + m\eta + n\zeta}{r^3}$$

$$\frac{\xi^2(2L-M-N) + \eta^2(2M-N-L) + \zeta^2(2N-L-M) + 3(P\eta\zeta + Q\zeta\xi + R\xi\eta)}{r^5}. \quad (9)$$

This expression may also be regarded as the potential energy of the unit pole in presence of the magnet, or more simply as the potential at the point ξ, η, ζ due to the magnet.

On the Centre of a Magnet and its Primary and Secondary Axes.

392.] This expression may be simplified by altering the directions of the coordinates and the position of the origin. In the first place, we shall make the direction of the axis of x parallel to the axis of the magnet. This is equivalent to making

$$l = 1, \quad m = 0, \quad n = 0. \quad (10)$$

If we change the origin of coordinates to the point (x', y', z'), the directions of the axes remaining unchanged, the volume-integrals lK, mK and nK will remain unchanged, but the others will be altered as follows:

$$L' = L - lKx', \qquad M' = M - mKy', \qquad N' = N - nKz'; \quad (11)$$

$$P' = P - K(mz' + ny'), \quad Q' = Q - K(nx' + lz'), \quad R' = R - K(ly' + mx'). \quad (12)$$

If we now make the direction of the axis of x parallel to the axis of the magnet, and put

$$x' = \frac{2L - M - N}{2K}, \qquad y' = \frac{R}{K}, \qquad z' = \frac{Q}{K}, \quad (13)$$

then for the new axes M and N have their values unchanged, and the value of L' becomes $\frac{1}{2}(M+N)$. P remains unchanged, and Q and R vanish. We may therefore write the potential thus,

$$K\frac{\xi}{r^3} + \frac{\frac{3}{2}(\eta^2 - \zeta^2)(M-N) + 3P\eta\zeta}{r}. \quad (14)$$

We have thus found a point, fixed with respect to the magnet, such that the second term of the potential assumes the most simple form when this point is taken as origin of coordinates. This point we therefore define as the centre of the magnet, and the axis drawn through it in the direction formerly defined as the direction of the magnetic axis may be defined as the principal axis of the magnet.

We may simplify the result still more by turning the axes of y and z round that of x through half the angle whose tangent is $\frac{P}{M-N}$. This will cause P to become zero, and the final form of the potential may be written

$$K\frac{\xi}{r^3} + \tfrac{3}{2}\frac{(\eta^2-\zeta^2)(M-N)}{r^5}. \tag{15}$$

This is the simplest form of the first two terms of the potential of a magnet. When the axes of y and z are thus placed they may be called the Secondary axes of the magnet.

We may also determine the centre of a magnet by finding the position of the origin of coordinates, for which the surface-integral of the square of the second term of the potential, extended over a sphere of unit radius, is a minimum.

The quantity which is to be made a minimum is, by Art. 141,

$$4(L^2+M^2+N^2-MN-NL-LM)+3(P^2+Q^2+R^2). \tag{16}$$

The changes in the values of this quantity due to a change of position of the origin may be deduced from equations (11) and (12). Hence the conditions of a minimum are

$$\left.\begin{aligned} 2l(2L-M-N)+3nQ+3mR &= 0,\\ 2m(2M-N-L)+3lR+3nP &= 0,\\ 2n(2N-L-M)+3mP+3lQ &= 0. \end{aligned}\right\} \tag{17}$$

If we assume $l=1$, $m=0$, $n=0$, these conditions become

$$2L-M-N=0, \quad Q=0, \quad R=0, \tag{18}$$

which are the conditions made use of in the previous investigation.

This investigation may be compared with that by which the potential of a system of gravitating matter is expanded. In the latter case, the most convenient point to assume as the origin is the centre of gravity of the system, and the most convenient axes are the principal axes of inertia through that point.

In the case of the magnet, the point corresponding to the centre of gravity is at an infinite distance in the direction of the axis, and the point which we call the centre of the magnet is a point having different properties from those of the centre of gravity. The quantities L, M, N correspond to the moments of inertia, and P, Q, R to the products of inertia of a material body, except that L, M and N are not necessarily positive quantities.

When the centre of the magnet is taken as the origin, the spherical harmonic of the second order is of the sectorial form, having its axis coinciding with that of the magnet, and this is true of no other point.

When the magnet is symmetrical on all sides of this axis, as in the case of a figure of revolution, the term involving the harmonic of the second order disappears entirely.

393.] At all parts of the earth's surface, except some parts of the Polar regions, one end of a magnet points towards the north, or at least in a northerly direction, and the other in a southerly direction. In speaking of the ends of a magnet we shall adopt the popular method of calling the end which points to the north the north end of the magnet. When, however, we speak in the language of the theory of magnetic fluids we shall use the words Boreal and Austral. Boreal magnetism is an imaginary kind of matter supposed to be most abundant in the northern parts of the earth, and Austral magnetism is the imaginary magnetic matter which prevails in the southern regions of the earth. The magnetism of the north end of a magnet is Austral, and that of the south end is Boreal. When therefore we speak of the north and south ends of a magnet we do not compare the magnet with the earth as the great magnet, but merely express the position which the magnet endeavours to take up when free to move. When, on the other hand, we wish to compare the distribution of imaginary magnetic fluid in the magnet with that in the earth we shall use the more grandiloquent words Boreal and Austral magnetism.

394.] In speaking of a field of magnetic force we shall use the phrase Magnetic North to indicate the direction in which the north end of a compass needle would point if placed in the field of force.

In speaking of a line of magnetic force we shall always suppose it to be traced from magnetic south to magnetic north, and shall call this direction positive. In the same way the direction of magnetization of a magnet is indicated by a line drawn from the south end of the magnet towards the north end, and the end of the magnet which points north is reckoned the positive end.

We shall consider Austral magnetism, that is, the magnetism of that end of a magnet which points north, as positive. If we denote its numerical value by m, then the magnetic potential

$$V = \Sigma\left(\frac{m}{r}\right),$$

and the positive direction of a line of force is that in which V diminishes.

CHAPTER II.

MAGNETIC FORCE AND MAGNETIC INDUCTION.

395.] We have already (Art. 386) determined the magnetic potential at a given point due to a magnet, the magnetization of which is given at every point of its substance, and we have shewn that the mathematical result may be expressed either in terms of the actual magnetization of every element of the magnet, or in terms of an imaginary distribution of 'magnetic matter,' partly condensed on the surface of the magnet and partly diffused throughout its substance.

The magnetic potential, as thus defined, is found by the same mathematical process, whether the given point is outside the magnet or within it. The force exerted on a unit magnetic pole placed at any point outside the magnet is deduced from the potential by the same process of differentiation as in the corresponding electrical problem. If the components of this force are a, β, γ,

$$a = -\frac{dV}{dx}, \quad \beta = -\frac{dV}{dy}, \quad \gamma = -\frac{dV}{dz}. \tag{1}$$

To determine by experiment the magnetic force at a point within the magnet we must begin by removing part of the magnetized substance, so as to form a cavity within which we are to place the magnetic pole. The force acting on the pole will depend, in general, in the form of this cavity, and on the inclination of the walls of the cavity to the direction of magnetization. Hence it is necessary, in order to avoid ambiguity in speaking of the magnetic force within a magnet, to specify the form and position of the cavity within which the force is to be measured. It is manifest that when the form and position of the cavity is specified, the point within it at which the magnetic pole is placed must be regarded as no longer within the substance of the magnet, and therefore the ordinary methods of determining the force become at once applicable.

396.] Let us now consider a portion of a magnet in which the direction and intensity of the magnetization are uniform. Within this portion let a cavity be hollowed out in the form of a cylinder, the axis of which is parallel to the direction of magnetization, and let a magnetic pole of unit strength be placed at the middle point of the axis.

Since the generating lines of this cylinder are in the direction of magnetization, there will be no superficial distribution of magnetism on the curved surface, and since the circular ends of the cylinder are perpendicular to the direction of magnetization, there will be a uniform superficial distribution, of which the surface-density is I for the negative end, and $-I$ for the positive end.

Let the length of the axis of the cylinder be $2b$, and its radius a. Then the force arising from this superficial distribution on a magnetic pole placed at the middle point of the axis is that due to the attraction of the disk on the positive side, and the repulsion of the disk on the negative side. These two forces are equal and in the same direction, and their sum is

$$R = 4\pi I\left(1 - \frac{b}{\sqrt{a^2+b^2}}\right). \tag{2}$$

From this expression it appears that the force depends, not on the absolute dimensions of the cavity, but on the ratio of the length to the diameter of the cylinder. Hence, however small we make the cavity, the force arising from the surface distribution on its walls will remain, in general, finite.

397.] We have hitherto supposed the magnetization to be uniform and in the same direction throughout the whole of the portion of the magnet from which the cylinder is hollowed out. When the magnetization is not thus restricted, there will in general be a distribution of imaginary magnetic matter through the substance of the magnet. The cutting out of the cylinder will remove part of this distribution, but since in similar solid figures the forces at corresponding points are proportional to the linear dimensions of the figures, the alteration of the force on the magnetic pole due to the volume-density of magnetic matter will diminish indefinitely as the size of the cavity is diminished, while the effect due to the surface-density on the walls of the cavity remains, in general, finite.

If, therefore, we assume the dimensions of the cylinder so small that the magnetization of the part removed may be regarded as everywhere parallel to the axis of the cylinder, and of constant magnitude I, the force on a magnetic pole placed at the middle point of the axis of the cylindrical hollow will be compounded of two forces. The first of these is that due to the distribution of magnetic matter on the outer surface of the magnet, and throughout its interior, exclusive of the portion hollowed out. The components of this force are α, β and γ, derived from the potential by equations (1). The second is the force R, acting along the axis of the cylinder in the direction of magnetization. The value of this force depends on the ratio of the length to the diameter of the cylindric cavity.

398.] *Case I.* Let this ratio be very great, or let the diameter of the cylinder be small compared with its length. Expanding the expression for R in terms of $\frac{a}{b}$, it becomes

$$R = 4\pi I\left\{\frac{1}{2}\frac{a^2}{b^2} - \frac{3}{8}\frac{a^4}{b^4} + \&c.\right\}, \tag{3}$$

a quantity which vanishes when the ratio of b to a is made infinite. Hence, when the cavity is a very narrow cylinder with its axis parallel to the direction of magnetization, the magnetic force within the cavity is not affected by the surface distribution on the ends of the cylinder, and the components of this force are simply α, β, γ, where

$$\alpha = -\frac{dV}{dx}, \quad \beta = -\frac{dV}{dy}, \quad \gamma = -\frac{dV}{dz}. \tag{4}$$

We shall define the force within a cavity of this form as the magnetic force within the magnet. Sir William Thomson has called this the Polar definition of magnetic force. When we have occasion to consider this force as a vector we shall denote it by $\mathfrak{H}$.

399.] *Case II.* Let the length of the cylinder be very small compared with its diameter, so that the cylinder becomes a thin disk. Expanding the expression for R in terms of $\frac{b}{a}$, it becomes

$$R = 4\pi I\left\{1 - \frac{b}{a} + \frac{1}{2}\frac{b^3}{a^3} - \&c.\right\}, \tag{5}$$

the ultimate value of which, when the ratio of a to b is made infinite, is $4\pi I$.

Hence, when the cavity is in the form of a thin disk, whose plane is normal to the direction of magnetization, a unit magnetic pole placed at the middle of the axis experiences a force $4\pi I$ in the direction of magnetization arising from the superficial magnetism on the circular surfaces of the disk *.

Since the components of I are A, B and C, the components of this force are $4\pi A$, $4\pi B$ and $4\pi C$. This must be compounded with the force whose components are α, β, γ.

400.] Let the actual force on the unit pole be denoted by the vector $\mathfrak{B}$, and its components by a, b and c, then

$$\left.\begin{aligned} a &= \alpha+4\pi A,\\ b &= \beta+4\pi B,\\ c &= \gamma+4\pi C. \end{aligned}\right\} \qquad (6)$$

We shall define the force within a hollow disk, whose plane sides are normal to the direction of magnetization, as the Magnetic Induction within the magnet. Sir William Thomson has called this the Electromagnetic definition of magnetic force.

The three vectors, the magnetization $\mathfrak{I}$, the magnetic force $\mathfrak{H}$, and the magnetic induction $\mathfrak{B}$ are connected by the vector equation

$$\mathfrak{B} = \mathfrak{H}+4\pi\mathfrak{I}. \qquad (7)$$

Line-Integral of Magnetic Force.

401.] Since the magnetic force, as defined in Art. 398, is that due to the distribution of free magnetism on the surface and through the interior of the magnet, and is not affected by the surface-magnetism of the cavity, it may be derived directly from the general expression for the potential of the magnet, and the line-integral of the magnetic force taken along any curve from the point A to the point B is

$$\int_A^B \left(\alpha\frac{dx}{ds}+\beta\frac{dy}{ds}+\gamma\frac{dz}{ds}\right)ds = V_A - V_B, \qquad (8)$$

where V_A and V_B denote the potentials at A and B respectively.

* *On the force within cavities of other forms.*

1. Any narrow crevasse. The force arising from the surface-magnetism is $4\pi I\cos\epsilon$ in the direction of the normal to the plane of the crevasse, where ϵ is the angle between this normal and the direction of magnetization. When the crevasse is parallel to the direction of magnetization the force is the magnetic force $\mathfrak{H}$; when the crevasse is perpendicular to the direction of magnetization the force is the magnetic induction $\mathfrak{B}$.

2. In an elongated cylinder, the axis of which makes an angle ϵ with the direction of magnetization, the force arising from the surface-magnetism is $2\pi I\sin\epsilon$, perpendicular to the axis in the plane containing the axis and the direction of magnetization.

3. In a sphere the force arising from surface-magnetism is $\frac{4}{3}\pi I$ in the direction of magnetization.

Surface-Integral of Magnetic Induction.

402.] The magnetic induction through the surface S is defined as the value of the integral

$$Q = \iint \mathfrak{B} \cos \epsilon \, dS, \tag{9}$$

where $\mathfrak{B}$ denotes the magnitude of the magnetic induction at the element of surface dS, and ϵ the angle between the direction of the induction and the normal to the element of surface, and the integration is to be extended over the whole surface, which may be either closed or bounded by a closed curve.

If a, b, c denote the components of the magnetic induction, and l, m, n the direction-cosines of the normal, the surface-integral may be written

$$Q = \iint (la + mb + nc)\, dS. \tag{10}$$

If we substitute for the components of the magnetic induction their values in terms of those of the magnetic force, and the magnetization as given in Art. 400, we find

$$Q = \iint (l\alpha + m\beta + n\gamma)\, dS + 4\pi \iint (lA + mB + nC)\, dS. \tag{11}$$

We shall now suppose that the surface over which the integration extends is a closed one, and we shall investigate the value of the two terms on the right-hand side of this equation.

Since the mathematical form of the relation between magnetic force and free magnetism is the same as that between electric force and free electricity, we may apply the result given in Art. 77 to the first term in the value of Q by substituting α, β, γ, the components of magnetic force, for X, Y, Z, the components of electric force in Art. 77, and M, the algebraic sum of the free magnetism within the closed surface, for e, the algebraic sum of the free electricity.

We thus obtain the equation

$$\iint (l\alpha + m\beta + n\gamma)\, dS = 4\pi M. \tag{12}$$

Since every magnetic particle has two poles, which are equal in numerical magnitude but of opposite signs, the algebraic sum of the magnetism of the particle is zero. Hence, those particles which are entirely within the closed surface S can contribute nothing to the algebraic sum of the magnetism within S. The value of M must therefore depend only on those magnetic particles which are cut by the surface S.

Consider a small element of the magnet of length s and transverse section k^2, magnetized in the direction of its length, so that the strength of its poles is m. The moment of this small magnet will be ms, and the intensity of its magnetization, being the ratio of the magnetic moment to the volume, will be

$$I = \frac{m}{k^2}. \tag{13}$$

Let this small magnet be cut by the surface S, so that the direction of magnetization makes an angle ϵ' with the normal drawn outwards from the surface, then if dS denotes the area of the section,

$$k^2 = dS \cos \epsilon'. \tag{14}$$

The negative pole $-m$ of this magnet lies within the surface S.

Hence, if we denote by dM the part of the free magnetism within S which is contributed by this little magnet,

$$dM = -m = -I k^2,$$
$$= -I \cos \epsilon' \, dS. \tag{15}$$

To find M, the algebraic sum of the free magnetism within the closed surface S, we must integrate this expression over the closed surface, so that

$$M = -\iint I \cos \epsilon' \, dS,$$

or writing A, B, C for the components of magnetization, and l, m, n for the direction-cosines of the normal drawn outwards,

$$M = -\iint (lA + mB + nC) \, dS. \tag{16}$$

This gives us the value of the integral in the second term of equation (11). The value of Q in that equation may therefore be found in terms of equations (12) and (16),

$$Q = 4\pi M - 4\pi M = 0, \tag{17}$$

or, *the surface-integral of the magnetic induction through any closed surface is zero.*

403.] If we assume as the closed surface that of the differential element of volume $dx\,dy\,dz$, we obtain the equation

$$\frac{da}{dx} + \frac{db}{dy} + \frac{dc}{dz} = 0. \tag{18}$$

This is the solenoidal condition which is always satisfied by the components of the magnetic induction.

Since the distribution of magnetic induction is solenoidal, the induction through any surface bounded by a closed curve depends only on the form and position of the closed curve, and not on that of the surface itself.

404.] Surfaces at every point of which

$$la+mb+nc=0 \tag{19}$$

are called Surfaces of no induction, and the intersection of two such surfaces is called a Line of induction. The conditions that a curve, s, may be a line of induction are

$$\frac{1}{a}\frac{dx}{ds}=\frac{1}{b}\frac{dy}{ds}=\frac{1}{c}\frac{dz}{ds}. \tag{20}$$

A system of lines of induction drawn through every point of a closed curve forms a tubular surface called a Tube of induction.

The induction across any section of such a tube is the same. If the induction is unity the tube is called a Unit tube of induction.

All that Faraday* says about lines of magnetic force and magnetic sphondyloids is mathematically true, if understood of the lines and tubes of magnetic induction.

The magnetic force and the magnetic induction are identical outside the magnet, but within the substance of the magnet they must be carefully distinguished. In a straight uniformly magnetized bar the magnetic force due to the magnet itself is from the end which points north, which we call the positive pole, towards the south end or negative pole, both within the magnet and in the space without.

The magnetic induction, on the other hand, is from the positive pole to the negative outside the magnet, and from the negative pole to the positive within the magnet, so that the lines and tubes of induction are re-entering or cyclic figures.

The importance of the magnetic induction as a physical quantity will be more clearly seen when we study electromagnetic phenomena. When the magnetic field is explored by a moving wire, as in Faraday's *Exp. Res.* 3076, it is the magnetic induction and not the magnetic force which is directly measured.

The Vector-Potential of Magnetic Induction.

405.] Since, as we have shewn in Art. 403, the magnetic induction through a surface bounded by a closed curve depends on

* *Exp. Res.*, series xxviii.

the closed curve, and not on the form of the surface which is bounded by it, it must be possible to determine the induction through a closed curve by a process depending only on the nature of that curve, and not involving the construction of a surface forming a diaphragm of the curve.

This may be done by finding a vector $\mathfrak{A}$ related to $\mathfrak{B}$, the magnetic induction, in such a way that the line-integral of $\mathfrak{A}$, extended round the closed curve, is equal to the surface-integral of $\mathfrak{B}$, extended over a surface bounded by the closed curve.

If, in Art. 24, we write F, G, H for the components of $\mathfrak{A}$, and a, b, c for the components of $\mathfrak{B}$, we find for the relation between these components

$$a = \frac{dH}{dy} - \frac{dG}{dz}, \quad b = \frac{dF}{dz} - \frac{dH}{dx}, \quad c = \frac{dG}{dx} - \frac{dF}{dy}. \tag{21}$$

The vector $\mathfrak{A}$, whose components are F, G, H, is called the vector-potential of magnetic induction. The vector-potential at a given point, due to a magnetized particle placed at the origin, is numerically equal to the magnetic moment of the particle divided by the square of the radius vector and multiplied by the sine of the angle between the axis of magnetization and the radius vector, and the direction of the vector-potential is perpendicular to the plane of the axis of magnetization and the radius vector, and is such that to an eye looking in the positive direction along the axis of magnetization the vector-potential is drawn in the direction of rotation of the hands of a watch.

Hence, for a magnet of any form in which A, B, C are the components of magnetization at the point $x\,y\,z$, the components of the vector-potential at the point $\xi\,\eta\,\zeta$, are

$$\left.\begin{aligned} F &= \iiint\left(B\frac{dp}{dz} - C\frac{dp}{dy}\right)dx\,dy\,dz, \\ G &= \iiint\left(C\frac{dp}{dx} - A\frac{dp}{dz}\right)dx\,dy\,dz, \\ H &= \iiint\left(A\frac{dp}{dy} - B\frac{dp}{dx}\right)dx\,dy\,dz\,; \end{aligned}\right\} \tag{22}$$

where p is put, for conciseness, for the reciprocal of the distance between the points (ξ, η, ζ) and (x, y, z), and the integrations are extended over the space occupied by the magnet.

406.] The scalar, or ordinary, potential of magnetic force, Art. 386, becomes when expressed in the same notation,

$$V = \iiint\left(A\frac{dp}{dx} + B\frac{dp}{dy} + C\frac{dp}{dz}\right)dx\,dy\,dz. \tag{23}$$

Remembering that $\frac{dp}{dx} = -\frac{dp}{d\xi}$, and that the integral

$$\iiint A\left(\frac{d^2p}{dx^2} + \frac{d^2p}{dy^2} + \frac{d^2p}{dz^2}\right) dx\, dy\, dz$$

has the value $-4\pi(A)$ when the point (ξ, η, ζ) is included within the limits of integration, and is zero when it is not so included, (A) being the value of A at the point (ξ, η, ζ), we find for the value of the x-component of the magnetic induction,

$$\begin{aligned} a &= \frac{dH}{d\eta} - \frac{dG}{d\zeta} \\ &= \iiint \left\{ A\left(\frac{d^2p}{dy\, d\eta} + \frac{d^2p}{dz\, d\zeta}\right) - B\frac{d^2p}{dx\, d\eta} - C\frac{d^2p}{dx\, d\zeta} \right\} dx\, dy\, dz \\ &= -\frac{d}{d\xi}\iiint \left\{ A\frac{dp}{dx} + B\frac{dp}{dy} + C\frac{dp}{dz} \right\} dx\, dy\, dz \\ &\qquad - \iiint A\left(\frac{d^2p}{dx^2} + \frac{d^2p}{dy^2} + \frac{d^2p}{dz^2}\right) dx\, dy\, dz. \end{aligned} \tag{24}$$

The first term of this expression is evidently $-\frac{dV}{d\xi}$, or α, the component of the magnetic force.

The quantity under the integral sign in the second term is zero for every element of volume except that in which the point (ξ, η, ζ) is included. If the value of A at the point (ξ, η, ζ) is (A), the value of the second term is $4\pi(A)$, where (A) is evidently zero at all points outside the magnet.

We may now write the value of the x-component of the magnetic induction

$$a = \alpha + 4\pi(A), \tag{25}$$

an equation which is identical with the first of those given in Art. 400. The equations for b and c will also agree with those of Art. 400.

We have already seen that the magnetic force $\mathfrak{H}$ is derived from the scalar magnetic potential V by the application of Hamilton's operator ∇, so that we may write, as in Art. 17,

$$\mathfrak{H} = -\nabla V, \tag{26}$$

and that this equation is true both without and within the magnet.

It appears from the present investigation that the magnetic induction $\mathfrak{B}$ is derived from the vector-potential $\mathfrak{A}$ by the application of the same operator, and that the result is true within the magnet as well as without it.

The application of this operator to a vector-function produces, in general, a scalar quantity as well as a vector. The scalar part, however, which we have called the convergence of the vector-function, vanishes when the vector-function satisfies the solenoidal condition

$$\frac{dF}{d\xi} + \frac{dG}{d\eta} + \frac{dH}{d\zeta} = 0. \tag{27}$$

By differentiating the expressions for F, G, H in equations (22), we find that this equation is satisfied by these quantities.

We may therefore write the relation between the magnetic induction and its vector-potential

$$\mathfrak{B} = \nabla \mathfrak{A},$$

which may be expressed in words by saying that the magnetic induction is the curl of its vector-potential. See Art. 25.

CHAPTER III.

MAGNETIC SOLENOIDS AND SHELLS *.

On Particular Forms of Magnets.

407.] If a long narrow filament of magnetic matter like a wire is magnetized everywhere in a longitudinal direction, then the product of any transverse section of the filament into the mean intensity of the magnetization across it is called the strength of the magnet at that section. If the filament were cut in two at the section without altering the magnetization, the two surfaces, when separated, would be found to have equal and opposite quantities of superficial magnetization, each of which is numerically equal to the strength of the magnet at the section.

A filament of magnetic matter, so magnetized that its strength is the same at every section, at whatever part of its length the section be made, is called a Magnetic Solenoid.

If m is the strength of the solenoid, ds an element of its length, r the distance of that element from a given point, and ϵ the angle which r makes with the axis of magnetization of the element, the potential at the given point due to the element is

$$\frac{m\,ds\cos\epsilon}{r^2} = \frac{m}{r^2}\frac{dr}{ds}ds.$$

Integrating this expression with respect to s, so as to take into account all the elements of the solenoid, the potential is found to be

$$V = m\left(\frac{1}{r_1} - \frac{1}{r_2}\right),$$

r_1 being the distance of the positive end of the solenoid, and r_2 that of the negative end from the point where V exists.

Hence the potential due to a solenoid, and consequently all its magnetic effects, depend only on its strength and the position of

* See Sir W. Thomson's 'Mathematical Theory of Magnetism,' *Phil. Trans.*, 1850, or *Reprint*.

its ends, and not at all on its form, whether straight or curved, between these points.

Hence the ends of a solenoid may be called in a strict sense its poles.

If a solenoid forms a closed curve the potential due to it is zero at every point, so that such a solenoid can exert no magnetic action, nor can its magnetization be discovered without breaking it at some point and separating the ends.

If a magnet can be divided into solenoids, all of which either form closed curves or have their extremities in the outer surface of the magnet, the magnetization is said to be solenoidal, and, since the action of the magnet depends entirely upon that of the ends of the solenoids, the distribution of imaginary magnetic matter will be entirely superficial.

Hence the condition of the magnetization being solenoidal is

$$\frac{dA}{dx}+\frac{dB}{dy}+\frac{dC}{dz}=0,$$

where A, B, C are the components of the magnetization at any point of the magnet.

408.] A longitudinally magnetized filament, of which the strength varies at different parts of its length, may be conceived to be made up of a bundle of solenoids of different lengths, the sum of the strengths of all the solenoids which pass through a given section being the magnetic strength of the filament at that section. Hence any longitudinally magnetized filament may be called a Complex Solenoid.

If the strength of a complex solenoid at any section is m, then the potential due to its action is

$$V=\int\frac{m}{r^2}\frac{dr}{ds}\,ds \text{ where } m \text{ is variable,}$$

$$=\frac{m_1}{r_1}-\frac{m_2}{r_2}-\int\frac{1}{r}\frac{dm}{ds}\,ds.$$

This shews that besides the action of the two ends, which may in this case be of different strengths, there is an action due to the distribution of imaginary magnetic matter along the filament with a linear density

$$\lambda=-\frac{dm}{ds}.$$

Magnetic Shells.

409.] If a thin shell of magnetic matter is magnetized in a

direction everywhere normal to its surface, the intensity of the magnetization at any place multiplied by the thickness of the sheet at that place is called the Strength of the magnetic shell at that place.

If the strength of a shell is everywhere equal, it is called a Simple magnetic shell; if it varies from point to point it may be conceived to be made up of a number of simple shells superposed and overlapping each other. It is therefore called a Complex magnetic shell.

Let dS be an element of the surface of the shell at Q, and Φ the strength of the shell, then the potential at any point, P, due to the element of the shell, is

$$dV = \Phi \frac{1}{r^2} dS \cos \epsilon,$$

where ϵ is the angle between the vector QP, or r and the normal drawn from the positive side of the shell.

But if $d\omega$ is the solid angle subtended by dS at the point P

$$r^2 d\omega = dS \cos \epsilon,$$

whence

$$dV = \Phi d\omega,$$

and therefore in the case of a simple magnetic shell

$$V = \Phi \omega,$$

or, *the potential due to a magnetic shell at any point is the product of its strength into the solid angle subtended by its edge at the given point**.

410.] The same result may be obtained in a different way by supposing the magnetic shell placed in any field of magnetic force, and determining the potential energy due to the position of the shell.

If V is the potential at the element dS, then the energy due to this element is

$$\Phi \left(l \frac{dV}{dx} + m \frac{dV}{dy} + n \frac{dV}{dz}\right) dS,$$

or, *the product of the strength of the shell into the part of the surface-integral of V due to the element dS of the shell.*

Hence, integrating with respect to all such elements, the energy due to the position of the shell in the field is equal to the product of the strength of the shell and the surface-integral of the magnetic induction taken over the surface of the shell.

Since this surface-integral is the same for any two surfaces which

* This theorem is due to Gauss, *General Theory of Terrestrial Magnetism*, § 38.

have the same bounding edge and do not include between them any centre of force, the action of the magnetic shell depends only on the form of its edge.

Now suppose the field of force to be that due to a magnetic pole of strength m. We have seen (Art. 76, *Cor.*) that the surface-integral over a surface bounded by a given edge is the product of the strength of the pole and the solid angle subtended by the edge at the pole. Hence the energy due to the mutual action of the pole and the shell is

$$\Phi m \omega,$$

and this (by Green's theorem, Art. 100) is equal to the product of the strength of the pole into the potential due to the shell at the pole. The potential due to the shell is therefore $\Phi \omega$.

411.] If a magnetic pole m starts from a point on the negative surface of a magnetic shell, and travels along any path in space so as to come round the edge to a point close to where it started but on the positive side of the shell, the solid angle will vary continuously, and will increase by 4π during the process. The work done by the pole will be $4\pi\Phi m$, and the potential at any point on the positive side of the shell will exceed that at the neighbouring point on the negative side by $4\pi\Phi$.

If a magnetic shell forms a closed surface, the potential outside the shell is everywhere zero, and that in the space within is everywhere $4\pi\Phi$, being positive when the positive side of the shell is inward. Hence such a shell exerts no action on any magnet placed either outside or inside the shell.

412.] If a magnet can be divided into simple magnetic shells, either closed or having their edges on the surface of the magnet, the distribution of magnetism is called Lamellar. If ϕ is the sum of the strengths of all the shells traversed by a point in passing from a given point to a point xyz by a line drawn within the magnet, then the conditions of lamellar magnetization are

$$A = \frac{d\phi}{dx}, \quad B = \frac{d\phi}{dy}, \quad C = \frac{d\phi}{dz}.$$

The quantity, ϕ, which thus completely determines the magnetization at any point may be called the Potential of Magnetization. It must be carefully distinguished from the Magnetic Potential.

413.] A magnet which can be divided into complex magnetic shells is said to have a complex lamellar distribution of magnetism. The condition of such a distribution is that the lines of magnetization must be such that a system of surfaces can be drawn cutting them at right angles. This condition is expressed by the well-known equation

$$A\left(\frac{dC}{dy}-\frac{dB}{dz}\right)+B\left(\frac{dA}{dz}-\frac{dC}{dx}\right)+C\left(\frac{dB}{dx}-\frac{dA}{dy}\right)=0.$$

Forms of the Potentials of Solenoidal and Lamellar Magnets.

414.] The general expression for the scalar potential of a magnet is

$$V=\iiint\left(A\frac{dp}{dx}+B\frac{dp}{dy}+C\frac{dp}{dz}\right)dx\,dy\,dz,$$

where p denotes the potential at (x, y, z) due to a unit magnetic pole placed at ξ, η, ζ or in other words, the reciprocal of the distance between (ξ, η, ζ), the point at which the potential is measured, and (x, y, z), the position of the element of the magnet to which it is due.

This quantity may be integrated by parts, as in Arts. 96, 386.

$$V=\iint p\,(Al+Bm+Cn)\,dS-\iiint p\left(\frac{dA}{dx}+\frac{dB}{dy}+\frac{dC}{dz}\right)dx\,dy\,dz,$$

where l, m, n are the direction-cosines of the normal drawn outwards from dS, an element of the surface of the magnet.

When the magnet is solenoidal the expression under the integral sign in the second term is zero for every point within the magnet, so that the triple integral is zero, and the scalar potential at any point, whether outside or inside the magnet, is given by the surface-integral in the first term.

The scalar potential of a solenoidal magnet is therefore completely determined when the normal component of the magnetization at every point of the surface is known, and it is independent of the form of the solenoids within the magnet.

415.] In the case of a lamellar magnet the magnetization is determined by ϕ, the potential of magnetization, so that

$$A=\frac{d\phi}{dx},\qquad B=\frac{d\phi}{dy},\qquad C=\frac{d\phi}{dz}.$$

The expression for V may therefore be written

$$V=\iiint\left(\frac{d\phi}{dx}\frac{dp}{dx}+\frac{d\phi}{dy}\frac{dp}{dy}+\frac{d\phi}{dz}\frac{dp}{dz}\right)dx\,dy\,dz.$$

Integrating this expression by parts, we find

$$V=\iint\phi\left(l\frac{dp}{dx}+m\frac{dp}{dy}+n\frac{dp}{dz}\right)dS-\iiint\phi\left(\frac{d^2p}{dx^2}+\frac{d^2p}{dy^2}+\frac{d^2p}{dz^2}\right)dx\,dy\,dz.$$

The second term is zero unless the point (ξ, η, ζ) is included in the magnet, in which case it becomes $4\pi(\phi)$ where (ϕ) is the value of ϕ at the point ξ, η, ζ. The surface-integral may be expressed in terms of r, the line drawn from (x, y, z) to (ξ, η, ζ), and θ the angle which this line makes with the normal drawn outwards from dS, so that the potential may be written

$$V = \iint \frac{1}{r^2}\phi \cos\theta \, dS + 4\pi(\phi),$$

where the second term is of course zero when the point (ξ, η, ζ) is not included in the substance of the magnet.

The potential, V, expressed by this equation, is continuous even at the surface of the magnet, where ϕ becomes suddenly zero, for if we write

$$\Omega = \iint \frac{1}{r^2}\phi \cos\theta \, dS,$$

and if Ω_1 is the value of Ω at a point just within the surface, and Ω_2 that at a point close to the first but outside the surface,

$$\Omega_2 = \Omega_1 + 4\pi(\phi),$$

or

$$V_2 = V_1.$$

The quantity Ω is not continuous at the surface of the magnet.

The components of magnetic induction are related to Ω by the equations

$$a = -\frac{d\Omega}{dx}, \quad b = -\frac{d\Omega}{dy}, \quad c = -\frac{d\Omega}{dz}.$$

416.] In the case of a lamellar distribution of magnetism we may also simplify the vector-potential of magnetic induction.

Its x-component may be written

$$F = \iiint \left(\frac{d\phi}{dy}\frac{dp}{dz} - \frac{d\phi}{dz}\frac{dp}{dy}\right) dx\,dy\,dz.$$

By integration by parts we may put this in the form of the surface-integral

$$F = \iint \phi \left(m\frac{dp}{dz} - n\frac{dp}{dy}\right) dS,$$

or

$$F = \iint p \left(m\frac{d\phi}{dz} - n\frac{d\phi}{dy}\right) dS.$$

The other components of the vector-potential may be written down from these expressions by making the proper substitutions.

On Solid Angles.

417.] We have already proved that at any point P the potential due to a magnetic shell is equal to the solid angle subtended by the edge of the shell multiplied by the strength of the shell. As we shall have occasion to refer to solid angles in the theory of electric currents, we shall now explain how they may be measured.

Definition. The solid angle subtended at a given point by a closed curve is measured by the area of a spherical surface whose centre is the given point and whose radius is unity, the outline of which is traced by the intersection of the radius vector with the sphere as it traces the closed curve. This area is to be reckoned positive or negative according as it lies on the left or the right-hand of the path of the radius vector as seen from the given point.

Let (ξ, η, ζ) be the given point, and let (x, y, z) be a point on the closed curve. The coordinates x, y, z are functions of s, the length of the curve reckoned from a given point. They are periodic functions of s, recurring whenever s is increased by the whole length of the closed curve.

We may calculate the solid angle ω directly from the definition thus. Using spherical coordinates with centre at (ξ, η, ζ), and putting

$$x-\xi = r\sin\theta\cos\phi, \quad y-\eta = r\sin\theta\sin\phi, \quad z-\zeta = r\cos\theta,$$

we find the area of any curve on the sphere by integrating

$$\omega = \int (1-\cos\theta)\, d\phi,$$

or, using the rectangular coordinates,

$$\omega = \int d\phi - \int_0^s \frac{z-\zeta}{r^3}\left[(x-\xi)\frac{dy}{ds} - (y-\eta)\frac{dx}{ds}\right] ds,$$

the integration being extended round the curve s.

If the axis of z passes once through the closed curve the first term is 2π. If the axis of z does not pass through it this term is zero.

418.] This method of calculating a solid angle involves a choice of axes which is to some extent arbitrary, and it does not depend solely on the closed curve. Hence the following method, in which no surface is supposed to be constructed, may be stated for the sake of geometrical propriety.

As the radius vector from the given point traces out the closed curve, let a plane passing through the given point roll on the closed curve so as to be a tangent plane at each point of the curve in succession. Let a line of unit-length be drawn from the given point perpendicular to this plane. As the plane rolls round the closed curve the extremity of the perpendicular will trace a second closed curve. Let the length of the second closed curve be σ, then the solid angle subtended by the first closed curve is

$$\omega = 2\pi - \sigma.$$

This follows from the well-known theorem that the area of a closed curve on a sphere of unit radius, together with the circumference of the polar curve, is numerically equal to the circumference of a great circle of the sphere.

This construction is sometimes convenient for calculating the solid angle subtended by a rectilinear figure. For our own purpose, which is to form clear ideas of physical phenomena, the following method is to be preferred, as it employs no constructions which do not flow from the physical data of the problem.

419.] A closed curve s is given in space, and we have to find the solid angle subtended by s at a given point P.

If we consider the solid angle as the potential of a magnetic shell of unit strength whose edge coincides with the closed curve, we must define it as the work done by a unit magnetic pole against the magnetic force while it moves from an infinite distance to the point P. Hence, if σ is the path of the pole as it approaches the point P, the potential must be the result of a line-integration along this path. It must also be the result of a line-integration along the closed curve s. The proper form of the expression for the solid angle must therefore be that of a double integration with respect to the two curves s and σ.

When P is at an infinite distance, the solid angle is evidently zero. As the point P approaches, the closed curve, as seen from the moving point, appears to open out, and the whole solid angle may be conceived to be generated by the apparent motion of the different elements of the closed curve as the moving point approaches.

As the point P moves from P to P' over the element $d\sigma$, the element QQ' of the closed curve, which we denote by ds, will change its position relatively to P, and the line on the unit sphere corresponding to QQ' will sweep over an area on the spherical surface, which we may write

$$d\omega = \Pi\, ds\, d\sigma. \qquad (1)$$

To find Π let us suppose P fixed while the closed curve is moved parallel to itself through a distance $d\sigma$ equal to PP' but in the opposite direction. The relative motion of the point P will be the same as in the real case.

During this motion the element QQ' will generate an area in the form of a parallelogram whose sides are parallel and equal to QQ' and PP'. If we construct a pyramid on this parallelogram as base with its vertex at P, the solid angle of this pyramid will be the increment $d\omega$ which we are in search of.

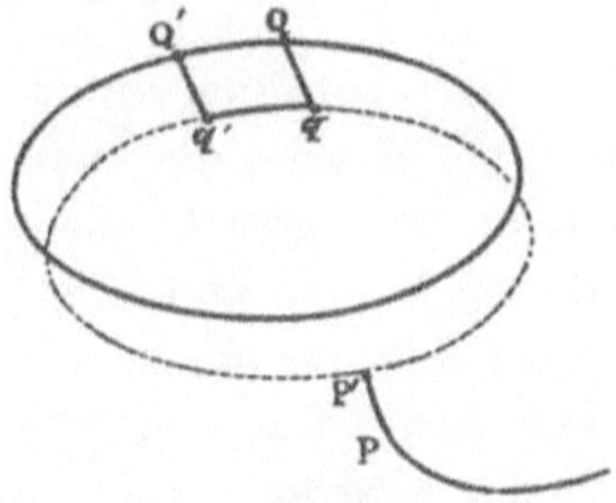

Fig. 3.

To determine the value of this solid angle, let θ and θ' be the angles which ds and $d\sigma$ make with PQ respectively, and let ϕ be the angle between the planes of these two angles, then the area of the projection of the parallelogram $ds . d\sigma$ on a plane perpendicular to PQ or r will be

$$ds\, d\sigma \sin\theta \sin\theta' \sin\phi,$$

and since this is equal to $r^2\, d\omega$, we find

$$d\omega = \Pi\, ds\, d\sigma = \frac{1}{r^2} \sin\theta \sin\theta' \sin\phi\, ds\, d\sigma. \tag{2}$$

Hence
$$\Pi = \frac{1}{r^2} \sin\theta \sin\theta' \sin\phi. \tag{3}$$

420.] We may express the angles θ, θ', and ϕ in terms of r, and its differential coefficients with respect to s and σ, for

$$\cos\theta = \frac{dr}{ds}, \quad \cos\theta' = \frac{dr}{d\sigma}, \quad \text{and} \quad \sin\theta \sin\theta' \cos\phi = r\frac{d^2r}{ds\, d\sigma}. \tag{4}$$

We thus find the following value for Π^2,

$$\Pi^2 = \frac{1}{r^4}\left[1 - \left(\frac{dr}{ds}\right)^2\right]\left[1 - \left(\frac{dr}{d\sigma}\right)^2\right] - \frac{1}{r^2}\left(\frac{d^2r}{ds\, d\sigma}\right)^2. \tag{5}$$

A third expression for Π in terms of rectangular coordinates may be deduced from the consideration that the volume of the pyramid whose solid angle is $d\omega$ and whose axis is r is

$$\tfrac{1}{3} r^3\, d\omega = \tfrac{1}{3} r^3\, \Pi\, ds\, d\sigma.$$

But the volume of this pyramid may also be expressed in terms of the projections of r, ds, and $d\sigma$ on the axis of x, y and z, as a determinant formed by these nine projections, of which we must take the third part. We thus find as the value of Π,

$$\Pi = \frac{1}{r^3} \begin{vmatrix} \xi - x, & \eta - y, & \zeta - z, \\ \frac{d\xi}{d\sigma}, & \frac{d\eta}{d\sigma}, & \frac{d\zeta}{d\sigma}, \\ \frac{dx}{ds}, & \frac{dy}{ds}, & \frac{dz}{ds}. \end{vmatrix} \tag{6}$$

This expression gives the value of Π free from the ambiguity of sign introduced by equation (5).

421.] The value of ω, the solid angle subtended by the closed curve at the point P, may now be written

$$\omega = \iint \Pi \, ds \, d\sigma + \omega_0, \tag{7}$$

where the integration with respect to s is to be extended completely round the closed curve, and that with respect to σ from A a fixed point on the curve to the point P. The constant ω_0 is the value of the solid angle at the point A. It is zero if A is at an infinite distance from the closed curve.

The value of ω at any point P is independent of the form of the curve between A and P provided that it does not pass through the magnetic shell itself. If the shell be supposed infinitely thin, and if P and P' are two points close together, but P on the positive and P' on the negative surface of the shell, then the curves AP and AP' must lie on opposite sides of the edge of the shell, so that PAP' is a line which with the infinitely short line $P'P$ forms a closed circuit embracing the edge. The value of ω at P exceeds that at P' by 4π, that is, by the surface of a sphere of radius unity.

Hence, if a closed curve be drawn so as to pass once through the shell, or in other words, if it be linked once with the edge of the shell, the value of the integral $\iint \Pi \, ds \, d\sigma$ extended round both curves will be 4π.

This integral therefore, considered as depending only on the closed curve s and the arbitrary curve AP, is an instance of a function of multiple values, since, if we pass from A to P along different paths the integral will have different values according to the number of times which the curve AP is twined round the curve s.

If one form of the curve between A and P can be transformed into another by continuous motion without intersecting the curve s, the integral will have the same value for both curves, but if during the transformation it intersects the closed curve n times the values of the integral will differ by $4\pi n$.

If s and σ are any two closed curves in space, then, if they are not linked together, the integral extended once round both is zero.

If they are intertwined n times in the same direction, the value of the integral is $4\pi n$. It is possible, however, for two curves to be intertwined alternately in opposite directions, so that they are inseparably linked together though the value of the integral is zero. See Fig. 4.

It was the discovery by Gauss of this very integral, expressing the work done on a magnetic pole while describing a closed curve in presence of a closed electric current, and indicating the geometrical connexion between the two closed curves, that led him to lament the small progress made in the Geometry of Position since the time of Leibnitz, Euler and Vandermonde. We have now, however, some progress to report, chiefly due to Riemann, Helmholtz and Listing.

Fig. 4.

422.] Let us now investigate the result of integrating with respect to s round the closed curve.

One of the terms of Π in equation (7) is

$$\frac{\xi - x}{r^3}\frac{d\eta}{d\sigma}\frac{dz}{ds} = \frac{d\eta}{d\sigma}\frac{d}{d\xi}\left(\frac{1}{r}\frac{dz}{ds}\right). \tag{8}$$

If we now write for brevity

$$F = \int \frac{1}{r}\frac{dx}{ds}\,ds, \quad G = \int \frac{1}{r}\frac{dy}{ds}\,ds, \quad H = \int \frac{1}{r}\frac{dz}{ds}\,ds, \tag{9}$$

the integrals being taken once round the closed curve s, this term of Π may be written

$$\frac{d\eta}{d\sigma}\frac{d^2 H}{d\xi\, ds},$$

and the corresponding term of $\int \Pi\, ds$ will be

$$\frac{d\eta}{d\sigma}\frac{dH}{d\xi}.$$

Collecting all the terms of Π, we may now write

$$-\frac{d\omega}{d\sigma} = -\int \Pi\, ds$$

$$= \left(\frac{dH}{d\eta} - \frac{dG}{d\zeta}\right)\frac{d\xi}{d\sigma} + \left(\frac{dF}{d\zeta} - \frac{dH}{d\eta}\right)\frac{d\eta}{d\sigma} + \left(\frac{dG}{d\xi} - \frac{dF}{d\eta}\right)\frac{d\zeta}{d\sigma}. \tag{10}$$

This quantity is evidently the rate of decrement of ω, the magnetic potential, in passing along the curve σ, or in other words, it is the magnetic force in the direction of $d\sigma$.

By assuming $d\sigma$ successively in the direction of the axes of x, y and z, we obtain for the values of the components of the magnetic force

$$\left.\begin{aligned}\alpha &= -\frac{d\omega}{d\xi} = \frac{dH}{d\eta} - \frac{dG}{d\zeta},\\ \beta &= -\frac{d\omega}{d\eta} = \frac{dF}{d\zeta} - \frac{dH}{d\xi},\\ \gamma &= -\frac{d\omega}{d\zeta} = \frac{dG}{d\xi} - \frac{dF}{d\eta}.\end{aligned}\right\} \tag{11}$$

The quantities F, G, H are the components of the vector-potential of the magnetic shell whose strength is unity, and whose edge is the curve s. They are not, like the scalar potential ω, functions having a series of values, but are perfectly determinate for every point in space.

The vector-potential at a point P due to a magnetic shell bounded by a closed curve may be found by the following geometrical construction:

Let a point Q travel round the closed curve with a velocity numerically equal to its distance from P, and let a second point R start from A and travel with a velocity the direction of which is always parallel to that of Q, but whose magnitude is unity. When Q has travelled once round the closed curve join AR, then the line AR represents in direction and in numerical magnitude the vector-potential due to the closed curve at P.

Potential Energy of a Magnetic Shell placed in a Magnetic Field.

423.] We have already shewn, in Art. 410, that the potential energy of a shell of strength ϕ placed in a magnetic field whose potential is V, is

$$M = \phi \iint \left(l\frac{dV}{dx} + m\frac{dV}{dy} + n\frac{dV}{dz} \right) dS, \tag{12}$$

where l, m, n are the direction-cosines of the normal to the shell drawn from the positive side, and the surface-integral is extended over the shell.

Now this surface-integral may be transformed into a line-integral by means of the vector-potential of the magnetic field, and we may write

$$M = -\phi \int \left(F\frac{dx}{ds} + G\frac{dy}{ds} + H\frac{dz}{ds} \right) ds, \tag{13}$$

where the integration is extended once round the closed curve s which forms the edge of the magnetic shell, the direction of ds being opposite to that of the hands of a watch when viewed from the positive side of the shell.

If we now suppose that the magnetic field is that due to a

second magnetic shell whose strength is ϕ', the values of F, G, H will be

$$F = \phi' \int \frac{1}{r} \frac{dx}{ds'} ds', \qquad G = \phi' \int \frac{1}{r} \frac{dy}{ds'} ds', \qquad H = \phi' \int \frac{1}{r} \frac{dz}{ds'} ds', \quad (14)$$

where the integrations are extended once round the curve s', which forms the edge of this shell.

Substituting these values in the expression for M we find

$$M = -\phi\phi' \iint \frac{1}{r} \left(\frac{dx}{ds} \frac{dx}{ds'} + \frac{dy}{ds} \frac{dy}{ds'} + \frac{dz}{ds} \frac{dz}{ds'} \right) ds\, ds', \qquad (15)$$

where the integration is extended once round s and once round s'. This expression gives the potential energy due to the mutual action of the two shells, and is, as it ought to be, the same when s and s' are interchanged. This expression with its sign reversed, when the strength of each shell is unity, is called the potential of the two closed curves s and s'. It is a quantity of great importance in the theory of electric currents. If we write ϵ for the angle between the directions of the elements ds and ds', the potential of s and s' may be written

$$\iint \frac{\cos \epsilon}{r} ds\, ds'. \qquad (16)$$

It is evidently a quantity of the dimension of a line.

CHAPTER IV.

INDUCED MAGNETIZATION.

424.] We have hitherto considered the actual distribution of magnetization in a magnet as given explicitly among the data of the investigation. We have not made any assumption as to whether this magnetization is permanent or temporary, except in those parts of our reasoning in which we have supposed the magnet broken up into small portions, or small portions removed from the magnet in such a way as not to alter the magnetization of any part.

We have now to consider the magnetization of bodies with respect to the mode in which it may be produced and changed. A bar of iron held parallel to the direction of the earth's magnetic force is found to become magnetic, with its poles turned the opposite way from those of the earth, or the same way as those of a compass needle in stable equilibrium.

Any piece of soft iron placed in a magnetic field is found to exhibit magnetic properties. If it be placed in a part of the field where the magnetic force is great, as between the poles of a horse-shoe magnet, the magnetism of the iron becomes intense. If the iron is removed from the magnetic field, its magnetic properties are greatly weakened or disappear entirely. If the magnetic properties of the iron depend entirely on the magnetic force of the field in which it is placed, and vanish when it is removed from the field, it is called Soft iron. Iron which is soft in the magnetic sense is also soft in the literal sense. It is easy to bend it and give it a permanent set, and difficult to break it.

Iron which retains its magnetic properties when removed from the magnetic field is called Hard iron. Such iron does not take up the magnetic state so readily as soft iron. The operation of hammering, or any other kind of vibration, allows hard iron under the influence of magnetic force to assume the magnetic state more readily, and to part with it more readily when the magnetizing force is removed. Iron which is magnetically hard is also more stiff to bend and more apt to break.

The processes of hammering, rolling, wire-drawing, and sudden cooling tend to harden iron, and that of annealing tends to soften it.

The magnetic as well as the mechanical differences between steel of hard and soft temper are much greater than those between hard and soft iron. Soft steel is almost as easily magnetized and demagnetized as iron, while the hardest steel is the best material for magnets which we wish to be permanent.

Cast iron, though it contains more carbon than steel, is not so retentive of magnetization.

If a magnet could be constructed so that the distribution of its magnetization is not altered by any magnetic force brought to act upon it, it might be called a rigidly magnetized body. The only known body which fulfils this condition is a conducting circuit round which a constant electric current is made to flow.

Such a circuit exhibits magnetic properties, and may therefore be called an electromagnet, but these magnetic properties are not affected by the other magnetic forces in the field. We shall return to this subject in Part IV.

All actual magnets, whether made of hardened steel or of loadstone, are found to be affected by any magnetic force which is brought to bear upon them.

It is convenient, for scientific purposes, to make a distinction between the permanent and the temporary magnetization, defining the permanent magnetization as that which exists independently of the magnetic force, and the temporary magnetization as that which depends on this force. We must observe, however, that this distinction is not founded on a knowledge of the intimate nature of magnetizable substances: it is only the expression of an hypothesis introduced for the sake of bringing calculation to bear on the phenomena. We shall return to the physical theory of magnetization in Chapter VI.

425.] At present we shall investigate the temporary magnetization on the assumption that the magnetization of any particle of the substance depends solely on the magnetic force acting on that particle. This magnetic force may arise partly from external causes, and partly from the temporary magnetization of neighbouring particles.

A body thus magnetized in virtue of the action of magnetic force, is said to be magnetized by induction, and the magnetization is said to be induced by the magnetizing force.

The magnetization induced by a given magnetizing force differs in different substances. It is greatest in the purest and softest iron, in which the ratio of the magnetization to the magnetic force may reach the value 32, or even 45 *.

Other substances, such as the metals nickel and cobalt, are capable of an inferior degree of magnetization, and all substances when subjected to a sufficiently strong magnetic force, are found to give indications of polarity.

When the magnetization is in the same direction as the magnetic force, as in iron, nickel, cobalt, &c., the substance is called Paramagnetic, Ferromagnetic, or more simply Magnetic. When the induced magnetization is in the direction opposite to the magnetic force, as in bismuth, &c., the substance is said to be Diamagnetic.

In all these substances the ratio of the magnetization to the magnetic force which produces it is exceedingly small, being only about $-\frac{1}{400000}$ in the case of bismuth, which is the most highly diamagnetic substance known.

In crystallized, strained, and organized substances the direction of the magnetization does not always coincide with that of the magnetic force which produces it. The relation between the components of magnetization, referred to axes fixed in the body, and those of the magnetic force, may be expressed by a system of three linear equations. Of the nine coefficients involved in these equations we shall shew that only six are independent. The phenomena of bodies of this kind are classed under the name of Magnecrystallic phenomena.

When placed in a field of magnetic force, crystals tend to set themselves so that the axis of greatest paramagnetic, or of least diamagnetic, induction is parallel to the lines of magnetic force. See Art. 435.

In soft iron, the direction of the magnetization coincides with that of the magnetic force at the point, and for small values of the magnetic force the magnetization is nearly proportional to it.

* Thalén, *Nova Acta*, *Reg. Soc. Sc.*, Upsal., 1863.

As the magnetic force increases, however, the magnetization increases more slowly, and it would appear from experiments described in Chap. VI, that there is a limiting value of the magnetization, beyond which it cannot pass, whatever be the value of the magnetic force.

In the following outline of the theory of induced magnetism, we shall begin by supposing the magnetization proportional to the magnetic force, and in the same line with it.

Definition of the Coefficient of Induced Magnetization.

426.] Let $\mathfrak{H}$ be the magnetic force, defined as in Art. 398, at any point of the body, and let $\mathfrak{I}$ be the magnetization at that point, then the ratio of $\mathfrak{I}$ to $\mathfrak{H}$ is called the Coefficient of Induced Magnetization.

Denoting this coefficient by κ, the fundamental equation of induced magnetism is

$$\mathfrak{I} = \kappa \mathfrak{H}. \tag{1}$$

The coefficient κ is positive for iron and paramagnetic substances, and negative for bismuth and diamagnetic substances. It reaches the value 32 in iron, and it is said to be large in the case of nickel and cobalt, but in all other cases it is a very small quantity, not greater than 0.00001.

The force $\mathfrak{H}$ arises partly from the action of magnets external to the body magnetized by induction, and partly from the induced magnetization of the body itself. Both parts satisfy the condition of having a potential.

427.] Let V be the potential due to magnetism external to the body, let Ω be that due to the induced magnetization, then if U is the actual potential due to both causes

$$U = V + \Omega. \tag{2}$$

Let the components of the magnetic force $\mathfrak{H}$, resolved in the directions of x, y, z, be α, β, γ, and let those of the magnetization $\mathfrak{I}$ be A, B, C, then by equation (1),

$$\left.\begin{aligned} A &= \kappa \alpha, \\ B &= \kappa \beta, \\ C &= \kappa \gamma. \end{aligned}\right\} \tag{3}$$

Multiplying these equations by dx, dy, dz respectively, and adding, we find

$$A\,dx + B\,dy + C\,dz = \kappa(\alpha\,dx + \beta\,dy + \gamma\,dz).$$

But since α, β and γ are derived from the potential U, we may write the second member $-\kappa\, dU$.

Hence, if κ is constant throughout the substance, the first member must also be a complete differential of a function of x, y and z, which we shall call ϕ, and the equation becomes

$$d\phi = -\kappa\, dU, \tag{4}$$

where $$A = \frac{d\phi}{dx}, \quad B = \frac{d\phi}{dy}, \quad C = \frac{d\phi}{dz}. \tag{5}$$

The magnetization is therefore lamellar, as defined in Art. 412.

It was shewn in Art. 386 that if ρ is the volume-density of free magnetism,

$$\rho = -\left(\frac{dA}{dx} + \frac{dB}{dy} + \frac{dC}{dz}\right),$$

which becomes in virtue of equations (3),

$$\rho = -\kappa\left(\frac{d\alpha}{dx} + \frac{d\beta}{dy} + \frac{d\gamma}{dz}\right).$$

But, by Art. 77,

$$\frac{d\alpha}{dx} + \frac{d\beta}{dy} + \frac{d\gamma}{dz} = 4\pi\rho.$$

Hence $$(1 + 4\pi\kappa)\rho = 0,$$

whence $$\rho = 0 \tag{6}$$

throughout the substance, and the magnetization is therefore solenoidal as well as lamellar. See Art. 407.

There is therefore no free magnetism except on the bounding surface of the body. If ν be the normal drawn inwards from the surface, the magnetic surface-density is

$$\sigma = -\frac{d\phi}{d\nu}. \tag{7}$$

The potential Ω due to this magnetization at any point may therefore be found from the surface-integral

$$\Omega = \iint \frac{\sigma}{r}\, dS. \tag{8}$$

The value of Ω will be finite and continuous everywhere, and will satisfy Laplace's equation at every point both within and without the surface. If we distinguish by an accent the value of Ω outside the surface, and if ν' be the normal drawn outwards, we have at the surface

$$\Omega' = \Omega; \tag{9}$$

$$\frac{d\Omega}{d\nu} + \frac{d\Omega'}{d\nu'} = -4\pi\sigma, \text{ by Art. 78,}$$

$$= 4\pi\frac{d\phi}{d\nu}, \text{ by (7),}$$

$$= -4\pi\kappa\frac{dU}{d\nu}, \text{ by (4),}$$

$$= -4\pi\kappa\left(\frac{dV}{d\nu} + \frac{d\Omega}{d\nu}\right), \text{ by (2).}$$

We may therefore write the surface-condition

$$(1+4\pi\kappa)\frac{d\Omega}{d\nu} + \frac{d\Omega'}{d\nu'} + 4\pi\kappa\frac{dV}{d\nu} = 0. \tag{10}$$

Hence the determination of the magnetism induced in a homogeneous isotropic body, bounded by a surface S, and acted upon by external magnetic forces whose potential is V, may be reduced to the following mathematical problem.

We must find two functions Ω and Ω' satisfying the following conditions:

Within the surface S, Ω must be finite and continuous, and must satisfy Laplace's equation.

Outside the surface S, Ω' must be finite and continuous, it must vanish at an infinite distance, and must satisfy Laplace's equation.

At every point of the surface itself, $\Omega = \Omega'$, and the derivatives of Ω, Ω' and V with respect to the normal must satisfy equation (10).

This method of treating the problem of induced magnetism is due to Poisson. The quantity k which he uses in his memoirs is not the same as κ, but is related to it as follows:

$$4\pi\kappa(k-1) + 3k = 0. \tag{11}$$

The coefficient κ which we have here used was introduced by J. Neumann.

428.] The problem of induced magnetism may be treated in a different manner by introducing the quantity which we have called, with Faraday, the Magnetic Induction.

The relation between $\mathfrak{B}$, the magnetic induction, $\mathfrak{H}$, the magnetic force, and $\mathfrak{I}$, the magnetization, is expressed by the equation

$$\mathfrak{B} = \mathfrak{H} + 4\pi\mathfrak{I}. \tag{12}$$

The equation which expresses the induced magnetization in terms of the magnetic force is

$$\mathfrak{I} = \kappa\mathfrak{H}. \tag{13}$$

Hence, eliminating $\mathfrak{I}$, we find

$$\mathfrak{B} = (1 + 4\pi\kappa)\mathfrak{H} \tag{14}$$

as the relation between the magnetic induction and the magnetic force in substances whose magnetization is induced by magnetic force.

In the most general case κ may be a function, not only of the position of the point in the substance, but of the direction of the vector $\mathfrak{H}$, but in the case which we are now considering κ is a numerical quantity.

If we next write

$$\mu = 1 + 4\pi\kappa, \tag{15}$$

we may define μ as the ratio of the magnetic induction to the magnetic force, and we may call this ratio the magnetic inductive capacity of the substance, thus distinguishing it from κ, the coefficient of induced magnetization.

If we write U for the total magnetic potential compounded of V, the potential due to external causes, and Ω for that due to the induced magnetization, we may express a, b, c, the components of magnetic induction, and α, β, γ, the components of magnetic force, as follows:

$$\left.\begin{aligned} a &= \mu\alpha = -\mu\frac{dU}{dx}, \\ b &= \mu\beta = -\mu\frac{dU}{dy}, \\ c &= \mu\gamma = -\mu\frac{dU}{dz}. \end{aligned}\right\} \tag{16}$$

The components a, b, c satisfy the solenoidal condition

$$\frac{da}{dx} + \frac{db}{dy} + \frac{dc}{dz} = 0. \tag{17}$$

Hence, the potential U must satisfy Laplace's equation

$$\frac{d^2U}{dx^2} + \frac{d^2U}{dy^2} + \frac{d^2U}{dz^2} = 0 \tag{18}$$

at every point where μ is constant, that is, at every point within the homogeneous substance, or in empty space.

At the surface itself, if ν is a normal drawn towards the magnetic substance, and ν' one drawn outwards, and if the symbols of quantities outside the substance are distinguished by accents, the condition of continuity of the magnetic induction is

$$a\frac{d\nu}{dx} + b\frac{d\nu}{dy} + c\frac{d\nu}{dz} + a'\frac{d\nu'}{dx} + b'\frac{d\nu'}{dy} + c'\frac{d\nu'}{dz} = 0; \tag{19}$$

or, by equations (16),

$$\mu \frac{dU}{d\nu} + \mu' \frac{dU'}{d\nu'} = 0. \tag{20}$$

μ', the coefficient of induction outside the magnet, will be unity unless the surrounding medium be magnetic or diamagnetic.

If we substitute for U its value in terms of V and Ω, and for μ its value in terms of κ, we obtain the same equation (10) as we arrived at by Poisson's method.

The problem of induced magnetism, when considered with respect to the relation between magnetic induction and magnetic force, corresponds exactly with the problem of the conduction of electric currents through heterogeneous media, as given in Art. 309.

The magnetic force is derived from the magnetic potential, precisely as the electric force is derived from the electric potential.

The magnetic induction is a quantity of the nature of a flux, and satisfies the same conditions of continuity as the electric current does.

In isotropic media the magnetic induction depends on the magnetic force in a manner which exactly corresponds with that in which the electric current depends on the electromotive force.

The specific magnetic inductive capacity in the one problem corresponds to the specific conductivity in the other. Hence Thomson, in his *Theory of Induced Magnetism* (*Reprint*, 1872, p. 484), has called this quantity the *permeability* of the medium.

We are now prepared to consider the theory of induced magnetism from what I conceive to be Faraday's point of view.

When magnetic force acts on any medium, whether magnetic or diamagnetic, or neutral, it produces within it a phenomenon called Magnetic Induction.

Magnetic induction is a directed quantity of the nature of a flux, and it satisfies the same conditions of continuity as electric currents and other fluxes do.

In isotropic media the magnetic force and the magnetic induction are in the same direction, and the magnetic induction is the product of the magnetic force into a quantity called the coefficient of induction, which we have expressed by μ.

In empty space the coefficient of induction is unity. In bodies capable of induced magnetization the coefficient of induction is $1 + 4\pi\kappa = \mu$, where κ is the quantity already defined as the coefficient of induced magnetization.

429.] Let μ, μ' be the values of μ on opposite sides of a surface separating two media, then if V, V' are the potentials in the two media, the magnetic forces towards the surface in the two media are $\frac{dV}{d\nu}$ and $\frac{dV'}{d\nu'}$.

The quantities of magnetic induction through the element of surface dS are $\mu \frac{dV}{d\nu} dS$ and $\mu' \frac{dV'}{d\nu'} dS$ in the two media respectively reckoned towards dS.

Since the total flux towards dS is zero,

$$\mu \frac{dV}{d\nu} + \mu' \frac{dV'}{d\nu'} = 0.$$

But by the theory of the potential near a surface of density σ,

$$\frac{dV}{d\nu} + \frac{dV'}{d\nu'} + 4\pi\sigma = 0.$$

Hence
$$\frac{dV}{d\nu}\left(1 - \frac{\mu}{\mu'}\right) + 4\pi\sigma = 0.$$

If κ_1 is the ratio of the superficial magnetization to the normal force in the first medium whose coefficient is μ, we have

$$4\pi\kappa_1 = \frac{\mu - \mu'}{\mu'}.$$

Hence κ_1 will be positive or negative according as μ is greater or less than μ'. If we put $\mu = 4\pi\kappa + 1$ and $\mu' = 4\pi\kappa' + 1$,

$$\kappa_1 = \frac{\kappa - \kappa'}{4\pi\kappa' + 1}.$$

In this expression κ and κ' are the coefficients of induced magnetization of the first and second medium deduced from experiments made in air, and κ_1 is the coefficient of induced magnetization of the first medium when surrounded by the second medium.

If κ' is greater than κ, then κ_1 is negative, or the apparent magnetization of the first medium is in the opposite direction from the magnetizing force.

Thus, if a vessel containing a weak aqueous solution of a paramagnetic salt of iron is suspended in a stronger solution of the same salt, and acted on by a magnet, the vessel moves as if it were magnetized in the opposite direction from that in which a magnet would set itself if suspended in the same place.

This may be explained by the hypothesis that the solution in the vessel is really magnetized in the same direction as the magnetic force, but that the solution which surrounds the vessel is magnetized more strongly in the same direction. Hence the vessel is like a weak magnet placed between two strong ones all magnetized in the same direction, so that opposite poles are in contact. The north pole of the weak magnet points in the same direction as those of the strong ones, but since it is in contact with the south pole of a stronger magnet, there is an excess of south magnetism in the neighbourhood of its north pole, which causes the small magnet to appear oppositely magnetized.

In some substances, however, the apparent magnetization is negative even when they are suspended in what is called a vacuum.

If we assume $\kappa = 0$ for a vacuum, it will be negative for these substances. No substance, however, has been discovered for which κ has a negative value numerically greater than $\frac{1}{4\pi}$, and therefore for all known substances μ is positive.

Substances for which κ is negative, and therefore μ less than unity, are called Diamagnetic substances. Those for which κ is positive, and μ greater than unity, are called Paramagnetic, Ferromagnetic, or simply magnetic, substances.

We shall consider the physical theory of the diamagnetic and paramagnetic properties when we come to electromagnetism, Arts. 831–845.

430.] The mathematical theory of magnetic induction was first given by Poisson *. The physical hypothesis on which he founded his theory was that of two magnetic fluids, an hypothesis which has the same mathematical advantages and physical difficulties as the theory of two electric fluids. In order, however, to explain the fact that, though a piece of soft iron can be magnetized by induction, it cannot be charged with unequal quantities of the two kinds of magnetism, he supposes that the substance in general is a non-conductor of these fluids, and that only certain small portions of the substance contain the fluids under circumstances in which they are free to obey the forces which act on them. These small magnetic elements of the substance contain each precisely equal quantities of the two fluids, and within each element the fluids move with perfect freedom, but the fluids can never pass from one magnetic element to another.

The problem therefore is of the same kind as that relating to a number of small conductors of electricity disseminated through a dielectric insulating medium. The conductors may be of any form provided they are small and do not touch each other.

If they are elongated bodies all turned in the same general

* *Mémoires de l'Institut*, 1824.

direction, or if they are crowded more in one direction than another, the medium, as Poisson himself shews, will not be isotropic. Poisson therefore, to avoid useless intricacy, examines the case in which each magnetic element is spherical, and the elements are disseminated without regard to axes. He supposes that the whole volume of all the magnetic elements in unit of volume of the substance is k.

We have already considered in Art. 314 the electric conductivity of a medium in which small spheres of another medium are distributed.

If the conductivity of the medium is μ_1, and that of the spheres μ_2, we have found that the conductivity of the composite system is

$$\mu = \mu_1 \frac{2\mu_1 + \mu_2 + 2k(\mu_2 - \mu_1)}{2\mu_1 + \mu_2 - k(\mu_2 - \mu_1)}.$$

Putting $\mu_1 = 1$ and $\mu_2 = \infty$, this becomes

$$\mu = \frac{1+2k}{1-k}.$$

This quantity μ is the electric conductivity of a medium consisting of perfectly conducting spheres disseminated through a medium of conductivity unity, the aggregate volume of the spheres in unit of volume being k.

The symbol μ also represents the coefficient of magnetic induction of a medium, consisting of spheres for which the permeability is infinite, disseminated through a medium for which it is unity.

The symbol k, which we shall call Poisson's Magnetic Coefficient, represents the ratio of the volume of the magnetic elements to the whole volume of the substance.

The symbol κ is known as Neumann's Coefficient of Magnetization by Induction. It is more convenient than Poisson's.

The symbol μ we shall call the Coefficient of Magnetic Induction. Its advantage is that it facilitates the transformation of magnetic problems into problems relating to electricity and heat.

The relations of these three symbols are as follows:

$$k = \frac{4\pi\kappa}{4\pi\kappa + 3}, \qquad k = \frac{\mu - 1}{\mu + 2},$$

$$\kappa = \frac{\mu - 1}{4\pi}, \qquad \kappa = \frac{3k}{4\pi(1-k)},$$

$$\mu = \frac{1+2k}{1-k}, \qquad \mu = 4\pi\kappa + 1.$$

If we put $\kappa = 32$, the value given by Thalén's * experiments on

* *Recherches sur les Propriétés Magnétiques du fer, Nova Acta*, Upsal, 1863.

soft iron, we find $k = \frac{134}{135}$. This, according to Poisson's theory, is the ratio of the volume of the magnetic molecules to the whole volume of the iron. It is impossible to pack a space with equal spheres so that the ratio of their volume to the whole space shall be so nearly unity, and it is exceedingly improbable that so large a proportion of the volume of iron is occupied by solid molecules whatever be their form. This is one reason why we must abandon Poisson's hypothesis. Others will be stated in Chapter VI. Of course the value of Poisson's mathematical investigations remains unimpaired, as they do not rest on his hypothesis, but on the experimental fact of induced magnetization.

CHAPTER V.

PARTICULAR PROBLEMS IN MAGNETIC INDUCTION.

A Hollow Spherical Shell.

481.] THE first example of the complete solution of a problem in magnetic induction was that given by Poisson for the case of a hollow spherical shell acted on by any magnetic forces whatever.

For simplicity we shall suppose the origin of the magnetic forces to be in the space outside the shell.

If V denotes the potential due to the external magnetic system, we may expand V in a series of solid harmonics of the form

$$V = C_0 S_0 + C_1 S_1 r + \&c. + C_i S_i r^i, \tag{1}$$

where r is the distance from the centre of the shell, S_i is a surface harmonic of order i, and C_i is a coefficient.

This series will be convergent provided r is less than the distance of the nearest magnet of the system which produces this potential. Hence, for the hollow spherical shell and the space within it, this expansion is convergent.

Let the external radius of the shell be a_2 and the inner radius a_1, and let the potential due to its induced magnetism be Ω. The form of the function Ω will in general be different in the hollow space, in the substance of the shell, and in the space beyond. If we expand these functions in harmonic series, then, confining our attention to those terms which involve the surface harmonic S_i, we shall find that if Ω_1 is that which corresponds to the hollow space within the shell, the expansion of Ω_1 must be in positive harmonics of the form $A_1 S_i r^i$, because the potential must not become infinite within the sphere whose radius is a_1.

In the substance of the shell, where r_1 lies between a_1 and a_2, the series may contain both positive and negative powers of r, of the form

$$A_2 S_i r^i + B_2 S_i r^{-(i+1)}.$$

Outside the shell, where r is greater than a_2, since the series must be convergent however great r may be, we must have only negative powers of r, of the form

$$B_3 S_i r^{-(i+1)}.$$

The conditions which must be satisfied by the function Ω are: It must be (1) finite, and (2) continuous, and (3) must vanish at an infinite distance, and it must (4) everywhere satisfy Laplace's equation.

On account of (1) $B_1 = 0$.

On account of (2) when $r = a_1$,

$$(A_1 - A_2)\, a_1^{2i+1} - B_2 = 0, \tag{2}$$

and when $r = a_2$,

$$(A_2 - A_3)\, a_2^{2i+1} + B_2 - B_3 = 0. \tag{3}$$

On account of (3) $A_3 = 0$, and the condition (4) is satisfied everywhere, since the functions are harmonic.

But, besides these, there are other conditions to be satisfied at the inner and outer surface in virtue of equation (10), Art. 427.

At the inner surface where $r = a_1$,

$$(1 + 4\pi\kappa)\frac{d\Omega_2}{dr} - \frac{d\Omega_1}{dr} + 4\pi\kappa\frac{dV}{dr} = 0, \tag{4}$$

and at the outer surface where $r = a_2$,

$$-(1 + 4\pi\kappa)\frac{d\Omega_2}{dr} + \frac{d\Omega_3}{dr} - 4\pi\kappa\frac{dV}{dr} = 0. \tag{5}$$

From these conditions we obtain the equations

$$(1+4\pi\kappa)\left(iA_2 a_1^{2i+1} - (i+1)B_2\right) - iA_1 a_1^{2i+1} + 4\pi\kappa i C_i a_1^{2i+1} = 0, \tag{6}$$

$$(1+4\pi\kappa)\left(iA_2 a_2^{2i+1} - (i+1)B_2\right) + (i+1)B_3 + 4\pi\kappa i C_i a_2^{2i+1} = 0; \tag{7}$$

and if we put

$$N_i = \frac{1}{(1+4\pi\kappa)(2i+1)^2 + (4\pi\kappa)^2 i(i+1)\left(1 - \left(\frac{a_1}{a_2}\right)^{2i+1}\right)}, \tag{8}$$

we find

$$A_1 = -(4\pi\kappa)^2 i(i+1)\left(1 - \left(\frac{a_1}{a_2}\right)^{2i+1}\right) N_i C_i, \tag{9}$$

$$A_2 = -4\pi\kappa i\left[2i+1+4\pi\kappa(i+1)\left(1 - \left(\frac{a_1}{a_2}\right)^{2i+1}\right)\right] N_i C_i, \tag{10}$$

$$B_2 = 4\pi\kappa i(2i+1)a_1^{2i+1} N_i C_i, \tag{11}$$

$$B_3 = 4\pi\kappa i\left(2i+1+4\pi\kappa(i+1)\right)\left(a_2^{2i+1} - a_1^{2i+1}\right) N_i C_i. \tag{12}$$

These quantities being substituted in the harmonic expansions give the part of the potential due to the magnetization of the shell. The quantity N_i is always positive, since $1 + 4\pi\kappa$ can never be negative. Hence A_1 is always negative, or in other words, the action of the magnetized shell on a point within it is always opposed to that of the external magnetic force whether the shell be paramagnetic or diamagnetic. The actual value of the resultant potential within the shell is

$$(C_i + A_1) S_i r^i,$$

or
$$(1+4\pi\kappa)(2i+1)^2 N_i C_i S_i r. \tag{13}$$

432.] When κ is a large number, as it is in the case of soft iron, then, unless the shell is very thin, the magnetic force within it is but a small fraction of the external force.

In this way Sir W. Thomson has rendered his marine galvanometer independent of external magnetic force by enclosing it in a tube of soft iron.

433.] The case of greatest practical importance is that in which $i = 1$. In this case

$$N_1 = \frac{1}{9(1+4\pi\kappa)+2(4\pi\kappa)^2\left(1-\left(\frac{a_1}{a_2}\right)^3\right)}, \tag{14}$$

$$\left.\begin{aligned}
A_1 &= -2(4\pi\kappa)^2\left(1-\left(\frac{a_1}{a_2}\right)^3\right) N_1 C_1,\\
A_2 &= -4\pi\kappa\left[3+8\pi\kappa\left(1-\left(\frac{a_1}{a_2}\right)^3\right)\right] N_1 C_1,\\
B_2 &= 12\pi\kappa a_1^3 N_1 C_1,\\
B_3 &= 4\pi\kappa(3+8\pi\kappa)(a_2^3-a_1^3) N_1 C_1.
\end{aligned}\right\} \tag{15}$$

The magnetic force within the hollow shell is in this case uniform and equal to

$$C_1 + A_1 = \frac{9(1+4\pi\kappa)}{9(1+4\pi\kappa)+2(4\pi\kappa)^2\left(1-\left(\frac{a_1}{a_2}\right)^3\right)} C_1. \tag{16}$$

If we wish to determine κ by measuring the magnetic force within a hollow shell and comparing it with the external magnetic force, the best value of the thickness of the shell may be found from the equation

$$1 - \frac{a_1^3}{a_2^3} = \frac{9}{2}\frac{1+4\pi\kappa}{(4\pi\kappa)^2}. \tag{17}$$

The magnetic force inside the shell is then half of its value outside.

Since, in the case of iron, κ is a number between 20 and 30, the thickness of the shell ought to be about the hundredth part of its radius. This method is applicable only when the value of κ is large. When it is very small the value of A_1 becomes insensible, since it depends on the square of κ.

For a nearly solid sphere with a very small spherical hollow,

$$\left.\begin{aligned} A_1 &= -\frac{2(4\pi\kappa)^2}{(3+4\pi\kappa)(3+8\pi\kappa)}C_1, \\ A_2 &= -\frac{4\pi\kappa}{3+4\pi\kappa}C_1, \\ B_3 &= \frac{4\pi\kappa}{3+4\pi\kappa}C_1 a_2^3. \end{aligned}\right\} \quad (18)$$

The whole of this investigation might have been deduced directly from that of conduction through a spherical shell, as given in Art. 312, by putting $k_1 = (1+4\pi\kappa)k_2$ in the expressions there given, remembering that A_1 and A_2 in the problem of conduction are equivalent to $C_1 + A_1$ and $C_1 + A_2$ in the problem of magnetic induction.

434.] The corresponding solution in two dimensions is graphically represented in Fig. XV, at the end of this volume. The lines of induction, which at a distance from the centre of the figure are nearly horizontal, are represented as disturbed by a cylindric rod magnetized transversely and placed in its position of stable equilibrium. The lines which cut this system at right angles represent the equipotential surfaces, one of which is a cylinder. The large dotted circle represents the section of a cylinder of a paramagnetic substance, and the dotted horizontal straight lines within it, which are continuous with the external lines of induction, represent the lines of induction within the substance. The dotted vertical lines represent the internal equipotential surfaces, and are continuous with the external system. It will be observed that the lines of induction are drawn nearer together within the substance, and the equipotential surfaces are separated farther apart by the paramagnetic cylinder, which, in the language of Faraday, conducts the lines of induction better than the surrounding medium.

If we consider the system of vertical lines as lines of induction, and the horizontal system as equipotential surfaces, we have, in the first place, the case of a cylinder magnetized transversely and placed in the position of unstable equilibrium among the lines of force, which it causes to diverge. In the second place, considering the large dotted circle as the section of a diamagnetic cylinder, the dotted straight lines within it, together with the lines external to it, represent the effect of a diamagnetic substance in separating the lines of induction and drawing together the equipotential surfaces, such a substance being a worse conductor of magnetic induction than the surrounding medium.

Case of a Sphere in which the Coefficients of Magnetization are Different in Different Directions.

435.] Let α, β, γ be the components of magnetic force, and A, B, C those of the magnetization at any point, then the most general linear relation between these quantities is given by the equations

$$\left.\begin{aligned} A &= r_1\alpha + p_3\beta + q_2\gamma, \\ B &= q_3\alpha + r_2\beta + p_1\gamma, \\ C &= p_2\alpha + q_1 h_2 + r_3\gamma, \end{aligned}\right\} \qquad (1)$$

where the coefficients r, p, q are the nine coefficients of magnetization.

Let us now suppose that these are the conditions of magnetization within a sphere of radius a, and that the magnetization at every point of the substance is uniform and in the same direction, having the components A, B, C.

Let us also suppose that the external magnetizing force is also uniform and parallel to one direction, and has for its components X, Y, Z.

The value of V is therefore

$$V = -(Xx + Yy + Zz), \qquad (2)$$

and that of Ω' the potential of the magnetization outside the sphere is

$$\Omega' = (Ax + By + Cz)\frac{4\pi a^3}{3r^3}. \qquad (3)$$

The value of Ω, the potential of the magnetization within the sphere, is

$$\Omega = \frac{4\pi}{3}(Ax + By + Cz). \qquad (4)$$

The actual potential within the sphere is $V + \Omega$, so that we shall have for the components of the magnetic force within the sphere

$$\left.\begin{aligned} \alpha &= X - \tfrac{4}{3}\pi A, \\ \beta &= Y - \tfrac{4}{3}\pi B, \\ \gamma &= Z - \tfrac{4}{3}\pi C. \end{aligned}\right\} \qquad (5)$$

Hence

$$\left.\begin{aligned} (1+\tfrac{4}{3}\pi r_1)A + \tfrac{4}{3}\pi p_3 B + \tfrac{4}{3}\pi q_2 C &= r_1 X + p_3 Y + q_2 Z, \\ \tfrac{4}{3}\pi q_3 A + (1+\tfrac{4}{3}\pi r_2)B + \tfrac{4}{3}\pi p_1 C &= q_3 X + r_2 Y + p_1 Z, \\ \tfrac{4}{3}\pi p_2 A + \tfrac{4}{3}\pi q_1 B + (1+\tfrac{4}{3}\pi r_3)C &= p_2 X + q_1 Y + r_3 Z. \end{aligned}\right\} \qquad (6)$$

Solving these equations, we find

$$\left.\begin{aligned} A &= r_1' X + p_3' Y + q_2' Z, \\ B &= q_3' X + r_2' Y + p_1' Z, \\ C &= p_2' X + q_1' Y + r_3' Z, \end{aligned}\right\} \qquad (7)$$

where
$$\left.\begin{aligned} D' r_1' &= r_1 + \tfrac{4}{3}\pi (r_3 r_1 - p_2 q_2 + r_1 r_2 - p_3 q_3) + (\tfrac{4}{3}\pi)^2 D, \\ D' p_1' &= p_1 - \tfrac{4}{3}\pi (q_2 q_3 - p_1 r_1), \\ D' q_1' &= q_1 - \tfrac{4}{3}\pi (p_2 p_3 - q_1 r_1), \\ &\text{\&c.,} \end{aligned}\right\} \qquad (8)$$
where D is the determinant of the coefficients on the right side of equations (6), and D' that of the coefficients on the left.

The new system of coefficients p', q', r' will be symmetrical only when the system p, q, r is symmetrical, that is, when the coefficients of the form p are equal to the corresponding ones of the form q.

436.] The moment of the couple tending to turn the sphere about the axis of x from y towards z is
$$\begin{aligned} L &= \tfrac{4}{3}\pi a^3 (ZB - YC) \\ &= \tfrac{4}{3}\pi a^3 \{p_1' Z^2 - q_1' Y^2 + (r_2' - r_3') YZ + X(q_3' Z - p_2' Y)\}. \end{aligned} \qquad (9)$$
If we make
$$X = 0, \qquad Y = F\cos\theta, \qquad Y = F\sin\theta,$$
this corresponds to a magnetic force F in the plane of yz, and inclined to y at an angle θ. If we now turn the sphere while this force remains constant the work done in turning the sphere will be $\int_0^{2\pi} L\,d\theta$ in each complete revolution. But this is equal to
$$\tfrac{4}{3}\pi a^3 F^2 (p_1' - q_1'). \qquad (10)$$
Hence, in order that the revolving sphere may not become an inexhaustible source of energy, $p_1' = q_1'$, and similarly $p_2' = q_2'$ and $p_3' = q_3'$.

These conditions shew that in the original equations the coefficient of B in the third equation is equal to that of C in the second, and so on. Hence, the system of equations is symmetrical, and the equations become when referred to the principal axes of magnetization,
$$\left.\begin{aligned} A &= \frac{r_1}{1 + \frac{4}{3}\pi r_1} X, \\ B &= \frac{r_2}{1 + \frac{4}{3}\pi r_2} Y, \\ C &= \frac{r_3}{1 + \frac{4}{3}\pi r_3} Z. \end{aligned}\right\} \qquad (11)$$

The moment of the couple tending to turn the sphere round the axis of x is
$$L = \tfrac{4}{3}\pi a^3 \frac{r_2 - r_3}{(1 + \frac{4}{3}\pi r_2)(1 + \frac{4}{3}\pi r_3)} YZ. \qquad (12)$$

In most cases the differences between the coefficients of magnetization in different directions are very small, so that we may put

$$L = \tfrac{4}{3}\pi a^3 \frac{r_2 - r_3}{(1+\frac{4}{3}\pi r)^2} F^2 \sin 2\theta. \tag{13}$$

This is the force tending to turn a crystalline sphere about the axis of x from y towards z. It always tends to place the axis of greatest magnetic coefficient (or least diamagnetic coefficient) parallel to the line of magnetic force.

The corresponding case in two dimensions is represented in Fig. XVI.

If we suppose the upper side of the figure to be towards the north, the figure represents the lines of force and equipotential surfaces as disturbed by a transversely magnetized cylinder placed with the north side eastwards. The resultant force tends to turn the cylinder from east to north. The large dotted circle represents a section of a cylinder of a crystalline substance which has a larger coefficient of induction along an axis from north-east to south-west than along an axis from north-west to south-east. The dotted lines within the circle represent the lines of induction and the equipotential surfaces, which in this case are not at right angles to each other. The resultant force on the cylinder is evidently to turn it from east to north.

437.] The case of an ellipsoid placed in a field of uniform and parallel magnetic force has been solved in a very ingenious manner by Poisson.

If V is the potential at the point (x, y, z), due to the gravitation of a body of any form of uniform density ρ, then $-\frac{dV}{dx}$ is the potential of the magnetism of the same body if uniformly magnetized in the direction of x with the intensity $I = \rho$.

For the value of $-\frac{dV}{dx}\delta x$ at any point is the excess of the value of V, the potential of the body, above V', the value of the potential when the body is moved $-\delta x$ in the direction of x.

If we supposed the body shifted through the distance $-\delta x$, and its density changed from ρ to $-\rho$ (that is to say, made of repulsive instead of attractive matter,) then $-\frac{dV}{dx}\delta x$ would be the potential due to the two bodies.

Now consider any elementary portion of the body containing a volume δv. Its quantity is $\rho\,\delta v$, and corresponding to it there is an element of the shifted body whose quantity is $-\rho\,\delta v$ at a distance $-\delta x$. The effect of these two elements is equivalent to that of a magnet of strength $\rho\,\delta r$ and length δx. The intensity of magnetization is found by dividing the magnetic moment of an element by its volume. The result is $\rho\,\delta x$.

Hence $-\dfrac{dV}{dx}\delta x$ is the magnetic potential of the body magnetized with the intensity $\rho\,\delta x$ in the direction of x, and $-\dfrac{dV}{dx}$ is that of the body magnetized with intensity ρ.

This potential may be also considered in another light. The body was shifted through the distance $-\delta x$ and made of density $-\rho$. Throughout that part of space common to the body in its two positions the density is zero, for, as far as attraction is concerned, the two equal and opposite densities annihilate each other. There remains therefore a shell of positive matter on one side and of negative matter on the other, and we may regard the resultant potential as due to these. The thickness of the shell at a point where the normal drawn outwards makes an angle ϵ with the axis of x is $\delta x \cos\epsilon$ and its density is ρ. The surface-density is therefore $\rho\,\delta x \cos\epsilon$, and, in the case in which the potential is $-\dfrac{dV}{dx}$, the surface-density is $\rho\cos\epsilon$.

In this way we can find the magnetic potential of any body uniformly magnetized parallel to a given direction. Now if this uniform magnetization is due to magnetic induction, the magnetizing force at all points within the body must also be uniform and parallel.

This force consists of two parts, one due to external causes, and the other due to the magnetization of the body. If therefore the external magnetic force is uniform and parallel, the magnetic force due to the magnetization must also be uniform and parallel for all points within the body.

Hence, in order that this method may lead to a solution of the problem of magnetic induction, $\dfrac{dV}{dx}$ must be a linear function of the coordinates x, y, z within the body, and therefore V must be a quadratic function of the coordinates.

Now the only cases with which we are acquainted in which V is a quadratic function of the coordinates within the body are those in which the body is bounded by a complete surface of the second degree, and the only case in which such a body is of finite dimensions is when it is an ellipsoid. We shall therefore apply the method to the case of an ellipsoid.

Let
$$\frac{x^2}{a^2}+\frac{y^2}{b^2}+\frac{z^2}{c^2}=1 \tag{1}$$
be the equation of the ellipsoid, and let Φ_0 denote the definite integral
$$\int_0^\infty \frac{d(\phi^2)}{\sqrt{(a^2+\phi^2)(b^2+\phi^2)(c^2+\phi^2)}}\,*. \tag{2}$$

Then if we make
$$L=2\pi abc\frac{d\Phi_0}{d(a^2)},\qquad M=2\pi abc\frac{d\Phi_0}{d(b^2)},\qquad N=2\pi abc\frac{d\Phi_0}{d(c^2)}, \tag{3}$$
the value of the potential within the ellipsoid will be
$$V_0=-\frac{\rho}{2}(Lx^2+My^2+Nz^2)+\text{const.} \tag{4}$$

If the ellipsoid is magnetized with uniform intensity I in a direction making angles whose cosines are l, m, n with the axes of x, y, z, so that the components of magnetization are
$$A=Il,\qquad B=Im,\qquad C=In,$$
the potential due to this magnetization within the ellipsoid will be
$$\Omega=-I(Llx+Mmy+Nnz). \tag{5}$$

If the external magnetizing force is $\mathfrak{H}$, and if its components are α, β, γ, its potential will be
$$V=Xx+Yy+Zz. \tag{6}$$

The components of the actual magnetizing force at any point within the body are therefore
$$X-AL,\qquad Y-BM,\qquad Z-CN. \tag{7}$$

The most general relations between the magnetization and the magnetizing force are given by three linear equations, involving nine coefficients. It is necessary, however, in order to fulfil the condition of the conservation of energy, that in the case of magnetic induction three of these should be equal respectively to other three, so that we should have
$$\left.\begin{aligned}A&=K_1(X-AL)+K'_3(Y-BM)+K'_2(Z-CN),\\ B&=K'_3(X-AL)+K_2(Y-BM)+K'_1(Z-CN),\\ C&=K'_2(X-AL)+K'_1(Y-BM)+K_3(Z-CN).\end{aligned}\right\} \tag{8}$$

From these equations we may determine A, B and C in terms of X, Y, Z, and this will give the most general solution of the problem.

The potential outside the ellipsoid will then be that due to the

* See Thomson and Tait's *Natural Philosophy*, § 522.

magnetization of the ellipsoid together with that due to the external magnetic force.

438.] The only case of practical importance is that in which

$$\kappa'_1 = \kappa'_2 = \kappa'_3 = 0. \tag{9}$$

We have then

$$\left.\begin{aligned} A &= \frac{\kappa_1}{1+\kappa_1 L} X, \\ B &= \frac{\kappa_2}{1+\kappa_2 M} Y, \\ C &= \frac{\kappa_3}{1+\kappa_3 N} Z. \end{aligned}\right\} \tag{10}$$

If the ellipsoid has two axes equal, and is of the planetary or flattened form,

$$b = c = \frac{a}{\sqrt{1-e^2}}; \tag{11}$$

$$\left.\begin{aligned} L &= 4\pi\left(\frac{1}{e^2} - \frac{\sqrt{1-e^2}}{e^3}\sin^{-1}e\right), \\ M = N &= 2\pi\left(\frac{\sqrt{1-e^2}}{e^3}\sin^{-1}e - \frac{1-e^2}{e^2}\right). \end{aligned}\right\} \tag{12}$$

If the ellipsoid is of the ovary or elongated form

$$a = b = \sqrt{1-e^2}\,c; \tag{13}$$

$$\left.\begin{aligned} L = M &= 2\pi\left(\frac{1}{e^2} - \frac{1-e^2}{2e^3}\log\frac{1+e}{1-e}\right), \\ N &= 4\pi\left(\frac{1}{e^2} - 1\right)\left(\frac{1}{2e}\log\frac{1+e}{1-e} - 1\right). \end{aligned}\right\} \tag{14}$$

In the case of a sphere, when $e = 0$,

$$L = M = N = \tfrac{4}{3}\pi. \tag{15}$$

In the case of a very flattened planetoid L becomes in the limit equal to 4π, and M and N become $\pi^2\frac{a}{c}$.

In the case of a very elongated ovoid L and M approximate to the value 2π, while N approximates to the form

$$4\pi\frac{a^2}{c^2}\left(\log\frac{2c}{a} - 1\right)$$

and vanishes when $e = 1$.

It appears from these results that—

(1) When κ, the coefficient of magnetization, is very small, whether positive or negative, the induced magnetization is nearly equal to the magnetizing force multiplied by κ, and is almost independent of the form of the body.

(2) When κ is a large positive quantity, the magnetization depends principally on the form of the body, and is almost independent of the precise value of κ, except in the case of a longitudinal force acting on an ovoid so elongated that $N\kappa$ is a small quantity though κ is large.

(3) If the value of κ could be negative and equal to $\frac{1}{4\pi}$ we should have an infinite value of the magnetization in the case of a magnetizing force acting normally to a flat plate or disk. The absurdity of this result confirms what we said in Art. 428.

Hence, experiments to determine the value of κ may be made on bodies of any form provided κ is very small, as it is in the case of all diamagnetic bodies, and all magnetic bodies except iron, nickel, and cobalt.

If, however, as in the case of iron, κ is a large number, experiments made on spheres or flattened figures are not suitable to determine κ; for instance, in the case of a sphere the ratio of the magnetization to the magnetizing force is as 1 to 4.22 if $\kappa = 30$, as it is in some kinds of iron, and if κ were infinite the ratio would be as 1 to 4.19, so that a very small error in the determination of the magnetization would introduce a very large one in the value of κ.

But if we make use of a piece of iron in the form of a very elongated ovoid, then, as long as $N\kappa$ is of moderate value compared with unity, we may deduce the value of κ from a determination of the magnetization, and the smaller the value of N the more accurate will be the value of κ.

In fact, if $N\kappa$ be made small enough, a small error in the value of N itself will not introduce much error, so that we may use any elongated body, such as a wire or long rod, instead of an ovoid.

We must remember, however, that it is only when the product $N\kappa$ is small compared with unity that this substitution is allowable. In fact the distribution of magnetism on a long cylinder with flat ends does not resemble that on a long ovoid, for the free magnetism is very much concentrated towards the ends of the cylinder, whereas it varies directly as the distance from the equator in the case of the ovoid.

The distribution of electricity on a cylinder, however, is really comparable with that on an ovoid, as we have already seen, Art. 152.

These results also enable us to understand why the magnetic moment of a permanent magnet can be made so much greater when the magnet has an elongated form. If we were to magnetize a disk with intensity I in a direction normal to its surface, and then leave it to itself, the interior particles would experience a constant demagnetizing force equal to $4\pi I$, and this, if not sufficient of itself to destroy part of the magnetization, would soon do so if aided by vibrations or changes of temperature.

If we were to magnetize a cylinder transversely the demagnetizing force would be only $2\pi I$.

If the magnet were a sphere the demagnetizing force would be $\frac{4}{3}\pi I$.

In a disk magnetized transversely the demagnetizing force is $\pi^2 \frac{a}{c} I$, and in an elongated ovoid magnetized longitudinally it is least of all, being $4\pi \frac{a^2}{c^2} I \log \frac{2c}{a}$.

Hence an elongated magnet is less likely to lose its magnetism than a short thick one.

The moment of the force acting on an ellipsoid having different magnetic coefficients for the three axes which tends to turn it about the axis of x, is

$$\tfrac{4}{3}\pi abc(BZ - CY) = \tfrac{4}{3}\pi abc YZ \frac{\kappa_3 - \kappa_2 + \kappa_2\kappa_3(M-N)}{(1-\kappa_2 M)(1-\kappa_3 N)}.$$

Hence, if κ_2 and κ_3 are small, this force will depend principally on the crystalline quality of the body and not on its shape, provided its dimensions are not very unequal, but if κ_2 and κ_3 are considerable, as in the case of iron, the force will depend principally on the shape of the body, and it will turn so as to set its longer axis parallel to the lines of force.

If a sufficiently strong, yet uniform, field of magnetic force could be obtained, an elongated isotropic diamagnetic body would also set itself with its longest dimension parallel to the lines of magnetic force.

439.] The question of the distribution of the magnetization of an ellipsoid of revolution under the action of any magnetic forces has been investigated by J. Neumann *. Kirchhoff † has extended the method to the case of a cylinder of infinite length acted on by any force.

* *Crelle*, bd. xxxvii (1848).
† *Crelle*, bd. xlviii (1854).

Green, in the 17th section of his Essay, has given an investigation of the distribution of magnetism in a cylinder of finite length acted on by a uniform external force parallel to its axis. Though some of the steps of this investigation are not very rigorous, it is probable that the result represents roughly the actual magnetization in this most important case. It certainly expresses very fairly the transition from the case of a cylinder for which κ is a large number to that in which it is very small, but it fails entirely in the case in which κ is negative, as in diamagnetic substances.

Green finds that the linear density of free magnetism at a distance x from the middle of a cylinder whose radius is a and whose length is $2l$, is

$$\lambda = \pi\kappa X p a \frac{e^{\frac{px}{a}} - e^{-\frac{px}{a}}}{e^{\frac{pl}{a}} + e^{-\frac{pl}{a}}},$$

where p is a numerical quantity to be found from the equation

$$0.231863 - 2\log_e p + 2p = \frac{1}{\pi\kappa p^2}.$$

The following are a few of the corresponding values of p and κ.

κ	p	κ	p
∞	0	11.802	0.07
336.4	0.01	9.137	0.08
62.02	0.02	7.517	0.09
48.416	0.03	6.319	0.10
29.475	0.04	0.1427	1.00
20.185	0.05	0.0002	10.00
14.794	0.06	0.0000	∞
		negative	imaginary.

When the length of the cylinder is great compared with its radius, the whole quantity of free magnetism on either side of the middle of the cylinder is, as it ought to be,

$$M = \pi^2 a \kappa X.$$

Of this $\frac{1}{2}pM$ is on the flat end of the cylinder, and the distance of the centre of gravity of the whole quantity M from the end of the cylinder is $\frac{a}{p}$.

When κ is very small p is large, and nearly the whole free magnetism is on the ends of the cylinder. As κ increases p diminishes, and the free magnetism is spread over a greater distance from the ends. When κ is infinite the free magnetism at any point of the cylinder is simply proportional to its distance from the middle point, the distribution being similar to that of free electricity on a conductor in a field of uniform force.

440.] In all substances except iron, nickel, and cobalt, the coefficient of magnetization is so small that the induced magnetization of the body produces only a very slight alteration of the forces in the magnetic field. We may therefore assume, as a first approximation, that the actual magnetic force within the body is the same as if the body had not been there. The superficial magnetization of the body is therefore, as a first approximation, $\kappa\frac{dV}{d\nu}$, where $\frac{dV}{d\nu}$ is the rate of increase of the magnetic potential due to the external magnet along a normal to the surface drawn inwards. If we now calculate the potential due to this superficial distribution, we may use it in proceeding to a second approximation.

To find the mechanical energy due to the distribution of magnetism on this first approximation we must find the surface-integral

$$E = \iint \kappa V \frac{dV}{d\nu} dS$$

taken over the whole surface of the body. Now we have shewn in Art. 100 that this is equal to the volume-integral

$$E = -\iiint \kappa \left(\overline{\frac{dV}{dx}}\Big|^2 + \overline{\frac{dV}{dy}}\Big|^2 + \overline{\frac{dV}{dz}}\Big|^2\right) dx\,dy\,dz$$

taken through the whole space occupied by the body, or, if R is the resultant magnetic force,

$$E = -\iiint \kappa R^2\, dx\,dy\,dz.$$

Now since the work done by the magnetic force on the body during a displacement δx is $X\delta x$ where X is the mechanical force in the direction of x, and since

$$\int X\delta x + E = \text{constant},$$

$$X = -\frac{dE}{dx} = \frac{d}{dx}\iiint \kappa R^2\, dx\,dy\,dz = \iiint \kappa \frac{d.R^2}{dx}\, dx\,dy\,dz,$$

which shews that the force acting on the body is as if every part of it tended to move from places where R^2 is less to places where it is greater with a force which on every unit of volume is

$$\kappa\frac{d.R^2}{dx}.$$

If κ is negative, as in diamagnetic bodies, this force is, as Faraday first shewed, from stronger to weaker parts of the magnetic field. Most of the actions observed in the case of diamagnetic bodies depend on this property.

Ship's Magnetism.

441.] Almost every part of magnetic science finds its use in navigation. The directive action of the earth's magnetism on the compass needle is the only method of ascertaining the ship's course when the sun and stars are hid. The declination of the needle from the true meridian seemed at first to be a hindrance to the application of the compass to navigation, but after this difficulty had been overcome by the construction of magnetic charts it appeared likely that the declination itself would assist the mariner in determining his ship's place.

The greatest difficulty in navigation had always been to ascertain the longitude; but since the declination is different at different points on the same parallel of latitude, an observation of the declination together with a knowledge of the latitude would enable the mariner to find his position on the magnetic chart.

But in recent times iron is so largely used in the construction of ships that it has become impossible to use the compass at all without taking into account the action of the ship, as a magnetic body, on the needle.

To determine the distribution of magnetism in a mass of iron of any form under the influence of the earth's magnetic force, even though not subjected to mechanical strain or other disturbances, is, as we have seen, a very difficult problem.

In this case, however, the problem is simplified by the following considerations.

The compass is supposed to be placed with its centre at a fixed point of the ship, and so far from any iron that the magnetism of the needle does not induce any perceptible magnetism in the ship. The size of the compass needle is supposed so small that we may regard the magnetic force at any point of the needle as the same.

The iron of the ship is supposed to be of two kinds only.

(1) Hard iron, magnetized in a constant manner.

(2) Soft iron, the magnetization of which is induced by the earth or other magnets.

In strictness we must admit that the hardest iron is not only capable of induction but that it may lose part of its so-called permanent magnetization in various ways.

The softest iron is capable of retaining what is called residual magnetization. The actual properties of iron cannot be accurately represented by supposing it compounded of the hard iron and the soft iron above defined. But it has been found that when a ship is acted on only by the earth's magnetic force, and not subjected to any extraordinary stress of weather, the supposition that the magnetism of the ship is due partly to permanent magnetization and partly to induction leads to sufficiently accurate results when applied to the correction of the compass.

The equations on which the theory of the variation of the compass is founded were given by Poisson in the fifth volume of the *Mémoires de l'Institut*, p. 533 (1824).

The only assumption relative to induced magnetism which is involved in these equations is, that if a magnetic force X due to external magnetism produces in the iron of the ship an induced magnetization, and if this induced magnetization exerts on the compass needle a disturbing force whose components are X', Y', Z', then, if the external magnetic force is altered in a given ratio, the components of the disturbing force will be altered in the same ratio.

It is true that when the magnetic force acting on iron is very great the induced magnetization is no longer proportional to the external magnetic force, but this want of proportionality is quite insensible for magnetic forces of the magnitude of those due to the earth's action.

Hence, in practice we may assume that if a magnetic force whose value is unity produces through the intervention of the iron of the ship a disturbing force at the compass needle whose components are a in the direction of x, d in that of y, and g in that of z, the components of the disturbing force due to a force X in the direction of x will be aX, dX, and gX.

If therefore we assume axes fixed in the ship, so that x is towards the ship's head, y to the starboard side, and z towards the keel, and if X, Y, Z represent the components of the earth's magnetic force in these directions, and X', Y', Z' the components of the combined magnetic force of the earth and ship on the compass needle,

$$\left.\begin{aligned} X' &= X + aX + bY + c\,Z + P, \\ Y' &= Y + dX + eY + f\,Z + Q, \\ Z' &= Z + gX + hY + k\,Z + R. \end{aligned}\right\} \qquad (1)$$

In these equations $a, b, c, d, e, f, g, h, k$ are nine constant coefficients depending on the amount, the arrangement, and the capacity for induction of the soft iron of the ship.

P, Q, and R are constant quantities depending on the permanent magnetization of the ship.

It is evident that these equations are sufficiently general if magnetic induction is a linear function of magnetic force, for they are neither more nor less than the most general expression of a vector as a linear function of another vector.

It may also be shewn that they are not too general, for, by a proper arrangement of iron, any one of the coefficients may be made to vary independently of the others.

Thus, a long thin rod of iron under the action of a longitudinal magnetic force acquires poles, the strength of each of which is numerically equal to the cross section of the rod multiplied by the magnetizing force and by the coefficient of induced magnetization. A magnetic force transverse to the rod produces a much feebler magnetization, the effect of which is almost insensible at a distance of a few diameters.

If a long iron rod be placed fore and aft with one end at a distance x from the compass needle, measured towards the ship's head, then, if the section of the rod is A, and its coefficient of magnetization κ, the strength of the pole will be $A\kappa X$, and, if $A = \frac{a x^2}{\kappa}$, the force exerted by this pole on the compass needle will be aX. The rod may be supposed so long that the effect of the other pole on the compass may be neglected.

We have thus obtained the means of giving any required value to the coefficient a.

If we place another rod of section B with one extremity at the same point, distant x from the compass toward the head of the vessel, and extending to starboard to such a distance that the distant pole produces no sensible effect on the compass, the disturbing force due to this rod will be in the direction of x, and equal to $\frac{B\kappa Y}{x^2}$, or if $B = \frac{b x^2}{\kappa}$, the force will be bY.

This rod therefore introduces the coefficient b.

A third rod extending downwards from the same point will introduce the coefficient c.

The coefficients d, e, f may be produced by three rods extending to head, to starboard, and downward from a point to starboard of the compass, and g, h, k by three rods in parallel directions from a point below the compass.

Hence each of the nine coefficients can be separately varied by means of iron rods properly placed.

The quantities P, Q, R are simply the components of the force on the compass arising from the permanent magnetization of the ship together with that part of the induced magnetization which is due to the action of this permanent magnetization.

A complete discussion of the equations (1), and of the relation between the true magnetic course of the ship and the course as indicated by the compass, is given by Mr. Archibald Smith in the Admiralty *Manual of the Deviation of the Compass.*

A valuable graphic method of investigating the problem is there given. Taking a fixed point as origin, a line is drawn from this point representing in direction and magnitude the horizontal part of the actual magnetic force on the compass-needle. As the ship is swung round so as to bring her head into different azimuths in succession, the extremity of this line describes a curve, each point of which corresponds to a particular azimuth.

Such a curve, by means of which the direction and magnitude of the force on the compass is given in terms of the magnetic course of the ship, is called a Dygogram.

There are two varieties of the Dygogram. In the first, the curve is traced on a plane fixed in space as the ship turns round. In the second kind, the curve is traced on a plane fixed with respect to the ship.

The dygogram of the first kind is the Limaçon of Pascal, that of the second kind is an ellipse. For the construction and use of these curves, and for many theorems as interesting to the mathematician as they are important to the navigator, the reader is referred to the Admiralty *Manual of the Deviation of the Compass.*

CHAPTER VI.

WEBER'S THEORY OF INDUCED MAGNETISM.

442.] We have seen that Poisson supposes the magnetization of iron to consist in a separation of the magnetic fluids within each magnetic molecule. If we wish to avoid the assumption of the existence of magnetic fluids, we may state the same theory in another form, by saying that each molecule of the iron, when the magnetizing force acts on it, becomes a magnet.

Weber's theory differs from this in assuming that the molecules of the iron are always magnets, even before the application of the magnetizing force, but that in ordinary iron the magnetic axes of the molecules are turned indifferently in every direction, so that the iron as a whole exhibits no magnetic properties.

When a magnetic force acts on the iron it tends to turn the axes of the molecules all in one direction, and so to cause the iron, as a whole, to become a magnet.

If the axes of all the molecules were set parallel to each other, the iron would exhibit the greatest intensity of magnetization of which it is capable. Hence Weber's theory implies the existence of a limiting intensity of magnetization, and the experimental evidence that such a limit exists is therefore necessary to the theory. Experiments shewing an approach to a limiting value of magnetization have been made by Joule * and by J. Müller †.

The experiments of Beetz ‡ on electrotype iron deposited under the action of magnetic force furnish the most complete evidence of this limit,—

A silver wire was varnished, and a very narrow line on the

* *Annals of Electricity*, iv. p. 131, 1839 ; *Phil. Mag.* [4] ii. p. 316.
† Pogg., *Ann.* lxxix. p. 337, 1850.
‡ Pogg. cxi. 1860.

metal was laid bare by making a fine longitudinal scratch on the varnish. The wire was then immersed in a solution of a salt of iron, and placed in a magnetic field with the scratch in the direction of a line of magnetic force. By making the wire the cathode of an electric current through the solution, iron was deposited on the narrow exposed surface of the wire, molecule by molecule. The filament of iron thus formed was then examined magnetically. Its magnetic moment was found to be very great for so small a mass of iron, and when a powerful magnetizing force was made to act in the same direction the increase of temporary magnetization was found to be very small, and the permanent magnetization was not altered. A magnetizing force in the reverse direction at once reduced the filament to the condition of iron magnetized in the ordinary way.

Weber's theory, which supposes that in this case the magnetizing force placed the axis of each molecule in the same direction during the instant of its deposition, agrees very well with what is observed.

Beetz found that when the electrolysis is continued under the action of the magnetizing force the intensity of magnetization of the subsequently deposited iron diminishes. The axes of the molecules are probably deflected from the line of magnetizing force when they are being laid down side by side with the molecules already deposited, so that an approximation to parallelism can be obtained only in the case of a very thin filament of iron.

If, as Weber supposes, the molecules of iron are already magnets, any magnetic force sufficient to render their axes parallel as they are electrolytically deposited will be sufficient to produce the highest intensity of magnetization in the deposited filament.

If, on the other hand, the molecules of iron are not magnets, but are only capable of magnetization, the magnetization of the deposited filament will depend on the magnetizing force in the same way in which that of soft iron in general depends on it. The experiments of Beetz leave no room for the latter hypothesis.

443.] We shall now assume, with Weber, that in every unit of volume of the iron there are n magnetic molecules, and that the magnetic moment of each is m. If the axes of all the molecules were placed parallel to one another, the magnetic moment of the unit of volume would be

$$M = nm,$$

and this would be the greatest intensity of magnetization of which the iron is capable.

In the unmagnetized state of ordinary iron Weber supposes the axes of its molecules to be placed indifferently in all directions.

To express this, we may suppose a sphere to be described, and a radius drawn from the centre parallel to the direction of the axis of each of the n molecules. The distribution of the extremities of these radii will express that of the axes of the molecules. In the case of ordinary iron these n points are equally distributed over every part of the surface of the sphere, so that the number of molecules whose axes make an angle less than a with the axis of x is

$$\frac{n}{2}(1-\cos a),$$

and the number of molecules whose axes make angles with that of x, between a and $a+da$ is therefore

$$\frac{n}{2}\sin a\, da.$$

This is the arrangement of the molecules in a piece of iron which has never been magnetized.

Let us now suppose that a magnetic force X is made to act on the iron in the direction of the axis of x, and let us consider a molecule whose axis was originally inclined a to the axis of x.

If this molecule is perfectly free to turn, it will place itself with its axis parallel to the axis of x, and if all the molecules did so, the very slightest magnetizing force would be found sufficient to develope the very highest degree of magnetization. This, however, is not the case.

The molecules do not turn with their axes parallel to x, and this is either because each molecule is acted on by a force tending to preserve it in its original direction, or because an equivalent effect is produced by the mutual action of the entire system of molecules.

Weber adopts the former of these suppositions as the simplest, and supposes that each molecule, when deflected, tends to return to its original position with a force which is the same as that which a magnetic force D, acting in the original direction of its axis, would produce.

The position which the axis actually assumes is therefore in the direction of the resultant of X and D.

Let APB represent a section of a sphere whose radius represents, on a certain scale, the force D.

Let the radius OP be parallel to the axis of a particular molecule in its original position.

Let SO represent on the same scale the magnetizing force X which is supposed to act from S towards O. Then, if the molecule is acted on by the force X in the direction SO, and by a force D in a direction parallel to OP, the original direction of its axis, its axis will set itself in the direction SP, that of the resultant of X and D.

Since the axes of the molecules are originally in all directions, P may be at any point of the sphere indifferently. In Fig. 5, in which X is less than D, SP, the final position of the axis, may be in any direction whatever, but not indifferently, for more of the molecules will have their axes turned towards A than towards B. In Fig. 6, in which X is greater than D, the axes of the molecules will be all confined within the cone STT' touching the sphere.

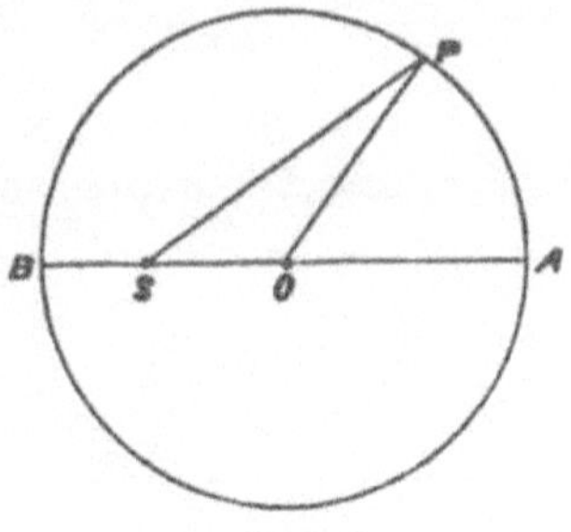

Fig. 5.

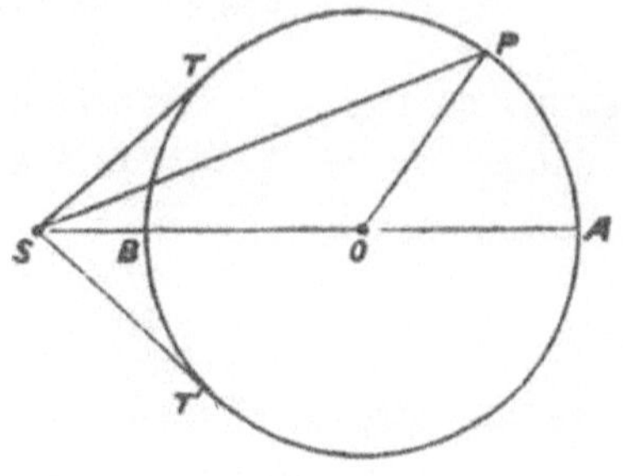

Fig. 6.

Hence there are two different cases according as X is less or greater than D.

Let $\alpha = AOP$, the original inclination of the axis of a molecule to the axis of x.

$\theta = ASP$, the inclination of the axis when deflected by the force X.

$\beta = SPO$, the angle of deflexion.

$SO = X$, the magnetizing force.

$OP = D$, the force tending towards the original position.

$SP = R$, the resultant of X and D.

m = magnetic moment of the molecule.

Then the moment of the statical couple due to X, tending to diminish the angle θ, is

$$mL = mX \sin\theta,$$

and the moment of the couple due to D, tending to increase θ, is

$$mL = mD \sin\beta.$$

Equating these values, and remembering that $\beta = \alpha - \theta$, we find

$$\tan\theta = \frac{D\sin\alpha}{X + D\cos\alpha} \tag{1}$$

to determine the direction of the axis after deflexion.

We have next to find the intensity of magnetization produced in the mass by the force X, and for this purpose we must resolve the magnetic moment of every molecule in the direction of x, and add all these resolved parts.

The resolved part of the moment of a molecule in the direction of x is

$$m\cos\theta.$$

The number of molecules whose original inclinations lay between α and $\alpha + d\alpha$ is

$$\frac{n}{2}\sin\alpha\, d\alpha.$$

We have therefore to integrate

$$I = \int_0^\pi \frac{mn}{2}\cos\theta\sin\alpha\, d\alpha, \tag{2}$$

remembering that θ is a function of α.

We may express both θ and α in terms of R, and the expression to be integrated becomes

$$\frac{mn}{4X^2D}(R^2 + X^2 - D^2)\,dR, \tag{3}$$

the general integral of which is

$$\frac{mnR}{12X^2D}(R^2 + 3X^2 - 3D^2) + C. \tag{4}$$

In the first case, that in which X is less than D, the limits of integration are $R = D + X$ and $R = D - X$. In the second case, in which X is greater than D, the limits are $R = X + D$ and $R = X - D$.

When X is less than D, $\qquad I = \frac{2}{3}\frac{mn}{D}X. \qquad (5)$

When X is equal to D, $\qquad I = \frac{2}{3}mn. \qquad (6)$

When X is greater than D, $\qquad I = mn\left(1 - \frac{1}{3}\frac{D^2}{X^2}\right); \qquad (7)$

and when X becomes infinite $\qquad I = mn. \qquad (8)$

According to this form of the theory, which is that adopted by Weber *, as the magnetizing force increases from 0 to D, the

* There is some mistake in the formula given by Weber (*Trans. Acad. Sax.* i. p. 572 (1852), or Pogg., *Ann.* lxxxvii. p. 167 (1852)) as the result of this integration, the steps of which are not given by him. His formula is

$$I = mn\frac{X}{\sqrt{X^2 + D^2}}\,\frac{X^4 + \frac{7}{6}X^2D^2 + \frac{2}{3}D^4}{X^4 + X^2D^2 + D^4}.$$

magnetization increases in the same proportion. When the magnetizing force attains the value D, the magnetization is two-thirds of its limiting value. When the magnetizing force is further increased, the magnetization, instead of increasing indefinitely, tends towards a finite limit.

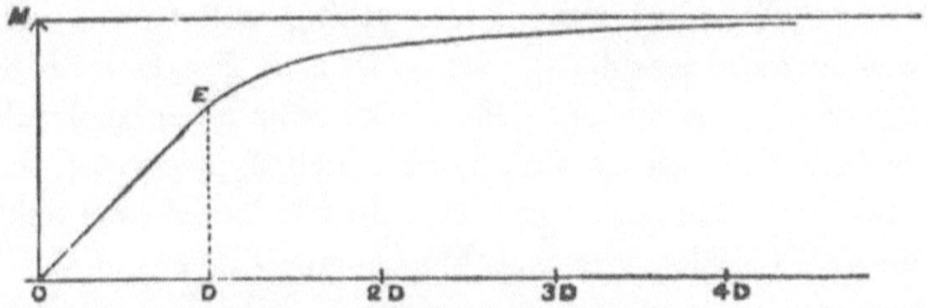

Fig. 7.

The law of magnetization is expressed in Fig. 7, where the magnetizing force is reckoned from O towards the right and the magnetization is expressed by the vertical ordinates. Weber's own experiments give results in satisfactory accordance with this law. It is probable, however, that the value of D is not the same for all the molecules of the same piece of iron, so that the transition from the straight line from O to E to the curve beyond E may not be so abrupt as is here represented.

444.] The theory in this form gives no account of the residual magnetization which is found to exist after the magnetizing force is removed. I have therefore thought it desirable to examine the results of making a further assumption relating to the conditions under which the position of equilibrium of a molecule may be permanently altered.

Let us suppose that the axis of a magnetic molecule, if deflected through any angle β less than β_0, will return to its original position when the deflecting force is removed, but that if the deflexion β exceeds β_0, then, when the deflecting force is removed, the axis will not return to its original position, but will be permanently deflected through an angle $\beta - \beta_0$, which may be called the permanent *set* of the molecule.

This assumption with respect to the law of molecular deflexion is not to be regarded as founded on any exact knowledge of the intimate structure of bodies, but is adopted, in our ignorance of the true state of the case, as an assistance to the imagination in following out the speculation suggested by Weber.

Let
$$L = D \sin \beta_0, \tag{9}$$

then, if the moment of the couple acting on a molecule is less than mL, there will be no permanent deflexion, but if it exceeds mL there will be a permanent change of the position of equilibrium.

To trace the results of this supposition, describe a sphere whose centre is O and radius $OL = L$.

As long as X is less than L everything will be the same as in the case already considered, but as soon as X exceeds L it will begin to produce a permanent deflexion of some of the molecules.

Let us take the case of Fig. 8, in which X is greater than L but less than D. Through S as vertex draw a double cone touching the sphere L. Let this cone meet the sphere D in P and Q. Then if the axis of a molecule in its original position lies between OA and OP, or between OB and OQ, it will be deflected through an angle less than β_0, and will not be permanently deflected. But if

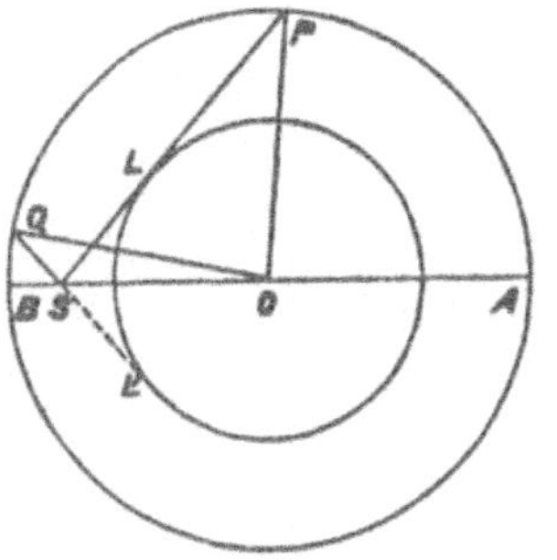

Fig. 8.

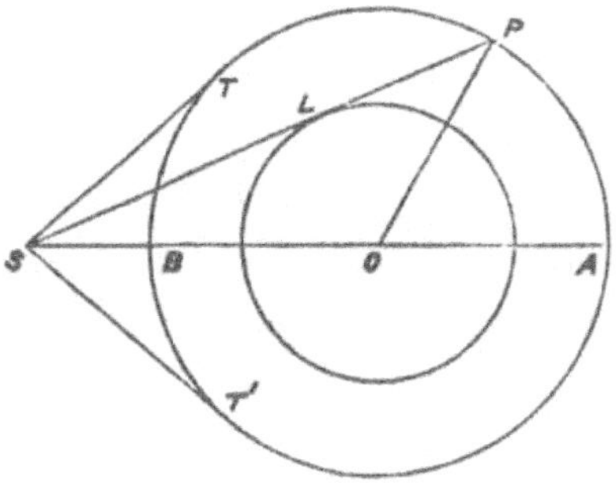

Fig. 9.

the axis of the molecule lies originally between OP and OQ, then a couple whose moment is greater than L will act upon it and will deflect it into the position SP, and when the force X ceases to act it will not resume its original direction, but will be permanently set in the direction OP.

Let us put

$$L = X \sin\theta_0 \quad \text{when} \quad \theta = PSA \text{ or } QSB,$$

then all those molecules whose axes, on the former hypotheses, would have values of θ between θ_0 and $\pi - \theta_0$ will be made to have the value θ_0 during the action of the force X.

During the action of the force X, therefore, those molecules whose axes when deflected lie within either sheet of the double cone whose semivertical angle is θ_0 will be arranged as in the former case, but all those whose axes on the former theory would lie outside of these sheets will be permanently deflected, so that their axes will form a dense fringe round that sheet of the cone which lies towards A.

As X increases, the number of molecules belonging to the cone about B continually diminishes, and when X becomes equal to D all the molecules have been wrenched out of their former positions of equilibrium, and have been forced into the fringe of the cone round A, so that when X becomes greater than D all the molecules form part of the cone round A or of its fringe.

When the force X is removed, then in the case in which X is less than L everything returns to its primitive state. When X is between L and D then there is a cone round A whose angle

$$AOP = \theta_0 + \beta_0,$$

and another cone round B whose angle

$$BOQ = \theta_0 - \beta_0.$$

Within these cones the axes of the molecules are distributed uniformly. But all the molecules, the original direction of whose axes lay outside of both these cones, have been wrenched from their primitive positions and form a fringe round the cone about A.

If X is greater than D, then the cone round B is completely dispersed, and all the molecules which formed it are converted into the fringe round A, and are inclined at the angle $\theta_0 + \beta_0$.

445.] Treating this case in the same way as before, we find for the intensity of the temporary magnetization during the action of the force X, which is supposed to act on iron which has never before been magnetized,

When X is less than L, $\quad I = \frac{2}{3} M \frac{X}{D}.$

When X is equal to L, $\quad I = \frac{2}{3} M \frac{L}{D}.$

When X is between L and D,

$$I = M\left\{\frac{2}{3}\frac{X}{D} + \left(1 - \frac{L^2}{X^2}\right)\left[\sqrt{1 - \frac{L^2}{D^2}} - \frac{2}{3}\sqrt{\frac{X^2}{D^2} - \frac{L^2}{D^2}}\right]\right\}.$$

When X is equal to D,

$$I = M\left\{\frac{2}{3} + \frac{1}{3}\left(1 - \frac{L^2}{D^2}\right)^{\frac{3}{2}}\right\}.$$

When X is greater than D,

$$I = M\left\{\frac{1}{3}\frac{X}{D} + \frac{1}{2} - \frac{1}{6}\frac{D}{X} + \frac{(D^2 - L^2)^{\frac{3}{2}}}{6X^2 D} - \frac{\sqrt{X^2 - L^2}}{6X^2 D}(2X^2 - 3XD + L^2)\right\}.$$

When X is infinite, $\quad I = M.$

When X is less than L the magnetization follows the former law, and is proportional to the magnetizing force. As soon as X exceeds L the magnetization assumes a more rapid rate of increase on account of the molecules beginning to be transferred from the one cone to the other. This rapid increase, however, soon comes to an end as the number of molecules forming the negative cone diminishes, and at last the magnetization reaches the limiting value M.

If we were to assume that the values of L and of D are different for different molecules, we should obtain a result in which the different stages of magnetization are not so distinctly marked.

The residual magnetization, I', produced by the magnetizing force X, and observed after the force has been removed, is as follows:

When X is less than L, No residual magnetization.

When X is between L and D,

$$I' = M\left(1 - \frac{L^2}{D^2}\right)\left(1 - \frac{L^2}{X^2}\right).$$

When X is equal to D,

$$I' = M\left(1 - \frac{L^2}{D^2}\right)^2.$$

When X is greater than D,

$$I' = \frac{1}{4} M \left\{1 - \frac{L^2}{XD} + \sqrt{1 - \frac{L^2}{D^2}}\sqrt{1 - \frac{L^2}{X^2}}\right\}^2.$$

When X is infinite,

$$I' = \frac{1}{4} M \left\{1 + \sqrt{1 - \frac{L^2}{D^2}}\right\}^2.$$

If we make

$$M = 1000, \qquad L = 3, \qquad D = 5,$$

we find the following values of the temporary and the residual magnetization:—

Magnetizing Force.	Temporary Magnetization.	Residual Magnetization.
X	I	I'
0	0	0
1	133	0
2	267	0
3	400	0
4	729	280
5	837	410
6	864	485
7	882	537
8	897	574
∞	1000	810

These results are laid down in Fig. 10.

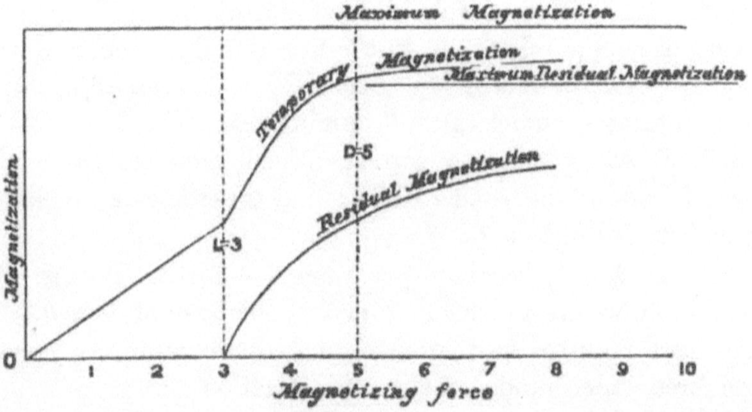

Fig. 10.

The curve of temporary magnetization is at first a straight line from $X = 0$ to $X = L$. It then rises more rapidly till $X = D$, and as X increases it approaches its horizontal asymptote.

The curve of residual magnetization begins when $X = L$, and approaches an asymptote at a distance $= .81\,M$.

It must be remembered that the residual magnetism thus found corresponds to the case in which, when the external force is removed, there is no demagnetizing force arising from the distribution of magnetism in the body itself. The calculations are therefore applicable only to very elongated bodies magnetized longitudinally. In the case of short, thick bodies the residual magnetism will be diminished by the reaction of the free magnetism in the same way as if an external reversed magnetizing force were made to act upon it.

446.] The scientific value of a theory of this kind, in which we make so many assumptions, and introduce so many adjustable constants, cannot be estimated merely by its numerical agreement with certain sets of experiments. If it has any value it is because it enables us to form a mental image of what takes place in a piece of iron during magnetization. To test the theory, we shall apply it to the case in which a piece of iron, after being subjected to a magnetizing force X_0, is again subjected to a magnetizing force X_1.

If the new force X_1 acts in the same direction in which X_0 acted, which we shall call the positive direction, then, if X_1 is less than X_0, it will produce no permanent set of the molecules, and when X_1 is removed the residual magnetization will be the same as that produced by X_0. If X_1 is greater than X_0, then it will produce exactly the same effect as if X_0 had not acted.

But let us suppose X_1 to act in the negative direction, and let us suppose $X_0 = L \operatorname{cosec} \theta_0$, and $X_1 = -L \operatorname{cosec} \theta_1$.

As X_1 increases numerically, θ_1 diminishes. The first molecules on which X_1 will produce a permanent deflexion are those which form the fringe of the cone round A, and these have an inclination when undeflected of $\theta_0 + \beta_0$.

As soon as $\theta_1 - \beta_0$ becomes less than $\theta_0 + \beta_0$ the process of demagnetization will commence. Since, at this instant, $\theta_1 = \theta_0 + 2\beta_0$, X_1, the force required to begin the demagnetization, is less than X_0, the force which produced the magnetization.

If the value of D and of L were the same for all the molecules, the slightest increase of X_1 would wrench the whole of the fringe of molecules whose axes have the inclination $\theta_0 + \beta_0$ into a position in which their axes are inclined $\theta_1 + \beta_0$ to the negative axis OB.

Though the demagnetization does not take place in a manner so sudden as this, it takes place so rapidly as to afford some confirmation of this mode of explaining the process.

Let us now suppose that by giving a proper value to the reverse force X_1 we have exactly demagnetized the piece of iron.

The axes of the molecules will not now be arranged indifferently in all directions, as in a piece of iron which has never been magnetized, but will form three groups.

(1) Within a cone of semiangle $\theta_1 - \beta_0$ surrounding the positive pole, the axes of the molecules remain in their primitive positions.

(2) The same is the case within a cone of semiangle $\theta_0 - \beta_0$ surrounding the negative pole.

(3) The directions of the axes of all the other molecules form a conical sheet surrounding the negative pole, and are at an inclination $\theta_1 + \beta_0$.

When X_0 is greater than D the second group is absent. When X_1 is greater than D the first group is also absent.

The state of the iron, therefore, though apparently demagnetized, is in a different state from that of a piece of iron which has never been magnetized.

To shew this, let us consider the effect of a magnetizing force X_2 acting in either the positive or the negative direction. The first permanent effect of such a force will be on the third group of molecules, whose axes make angles $= \theta_1 + \beta_0$ with the negative axis.

If the force X_2 acts in the negative direction it will begin to produce a permanent effect as soon as $\theta_2+\beta_0$ becomes less than $\theta_1+\beta_0$, that is, as soon as X_2 becomes greater than X_1. But if X_2 acts in the positive direction it will begin to remagnetize the iron as soon as $\theta_2-\beta$ becomes less than $\theta_1+\beta_0$, that is, when $\theta_2=\theta_1+2\beta_0$, or while X_2 is still much less than X_1.

It appears therefore from our hypothesis that—

When a piece of iron is magnetized by means of a force X_0, its magnetism cannot be increased without the application of a force greater than X_0. A reverse force, less than X_0, is sufficient to diminish its magnetization.

If the iron is exactly demagnetized by a reversed force X_1, then it cannot be magnetized in the reversed direction without the application of a force greater than X_1, but a positive force less than X_1 is sufficient to begin to remagnetize the iron in its original direction.

These results are consistent with what has been actually observed by Ritchie *, Jacobi †, Marianini ‡, and Joule §.

A very complete account of the relations of the magnetization of iron and steel to magnetic forces and to mechanical strains is given by Wiedemann in his *Galvanismus*. By a detailed comparison of the effects of magnetization with those of torsion, he shews that the ideas of elasticity and plasticity which we derive from experiments on the temporary and permanent torsion of wires can be applied with equal propriety to the temporary and permanent magnetization of iron and steel.

447.] Matteucci ‖ found that the extension of a hard iron bar during the action of the magnetizing force increases its temporary magnetism. This has been confirmed by Wertheim. In the case of soft bars the magnetism is diminished by extension.

The permanent magnetism of a bar increases when it is extended, and diminishes when it is compressed.

Hence, if a piece of iron is first magnetized in one direction, and then extended in another direction, the direction of magnetization will tend to approach the direction of extension. If it be compressed, the direction of magnetization will tend to become normal to the direction of compression.

This explains the result of an experiment of Wiedemann's. A

* *Phil. Mag.*, 1833. † Pog., *Ann.*, 1834.
‡ *Ann. de Chimie et de Physique*, 1846. § *Phil. Trans.*, 1855, p. 287.
‖ *Ann. de Chimie et de Physique*, 1858.

current was passed downward through a vertical wire. If, either during the passage of the current or after it has ceased, the wire be twisted in the direction of a right-handed screw, the lower end becomes a north pole.

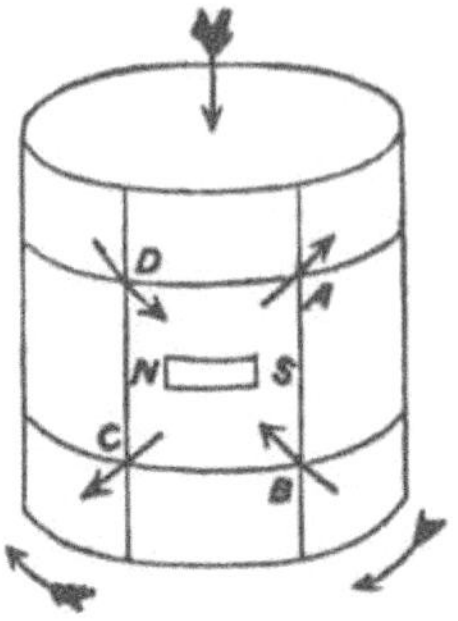

Fig. 11.

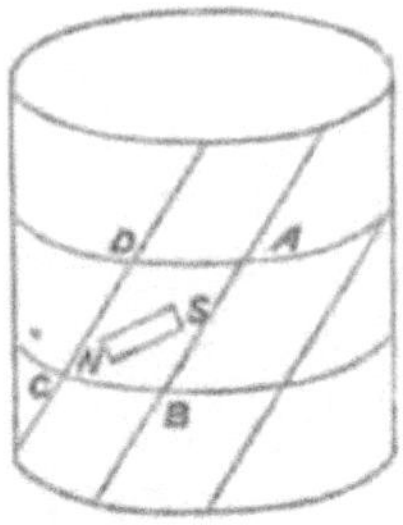

Fig. 12.

Here the downward current magnetizes every part of the wire in a tangential direction, as indicated by the letters *NS*.

The twisting of the wire in the direction of a right-handed screw causes the portion *ABCD* to be extended along the diagonal *AC* and compressed along the diagonal *BD*. The direction of magnetization therefore tends to approach *AC* and to recede from *BD*, and thus the lower end becomes a north pole and the upper end a south pole.

Effect of Magnetization on the Dimensions of the Magnet.

448.] Joule*, in 1842, found that an iron bar becomes lengthened when it is rendered magnetic by an electric current in a coil which surrounds it. He afterwards† shewed, by placing the bar in water within a glass tube, that the volume of the iron is not augmented by this magnetization, and concluded that its transverse dimensions were contracted.

Finally, he passed an electric current through the axis of an iron tube, and back outside the tube, so as to make the tube into a closed magnetic solenoid, the magnetization being at right angles to the axis of the tube. The length of the axis of the tube was found in this case to be shortened.

He found that an iron rod under longitudinal pressure is also elongated when it is magnetized. When, however, the rod is under considerable longitudinal tension, the effect of magnetization is to shorten it.

* Sturgeon's *Annals of Electricity*, vol. viii. p. 219.
† *Phil. Mag.*, 1847.

This was the case with a wire of a quarter of an inch diameter when the tension exceeded 600 pounds weight.

In the case of a hard steel wire the effect of the magnetizing force was in every case to shorten the wire, whether the wire was under tension or pressure. The change of length lasted only as long as the magnetizing force was in action, no alteration of length was observed due to the permanent magnetization of the steel.

Joule found the elongation of iron wires to be nearly proportional to the square of the actual magnetization, so that the first effect of a demagnetizing current was to shorten the wire.

On the other hand, he found that the shortening effect on wires under tension, and on steel, varied as the product of the magnetization and the magnetizing current.

Wiedemann found that if a vertical wire is magnetized with its north end uppermost, and if a current is then passed downwards through the wire, the lower end of the wire, if free, twists in the direction of the hands of a watch as seen from above, or, in other words, the wire becomes twisted like a right-handed screw.

In this case the magnetization due to the action of the current on the previously existing magnetization is in the direction of a left-handed screw round the wire. Hence the twisting would indicate that when the iron is magnetized it contracts in the direction of magnetization and expands in directions at right angles to the magnetization. This, however, seems not to agree with Joule's results.

For further developments of the theory of magnetization, see Arts. 832–845.

CHAPTER VII.

MAGNETIC MEASUREMENTS.

449.] THE principal magnetic measurements are the determination of the magnetic axis and magnetic moment of a magnet, and that of the direction and intensity of the magnetic force at a given place.

Since these measurements are made near the surface of the earth, the magnets are always acted on by gravity as well as by terrestrial magnetism, and since the magnets are made of steel their magnetism is partly permanent and partly induced. The permanent magnetism is altered by changes of temperature, by strong induction, and by violent blows; the induced magnetism varies with every variation of the external magnetic force.

The most convenient way of observing the force acting on a magnet is by making the magnet free to turn about a vertical axis. In ordinary compasses this is done by balancing the magnet on a vertical pivot. The finer the point of the pivot the smaller is the moment of the friction which interferes with the action of the magnetic force. For more refined observations the magnet is suspended by a thread composed of a silk fibre without twist, either single, or doubled on itself a sufficient number of times, and so formed into a thread of parallel fibres. each of which supports as nearly as possible an equal part of the weight. The force of torsion of such a thread is much less than that of a metal wire of equal strength, and it may be calculated in terms of the observed azimuth of the magnet, which is not the case with the force arising from the friction of a pivot.

The suspension fibre can be raised or lowered by turning a horizontal screw which works in a fixed nut. The fibre is wound round the thread of the screw, so that when the screw is turned the suspension fibre always hangs in the same vertical line.

The suspension fibre carries a small horizontal divided circle called the Torsion-circle, and a stirrup with an index, which can be placed so that the index coincides with any given division of the torsion circle. The stirrup is so shaped that the magnet bar can be fitted into it with its axis horizontal, and with any one of its four sides uppermost.

To ascertain the zero of torsion a non-magnetic body of the same weight as the magnet is placed in the stirrup, and the position of the torsion circle when in equilibrium ascertained.

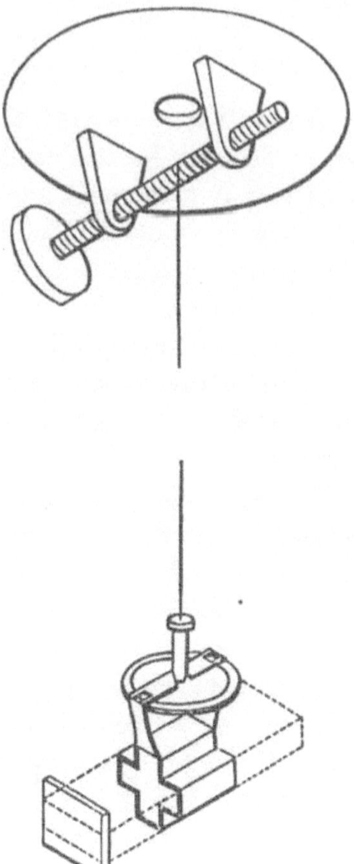
Fig. 13.

The magnet itself is a piece of hard-tempered steel. According to Gauss and Weber its length ought to be at least eight times its greatest transverse dimension. This is necessary when permanence of the direction of the magnetic axis within the magnet is the most important consideration. Where promptness of motion is required the magnet should be shorter, and it may even be advisable in observing sudden alterations in magnetic force to use a bar magnetized transversely and suspended with its longest dimension vertical *.

450.] The magnet is provided with an arrangement for ascertaining its angular position. For ordinary purposes its ends are pointed, and a divided circle is placed below the ends, by which their positions are read off by an eye placed in a plane through the suspension thread and the point of the needle.

For more accurate observations a plane mirror is fixed to the magnet, so that the normal to the mirror coincides as nearly as possible with the axis of magnetization. This is the method adopted by Gauss and Weber.

Another method is to attach to one end of the magnet a lens and to the other end a scale engraved on glass, the distance of the lens

* Joule, *Proc. Phil. Soc., Manchester*, Nov. 29, 1864.

from the scale being equal to the principal focal length of the lens. The straight line joining the zero of the scale with the optical centre of the lens ought to coincide as nearly as possible with the magnetic axis.

As these optical methods of ascertaining the angular position of suspended apparatus are of great importance in many physical researches, we shall here consider once for all their mathematical theory.

Theory of the Mirror Method.

We shall suppose that the apparatus whose angular position is to be determined is capable of revolving about a vertical axis. This axis is in general a fibre or wire by which it is suspended. The mirror should be truly plane, so that a scale of millimetres may be seen distinctly by reflexion at a distance of several metres from the mirror.

The normal through the middle of the mirror should pass through the axis of suspension, and should be accurately horizontal. We shall refer to this normal as the line of collimation of the apparatus.

Having roughly ascertained the mean direction of the line of collimation during the experiments which are to be made, a telescope is erected at a convenient distance in front of the mirror, and a little above the level of the mirror.

The telescope is capable of motion in a vertical plane, it is directed towards the suspension fibre just above the mirror, and a fixed mark is erected in the line of vision, at a horizontal distance from the object glass equal to twice the distance of the mirror from the object glass. The apparatus should, if possible, be so arranged that this mark is on a wall or other fixed object. In order to see the mark and the suspension fibre at the same time through the telescope, a cap may be placed over the object glass having a slit along a vertical diameter. This should be removed for the other observations. The telescope is then adjusted so that the mark is seen distinctly to coincide with the vertical wire at the focus of the telescope. A plumb-line is then adjusted so as to pass close in front of the optical centre of the object glass and to hang below the telescope. Below the telescope and just behind the plumb-line a scale of equal parts is placed so as to be bisected at right angles by the plane through the mark, the suspension-fibre, and the plumb-line. The sum of the heights of the scale and the object glass should be equal to twice the height of the mirror from the floor. The telescope being now directed towards the mirror will see in it the reflexion of the scale. If the part of the scale where the plumb-line crosses it appears to coincide with the vertical wire of the telescope, then the line of collimation of the mirror coincides with the plane through the mark and the optical centre of the object glass. If the vertical wire coincides with any other division of the scale, the angular position of the line of collimation is to be found as follows :—

Let the plane of the paper be horizontal, and let the various points be projected on this plane. Let O be the centre of the object glass of the telescope, P the fixed mark, P and the vertical wire of the telescope are conjugate foci with respect to the object glass. Let M be the point where OP cuts the plane of the mirror. Let MN be the normal to the mirror; then $OMN = \theta$ is the angle which the line of collimation makes with the fixed plane. Let MS be a line in the plane of OM and MN, such that $NMS = OMN$, then S will be the part of the scale which will be seen by reflexion to coincide with the vertical wire of the telescope. Now, since

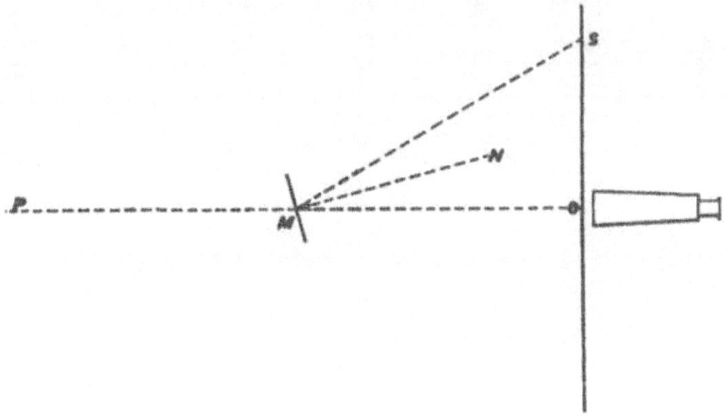

Fig. 14.

MN is horizontal, the projected angles OMN and NMS in the figure are equal, and $OMS = 2\theta$. Hence $OS = OM \tan 2\theta$.

We have therefore to measure OM in terms of the divisions of the scale; then, if s_0 is the division of the scale which coincides with the plumb-line, and s the observed division,

$$s - s_0 = OM \tan 2\theta,$$

whence θ may be found. In measuring OM we must remember that if the mirror is of glass, silvered at the back, the virtual image of the reflecting surface is at a distance behind the front surface of the glass $= \frac{t}{\mu}$, where t is the thickness of the glass, and μ is the index of refraction.

We must also remember that if the line of suspension does not pass through the point of reflexion, the position of M will alter with θ. Hence, when it is possible, it is advisable to make the centre of the mirror coincide with the line of suspension.

It is also advisable, especially when large angular motions have to be observed, to make the scale in the form of a concave cylindric surface, whose axis is the line of suspension. The angles are then observed at once in circular measure without reference to a table of tangents. The scale should be carefully adjusted, so that the axis of the cylinder coincides with the suspension fibre. The numbers on the scale should always run from the one end to the other in the same direction so as to avoid negative readings. Fig. 15

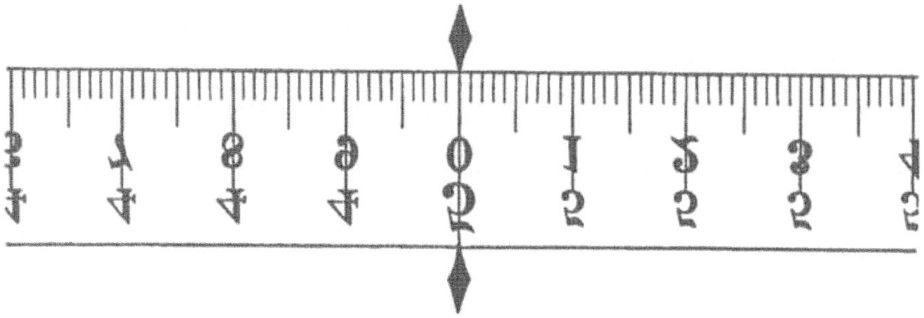

Fig. 15.

represents the middle portion of a scale to be used with a mirror and an inverting telescope.

This method of observation is the best when the motions are slow. The observer sits at the telescope and sees the image of the scale moving to right or to left past the vertical wire of the telescope. With a clock beside him he can note the instant at which a given division of the scale passes the wire, or the division of the scale which is passing at a given tick of the clock, and he can also record the extreme limits of each oscillation.

When the motion is more rapid it becomes impossible to read the divisions of the scale except at the instants of rest at the extremities of an oscillation. A conspicuous mark may be placed at a known division of the scale, and the instant of transit of this mark may be noted.

When the apparatus is very light, and the forces variable, the motion is so prompt and swift that observation through a telescope would be useless. In this case the observer looks at the scale directly, and observes the motions of the image of the vertical wire thrown on the scale by a lamp.

It is manifest that since the image of the scale reflected by the mirror and refracted by the object glass coincides with the vertical wire, the image of the vertical wire, if sufficiently illuminated, will coincide with the scale. To observe this the room is darkened, and the concentrated rays of a lamp are thrown on the vertical wire towards the object glass. A bright patch of light crossed by the shadow of the wire is seen on the scale. Its motions can be followed by the eye, and the division of the scale at which it comes to rest can be fixed on by the eye and read off at leisure. If it be desired to note the instant of the passage of the bright spot past a given point on the scale, a pin or a bright metal wire may be placed there so as to flash out at the time of passage.

By substituting a small hole in a diaphragm for the cross wire the image becomes a small illuminated dot moving to right or left on the scale, and by substituting for the scale a cylinder revolving by clock work about a horizontal axis and covered with photographic paper, the spot of light traces out a curve which can be afterwards rendered visible. Each abscissa of this curve corresponds to a particular time, and the ordinate indicates the angular position of the mirror at that time. In this way an automatic system of continuous registration of all the elements of terrestrial magnetism has been established at Kew and other observatories.

In some cases the telescope is dispensed with, a vertical wire is illuminated by a lamp placed behind it, and the mirror is a concave one, which forms the image of the wire on the scale as a dark line across a patch of light.

451.] In the Kew portable apparatus, the magnet is made in the form of a tube, having at one end a lens, and at the other a glass scale, so adjusted as to be at the principal focus of the lens. Light is admitted from behind the scale, and after passing through the lens it is viewed by means of a telescope.

Since the scale is at the principal focus of the lens, rays from any division of the scale emerge from the lens parallel, and if the telescope is adjusted for celestial objects, it will shew the scale in optical coincidence with the cross wires of the telescope. If a given division of the scale coincides with the intersection of the cross wires, then the line joining that division with the optical centre of the lens must be parallel to the line of collimation of the telescope. By fixing the magnet and moving the telescope, we may ascertain the angular value of the divisions of the scale, and then, when the magnet is suspended and the position of the telescope known, we may determine the position of the magnet at any instant by reading off the division of the scale which coincides with the cross wires.

The telescope is supported on an arm which is centred in the line of the suspension fibre, and the position of the telescope is read off by verniers on the azimuth circle of the instrument.

This arrangement is suitable for a small portable magnetometer in which the whole apparatus is supported on one tripod, and in which the oscillations due to accidental disturbances rapidly subside.

Determination of the Direction of the Axis of the Magnet, and of the Direction of Terrestrial Magnetism.

452.] Let a system of axes be drawn in the magnet, of which the axis of z is in the direction of the length of the bar, and x and y perpendicular to the sides of the bar supposed a parallelepiped.

Let l, m, n and λ, μ, ν be the angles which the magnetic axis and the line of collimation make with these axes respectively.

Let M be the magnetic moment of the magnet, let H be the horizontal component of terrestrial magnetism, let Z be the vertical component, and let δ be the azimuth in which H acts, reckoned from the north towards the west.

Let ζ be the observed azimuth of the line of collimation, let α be the azimuth of the stirrup, and β the reading of the index of the torsion circle, then $\alpha-\beta$ is the azimuth of the lower end of the suspension fibre.

Let γ be the value of $\alpha-\beta$ when there is no torsion, then the moment of the force of torsion tending to diminish α will be

$$\tau(\alpha-\beta-\gamma),$$

where τ is a coefficient of torsion depending on the nature of the fibre.

To determine λ, fix the stirrup so that y is vertical and upwards, z to the north and x to the west, and observe the azimuth ζ of the line of collimation. Then remove the magnet, turn it through an angle π about the axis of z and replace it in this inverted position, and observe the azimuth ζ' of the line of collimation when y is downwards and x to the east,

$$\zeta = a + \frac{\pi}{2} - \lambda, \tag{1}$$

$$\zeta' = a - \frac{\pi}{2} + \lambda. \tag{2}$$

Hence
$$\lambda = \frac{\pi}{2} + \frac{1}{2}(\zeta' - \zeta). \tag{3}$$

Next, hang the stirrup to the suspension fibre, and place the magnet in it, adjusting it carefully so that y may be vertical and upwards, then the moment of the force tending to increase a is

$$MH \sin m \sin\left(\delta - a - \frac{\pi}{2} + l\right) - \tau(a - \beta - \gamma). \tag{4}$$

But if ζ is the observed azimuth of the line of collimation

$$\zeta = a + \frac{\pi}{2} - \lambda, \tag{5}$$

so that the force may be written

$$MH \sin m \sin(\delta - \zeta + l - \lambda) - \tau\left(\zeta + \lambda - \frac{\pi}{2} - \beta - \gamma\right). \tag{6}$$

When the apparatus is in equilibrium this quantity is zero for a particular value of ζ.

When the apparatus never comes to rest, but must be observed in a state of vibration, the value of ζ corresponding to the position of equilibrium may be calculated by a method which will be described in Art. 735.

When the force of torsion is small compared with the moment of the magnetic force, we may put $\delta - \zeta + l - \lambda$ for the sine of that angle.

If we give to β, the reading of the torsion circle, two different values, β_1 and β_2, and if ζ_1 and ζ_2 are the corresponding values of ζ,

$$MH \sin m(\zeta_1 - \zeta_2) = \tau(\zeta_1 - \zeta_2 - \beta_1 + \beta_2), \tag{7}$$

or, if we put

$$\frac{\zeta_1 - \zeta_2}{\zeta_1 - \zeta_2 - \beta_1 + \beta_2} = \tau', \quad \text{then} \quad \tau = MH \sin m\, \tau', \tag{8}$$

and equation (7) becomes, dividing by $MH \sin m$,

$$\delta - \zeta + l - \lambda - \tau'\left(\zeta + \lambda - \frac{\pi}{2} - \beta - \gamma\right) = 0. \tag{9}$$

If we now reverse the magnet so that y is downwards, and adjust the apparatus till y is exactly vertical, and if ζ' is the new value of the azimuth, and δ' the corresponding declination,

$$\delta' - \zeta' - l + \lambda - \tau'\left(\zeta' - \lambda + \frac{\pi}{2} - \beta - \gamma\right) = 0, \tag{10}$$

whence
$$\frac{\delta + \delta'}{2} = \tfrac{1}{2}(\zeta + \zeta') + \tfrac{1}{2}\tau'(\zeta + \zeta' - 2(\beta + \gamma)). \tag{11}$$

The reading of the torsion circle should now be adjusted, so that the coefficient of τ' may be as nearly as possible zero. For this purpose we must determine γ, the value of $\alpha-\beta$ when there is no torsion. This may be done by placing a non-magnetic bar of the same weight as the magnet in the stirrup, and determining $\alpha-\beta$ when there is equilibrium. Since τ' is small, great accuracy is not required. Another method is to use a torsion bar of the same weight as the magnet, containing within it a very small magnet whose magnetic moment is $\frac{1}{n}$ of that of the principal magnet. Since τ remains the same, τ' will become $n\tau'$, and if ζ_1 and ζ_1' are the values of ζ as found by the torsion bar,

$$\delta = \tfrac{1}{2}(\zeta_1+\zeta_1')+\tfrac{1}{2}n\tau'(\zeta_1+\zeta_1'-2(\beta+\gamma)). \qquad (12)$$

Subtracting this equation from (11),

$$2(n-1)(\beta+\gamma) = \left(n+\frac{1}{\tau'}\right)(\zeta_1+\zeta_1')-\left(1+\frac{1}{\tau'}\right)(\zeta+\zeta'). \qquad (13)$$

Having found the value of $\beta+\gamma$ in this way, β, the reading of the torsion circle, should be altered till

$$\zeta+\zeta'-2(\beta+\gamma) = 0, \qquad (14)$$

as nearly as possible in the ordinary position of the apparatus.

Then, since τ' is a very small numerical quantity, and since its coefficient is very small, the value of the second term in the expression for δ will not vary much for small errors in the values of τ' and γ, which are the quantities whose values are least accurately known.

The value of δ, the magnetic declination, may be found in this way with considerable accuracy, provided it remains constant during the experiments, so that we may assume $\delta'=\delta$.

When great accuracy is required it is necessary to take account of the variations of δ during the experiment. For this purpose observations of another suspended magnet should be made at the same instants that the different values of ζ are observed, and if η, η' are the observed azimuths of the second magnet corresponding to ζ and ζ', and if δ and δ' are the corresponding values of δ, then

$$\delta'-\delta = \eta'-\eta. \qquad (15)$$

Hence, to find the value of δ we must add to (11) a correction

$$\tfrac{1}{2}(\eta-\eta').$$

The declination at the time of the first observation is therefore

$$\delta = \tfrac{1}{2}(\zeta+\zeta'+\eta-\eta')+\tfrac{1}{2}\tau'(\zeta+\zeta'-2\beta-2\gamma). \qquad (16)$$

To find the direction of the magnetic axis within the magnet subtract (10) from (9) and add (15),

$$l = \lambda + \tfrac{1}{2}(\zeta - \zeta') - \tfrac{1}{2}(\eta - \eta') + \tfrac{1}{2}\tau'(\zeta - \zeta' + 2\lambda - \pi). \tag{17}$$

By repeating the experiments with the bar on its two edges, so that the axis of x is vertically upwards and downwards, we can find the value of m. If the axis of collimation is capable of adjustment it ought to be made to coincide with the magnetic axis as nearly as possible, so that the error arising from the magnet not being exactly inverted may be as small as possible *.

On the Measurement of Magnetic Forces.

453.] The most important measurements of magnetic force are those which determine M, the magnetic moment of a magnet, and H, the intensity of the horizontal component of terrestrial magnetism. This is generally done by combining the results of two experiments, one of which determines the ratio and the other the product of these two quantities.

The intensity of the magnetic force due to an infinitely small magnet whose magnetic moment is M, at a point distant r from the centre of the magnet in the positive direction of the axis of the magnet, is

$$R = 2\frac{M}{r^3} \tag{1}$$

and is in the direction of r. If the magnet is of finite size but spherical, and magnetized uniformly in the direction of its axis, this value of the force will still be exact. If the magnet is a solenoidal bar magnet of length $2L$,

$$R = 2\frac{M}{r^3}\left(1 + 2\frac{L^2}{r^2} + 3\frac{L^4}{r^4} + \text{\&c.}\right). \tag{2}$$

If the magnet be of any kind, provided its dimensions are all small compared with r,

$$R = 2\frac{M}{r^3}\left(1 + A_1\frac{1}{r} + A_2\frac{1}{r^2}\right) + \text{\&c.}, \tag{3}$$

where A_1, A_2, &c. are coefficients depending on the distribution of the magnetization of the bar.

Let H be the intensity of the horizontal part of terrestrial magnetism at any place. H is directed towards magnetic north. Let r be measured towards magnetic west, then the magnetic force at the extremity of r will be H towards the north and R towards

* See a Paper on 'Imperfect Inversion,' by W. Swan. *Trans. R. S. Edin.*, vol. xxi (1855), p. 349.

the west. The resultant force will make an angle θ with the magnetic meridian, measured towards the west, and such that

$$R = H \tan\theta. \tag{4}$$

Hence, to determine $\frac{R}{H}$ we proceed as follows:—

The direction of the magnetic north having been ascertained, a magnet, whose dimensions should not be too great, is suspended as in the former experiments, and the deflecting magnet M is placed so that its centre is at a distance r from that of the suspended magnet, in the same horizontal plane, and due magnetic east.

The axis of M is carefully adjusted so as to be horizontal and in the direction of r.

The suspended magnet is observed before M is brought near and also after it is placed in position. If θ is the observed deflexion, we have, if we use the approximate formula (1),

$$\frac{M}{H} = \frac{r^3}{2}\tan\theta; \tag{5}$$

or, if we use the formula (3),

$$\frac{1}{2}\frac{H}{M}r^3\tan\theta = 1 + A_1\frac{1}{r} + A_2\frac{1}{r^2} + \&c. \tag{6}$$

Here we must bear in mind that though the deflexion θ can be observed with great accuracy, the distance r between the centres of the magnets is a quantity which cannot be precisely determined, unless both magnets are fixed and their centres defined by marks.

This difficulty is overcome thus:

The magnet M is placed on a divided scale which extends east and west on both sides of the suspended magnet. The middle point between the ends of M is reckoned the centre of the magnet. This point may be marked on the magnet and its position observed on the scale, or the positions of the ends may be observed and the arithmetic mean taken. Call this s_1, and let the line of the suspension fibre of the suspended magnet when produced cut the scale at s_0, then $r_1 = s_1 - s_0$, where s_1 is known accurately and s_0 approximately. Let θ_1 be the deflexion observed in this position of M.

Now reverse M, that is, place it on the scale with its ends reversed, then r_1 will be the same, but M and A_1, A_3, &c. will have their signs changed, so that if θ_2 is the deflexion,

$$-\frac{1}{2}\frac{H}{M}r_1^3\tan\theta_2 = 1 - A_1\frac{1}{r_1} + A_2\frac{1}{r_1^2} - \&c. \tag{7}$$

Taking the arithmetical mean of (6) and (7),

$$\frac{1}{4}\frac{H}{M}r_1^3(\tan\theta_1-\tan\theta_2) = 1+A_2\frac{1}{r_1^2}+A_4\frac{1}{r_1^4}+\&c. \qquad (8)$$

Now remove M to the west side of the suspended magnet, and place it with its centre at the point marked $2s_0-s$ on the scale. Let the deflexion when the axis is in the first position be θ_3, and when it is in the second θ_4, then, as before,

$$\frac{1}{4}\frac{H}{M}r_2^3(\tan\theta_3-\tan\theta_4) = 1+A_2\frac{1}{r_2^2}+A_4\frac{1}{r_2^4}+\&c. \qquad (9)$$

Let us suppose that the true position of the centre of the suspended magnet is not s_0 but $s_0+\sigma$, then

$$r_1 = r-\sigma, \qquad r_2 = r+\sigma, \qquad (10)$$

and
$$\frac{1}{2}(r_1^n+r_2^n) = r^n\left(1+\frac{n(n-1)}{2}\frac{\sigma^2}{r^2}+\&c.\right); \qquad (11)$$

and since $\frac{\sigma^2}{r^2}$ may be neglected if the measurements are carefully made, we are sure that we may take the arithmetical mean of r_1^n and r_2^n for r^n.

Hence, taking the arithmetical mean of (8) and (9),

$$\frac{1}{8}\frac{H}{M}r^3(\tan\theta_1-\tan\theta_2+\tan\theta_3-\tan\theta_4) = 1+A_2\frac{1}{r^2}+\&c., \quad (12)$$

or, making

$$\frac{1}{4}(\tan\theta_1-\tan\theta_2+\tan\theta_3-\tan\theta_4) = D, \qquad (13)$$

$$\frac{1}{2}\frac{H}{M}Dr^3 = 1+A_2\frac{1}{r^2}+\&c.$$

454.] We may now regard D and r as capable of exact determination.

The quantity A_2 can in no case exceed $2L^2$, where L is half the length of the magnet, so that when r is considerable compared with L we may neglect the term in A_2 and determine the ratio of H to M at once. We cannot, however, assume that A_2 is equal to $2L^2$, for it may be less, and may even be negative for a magnet whose largest dimensions are transverse to the axis. The term in A_4, and all higher terms, may safely be neglected.

To eliminate A_2, repeat the experiment, using distances r_1, r_2, r_3, &c., and let the values of D be D_1, D_2, D_3, &c., then

$$D_1 = \frac{2M}{H}\left(\frac{1}{r_1^3}+\frac{A_2}{r_1^5}\right),$$

$$D_2 = \frac{2M}{H}\left(\frac{1}{r_2^3}+\frac{A_2}{r_2^5}\right),$$

&c. &c.

If we suppose that the probable errors of these equations are equal, as they will be if they depend on the determination of D only, and if there is no uncertainty about r, then, by multiplying each equation by r^{-3} and adding the results, we obtain one equation, and by multiplying each equation by r^{-5} and adding we obtain another, according to the general rule in the theory of the combination of fallible measures when the probable error of each equation is supposed the same.

Let us write

$$\Sigma(Dr^{-3}) \text{ for } D_1 r_1^{-3} + D_2 r_2^{-3} + D_3 r_3^{-3} + \&c.,$$

and use similar expressions for the sums of other groups of symbols, then the two resultant equations may be written

$$\Sigma(Dr^{-3}) = \frac{2M}{H}\left(\Sigma(r^{-6}) + A_2 \Sigma(r^{-8})\right),$$

$$\Sigma(Dr^{-5}) = \frac{2M}{H}\left(\Sigma(r^{-8}) + A_2 \Sigma(r^{-10})\right),$$

whence

$$\frac{2M}{H}\{\Sigma(r^{-6})\Sigma(r^{-10}) - [\Sigma(r^{-8})]^2\} = \Sigma(Dr^{-3})\Sigma(r^{-10}) - \Sigma(Dr^{-5})\Sigma(r^{-8}),$$

and $A_2\{\Sigma(Dr^{-3})\Sigma(r^{-10}) - \Sigma(Dr^{-5})\Sigma(r^{-8})\}$

$$= \Sigma(Dr^{-5})\Sigma(r^{-6}) - \Sigma(Dr^{-3})\Sigma(r^{-8}).$$

The value of A_2 derived from these equations ought to be less than half the square of the length of the magnet M. If it is not we may suspect some error in the observations. This method of observation and reduction was given by Gauss in the 'First Report of the Magnetic Association.'

When the observer can make only two series of experiments at distances r_1 and r_2, the value of $\frac{2M}{H}$ derived from these experiments is

$$Q = \frac{2M}{H} = \frac{D_1 r_1^5 - D_2 r_2^5}{r_1^2 - r_2^2}, \qquad A_2 = \frac{D_2 r_2^3 - D_1 r_1^3}{r_1^2 - r_2^2} r_1^2 r_2^2.$$

If δD_1 and δD_2 are the actual errors of the observed deflexions D_1 and D_2, the actual error of the calculated result Q will be

$$\delta Q = \frac{r_1^5 \delta D_1 - r_2^5 \delta D_2}{r_1^2 - r_2^2}.$$

If we suppose the errors δD_1 and δD_2 to be independent, and that the probable value of either is δD, then the probable value of the error in the calculated value of Q will be δQ, where

$$(\delta Q)^2 = \frac{r_1^{10} + r_2^{10}}{(r_1^2 - r_2^2)^2}(\delta D)^2.$$

If we suppose that one of these distances, say the smaller, is given, the value of the greater distance may be determined so as to make δQ a minimum. This condition leads to an equation of the fifth degree in r_1^2, which has only one real root greater than r_2^2. From this the best value of r_1 is found to be $r_1 = 1.3189 r_2$*.

If one observation only is taken the best distance is when

$$\frac{\delta D}{D} = \sqrt{3}\frac{\delta r}{r},$$

where δD is the probable error of a measurement of deflexion, and δr is the probable error of a measurement of distance.

Method of Sines.

455.] The method which we have just described may be called the Method of Tangents, because the tangent of the deflexion is a measure of the magnetic force.

If the line r_1, instead of being measured east or west, is adjusted till it is at right angles with the axis of the deflected magnet, then R is the same as before, but in order that the suspended magnet may remain perpendicular to r, the resolved part of the force H in the direction of r must be equal and opposite to R. Hence, if θ is the deflexion, $R = H \sin\theta$.

This method is called the Method of Sines. It can be applied only when R is less than H.

In the Kew portable apparatus this method is employed. The suspended magnet hangs from a part of the apparatus which revolves along with the telescope and the arm for the deflecting magnet, and the rotation of the whole is measured on the azimuth circle.

The apparatus is first adjusted so that the axis of the telescope coincides with the mean position of the line of collimation of the magnet in its undisturbed state. If the magnet is vibrating, the true azimuth of magnetic north is found by observing the extremities of the oscillation of the transparent scale and making the proper correction of the reading of the azimuth circle.

The deflecting magnet is then placed upon a straight rod which passes through the axis of the revolving apparatus at right angles to the axis of the telescope, and is adjusted so that the axis of the deflecting magnet is in a line passing through the centre of the suspended magnet.

The whole of the revolving apparatus is then moved till the line

* See Airy's *Magnetism*.

of collimation of the suspended magnet again coincides with the axis of the telescope, and the new azimuth reading is corrected, if necessary, by the mean of the scale readings at the extremities of an oscillation.

The difference of the corrected azimuths gives the deflexion, after which we proceed as in the method of tangents, except that in the expression for D we put $\sin\theta$ instead of $\tan\theta$.

In this method there is no correction for the torsion of the suspending fibre, since the relative position of the fibre, telescope, and magnet is the same at every observation.

The axes of the two magnets remain always at right angles in this method, so that the correction for length can be more accurately made.

456.] Having thus measured the ratio of the moment of the deflecting magnet to the horizontal component of terrestrial magnetism, we have next to find the product of these quantities, by determining the moment of the couple with which terrestrial magnetism tends to turn the same magnet when its axis is deflected from the magnetic meridian.

There are two methods of making this measurement, the dynamical, in which the time of vibration of the magnet under the action of terrestrial magnetism is observed, and the statical, in which the magnet is kept in equilibrium between a measurable statical couple and the magnetic force.

The dynamical method requires simpler apparatus and is more accurate for absolute measurements, but takes up a considerable time, the statical method admits of almost instantaneous measurement, and is therefore useful in tracing the changes of the intensity of the magnetic force, but it requires more delicate apparatus, and is not so accurate for absolute measurement.

Method of Vibrations.

The magnet is suspended with its magnetic axis horizontal, and is set in vibration in small arcs. The vibrations are observed by means of any of the methods already described.

A point on the scale is chosen corresponding to the middle of the arc of vibration. The instant of passage through this point of the scale in the positive direction is observed. If there is sufficient time before the return of the magnet to the same point, the instant of passage through the point in the negative direction is also observed, and the process is continued till $n+1$ positive and

n negative passages have been observed. If the vibrations are too rapid to allow of every consecutive passage being observed, every third or every fifth passage is observed, care being taken that the observed passages are alternately positive and negative.

Let the observed times of passage be T_1, T_2, T_{2n+1}, then if we put

$$\frac{1}{n}(\tfrac{1}{2}T_1 + T_3 + T_5 + \&c. + T_{2n-1} + \tfrac{1}{2}T_{2n+1}) = T_{n+1},$$

$$\frac{1}{n}(T_2 + T_4 \;\&c. \qquad + T_{2n}) = T'_{n+1};$$

then T_{n+1} is the mean time of the positive passages, and ought to agree with T'_{n+1}, the mean time of the negative passages, if the point has been properly chosen. The mean of these results is to be taken as the mean time of the middle passage.

After a large number of vibrations have taken place, but before the vibrations have ceased to be distinct and regular, the observer makes another series of observations, from which he deduces the mean time of the middle passage of the second series.

By calculating the period of vibration either from the first series of observations or from the second, he ought to be able to be certain of the number of whole vibrations which have taken place in the interval between the time of middle passage in the two series. Dividing the interval between the mean times of middle passage in the two series by this number of vibrations, the mean time of vibration is obtained.

The observed time of vibration is then to be reduced to the time of vibration in infinitely small arcs by a formula of the same kind as that used in pendulum observations, and if the vibrations are found to diminish rapidly in amplitude, there is another correction for resistance, see Art. 740. These corrections, however, are very small when the magnet hangs by a fibre, and when the arc of vibration is only a few degrees.

The equation of motion of the magnet is

$$A\frac{d^2\theta}{dt^2} + MH\sin\theta + MH\tau'(\theta - \gamma) = 0$$

where θ is the angle between the magnetic axis and the direction of the force H, A is the moment of inertia of the magnet and suspended apparatus, M is the magnetic moment of the magnet, H the intensity of the horizontal magnetic force, and $MH\tau'$ the coefficient of torsion: τ' is determined as in Art. 452, and is a very small quantity. The value of θ for equilibrium is

$$\theta_0 = \frac{\tau'\gamma}{1+\tau'}, \quad \text{a very small angle,}$$

and the solution of the equation for small values of the amplitude, C is

$$\theta = C\cos\left(2\pi\frac{t}{T} + a\right) + \theta_0,$$

where T is the periodic time, and C the amplitude, and

$$T^2 = \frac{4\pi^2 A}{MH(1+\tau')};$$

whence we find the value of MH,

$$MH = \frac{4\pi^2 A}{T^2(1+\tau')}.$$

Here T is the time of a complete vibration determined from observation. A, the moment of inertia, is found once for all for the magnet, either by weighing and measuring it if it is of a regular figure, or by a dynamical process of comparison with a body whose moment of inertia is known.

Combining this value of MH with that of $\frac{M}{H}$ formerly obtained,

we get $$M^2 = (MH)\left(\frac{M}{H}\right) = \frac{2\pi^2 A}{T^2(1+\tau')}Dr^3,$$

and $$H^2 = (MH)\left(\frac{H}{M}\right) = \frac{8\pi^2 A}{T^2(1+\tau')Dr^3}.$$

457.] We have supposed that H and M continue constant during the two series of experiments. The fluctuations of H may be ascertained by simultaneous observations of the bifilar magnetometer to be presently described, and if the magnet has been in use for some time, and is not exposed during the experiments to changes of temperature or to concussion, the part of M which depends on permanent magnetism may be assumed to be constant. All steel magnets, however, are capable of induced magnetism depending on the action of external magnetic force.

Now the magnet when employed in the deflexion experiments is placed with its axis east and west, so that the action of terrestrial magnetism is transverse to the magnet, and does not tend to increase or diminish M. When the magnet is made to vibrate, its axis is north and south, so that the action of terrestrial magnetism tends to magnetize it in the direction of the axis, and therefore to increase its magnetic moment by a quantity kH, where k is a coefficient to be found by experiments on the magnet.

There are two ways in which this source of error may be avoided without calculating k, the experiments being arranged so that the magnet shall be in the same condition when employed in deflecting another magnet and when itself swinging.

We may place the deflecting magnet with its axis pointing north, at a distance r from the centre of the suspended magnet, the line r making an angle whose cosine is $\sqrt{\frac{1}{3}}$ with the magnetic meridian. The action of the deflecting magnet on the suspended one is then at right angles to its own direction, and is equal to

$$R = \sqrt{2}\,\frac{M}{r^3}.$$

Here M is the magnetic moment when the axis points north, as in the experiment of vibration, so that no correction has to be made for induction.

This method, however, is extremely difficult, owing to the large errors which would be introduced by a slight displacement of the deflecting magnet, and as the correction by reversing the deflecting magnet is not applicable here, this method is not to be followed except when the object is to determine the coefficient of induction.

The following method, in which the magnet while vibrating is freed from the inductive action of terrestrial magnetism, is due to Dr. J. P. Joule *.

Two magnets are prepared whose magnetic moments are as nearly equal as possible. In the deflexion experiments these magnets are used separately, or they may be placed simultaneously on opposite sides of the suspended magnet to produce a greater deflexion. In these experiments the inductive force of terrestrial magnetism is transverse to the axis.

Let one of these magnets be suspended, and let the other be placed parallel to it with its centre exactly below that of the suspended magnet, and with its axis in the same direction. The force which the fixed magnet exerts on the suspended one is in the opposite direction from that of terrestrial magnetism. If the fixed magnet be gradually brought nearer to the suspended one the time of vibration will increase, till at a certain point the equilibrium will cease to be stable, and beyond this point the suspended magnet will make oscillations in the reverse position. By experimenting in this way a position of the fixed magnet is found at which it exactly neutralizes the effect of terrestrial magnetism on the suspended one. The two magnets are fastened together so as to be parallel, with their axes turned the same way, and at the distance just found by experiment. They are then suspended in the usual way and made to vibrate together through small arcs.

* *Proc. Phil. S., Manchester*, March 19, 1867.

The lower magnet exactly neutralizes the effect of terrestrial magnetism on the upper one, and since the magnets are of equal moment, the upper one neutralizes the inductive action of the earth on the lower one.

The value of M is therefore the same in the experiment of vibration as in the experiment of deflexion, and no correction for induction is required.

458.] The most accurate method of ascertaining the intensity of the horizontal magnetic force is that which we have just described. The whole series of experiments, however, cannot be performed with sufficient accuracy in much less than an hour, so that any changes in the intensity which take place in periods of a few minutes would escape observation. Hence a different method is required for observing the intensity of the magnetic force at any instant.

The statical method consists in deflecting the magnet by means of a statical couple acting in a horizontal plane. If L be the moment of this couple, M the magnetic moment of the magnet, H the horizontal component of terrestrial magnetism, and θ the deflexion,

$$MH \sin\theta = L.$$

Hence, if L is known in terms of θ, MH can be found.

The couple L may be generated in two ways, by the torsional elasticity of a wire, as in the ordinary torsion balance, or by the weight of the suspended apparatus, as in the bifilar suspension.

In the torsion balance the magnet is fastened to the end of a vertical wire, the upper end of which can be turned round, and its rotation measured by means of a torsion circle.

We have then

$$L = \tau(a - a_0 - \theta) = MH \sin\theta.$$

Here a_0 is the value of the reading of the torsion circle when the axis of the magnet coincides with the magnetic meridian, and a is the actual reading. If the torsion circle is turned so as to bring the magnet nearly perpendicular to the magnetic meridian, so that

$$\theta = \frac{\pi}{2} - \theta', \quad \text{then} \quad \tau\left(a - a_0 - \frac{\pi}{2} + \theta'\right) = MH(1 - \tfrac{1}{2}\theta'^2),$$

$$\text{or} \quad MH = \tau(1 + \tfrac{1}{2}\theta'^2)\left(a - a_0 - \frac{\pi}{2} + \theta'\right).$$

By observing θ', the deflexion of the magnet when in equilibrium, we can calculate MH provided we know τ.

If we only wish to know the relative value of H at different times it is not necessary to know either M or τ.

We may easily determine τ in absolute measure by suspending a non-magnetic body from the same wire and observing its time of oscillation, then if A is the moment of inertia of this body, and T the time of a complete vibration,

$$\tau = \frac{4\pi^2 A}{T^2}.$$

The chief objection to the use of the torsion balance is that the zero-reading a_0 is liable to change. Under the constant twisting force, arising from the tendency of the magnet to turn to the north, the wire gradually acquires a permanent twist, so that it becomes necessary to determine the zero-reading of the torsion circle afresh at short intervals of time.

Bifilar Suspension.

459.] The method of suspending the magnet by two wires or fibres was introduced by Gauss and Weber. As the bifilar suspension is used in many electrical instruments, we shall investigate it more in detail. The general appearance of the suspension is shewn in Fig. 16, and Fig. 17 represents the projection of the wires on a horizontal plane.

AB and $A'B'$ are the projections of the two wires.

AA' and BB' are the lines joining the upper and the lower ends of the wires.

a and b are the lengths of these lines.

α and β their azimuths.

W and W' the vertical components of the tensions of the wires.

Q and Q' their horizontal components.

h the vertical distance between AA' and BB'.

The forces which act on the magnet are—its weight, the couple arising from terrestrial magnetism, the torsion of the wires (if any) and their tensions. Of these the effects of magnetism and of torsion are of the nature of couples. Hence the resultant of the tensions must consist of a vertical force, equal to the weight of the magnet, together with a couple. The resultant of the vertical components of the tensions is therefore along the line whose projection is O, the intersection of AA' and BB', and either of these lines is divided in O in the ratio of W' to W.

The horizontal components of the tensions form a couple, and are therefore equal in magnitude and parallel in direction. Calling either of them Q, the moment of the couple which they form is

$$L = Q \cdot PP', \tag{1}$$

where PP' is the distance between the parallel lines AB and $A'B'$.

To find the value of L we have the equations of moments

$$Qh = W.AB = W'.A'B', \tag{2}$$

and the geometrical equation

$$(AB+A'B')\,PP' = ab\sin(\alpha-\beta), \tag{3}$$

whence we obtain,

$$L = Q.PP' = \frac{ab}{h}\frac{WW'}{W+W'}\sin(\alpha-\beta). \tag{4}$$

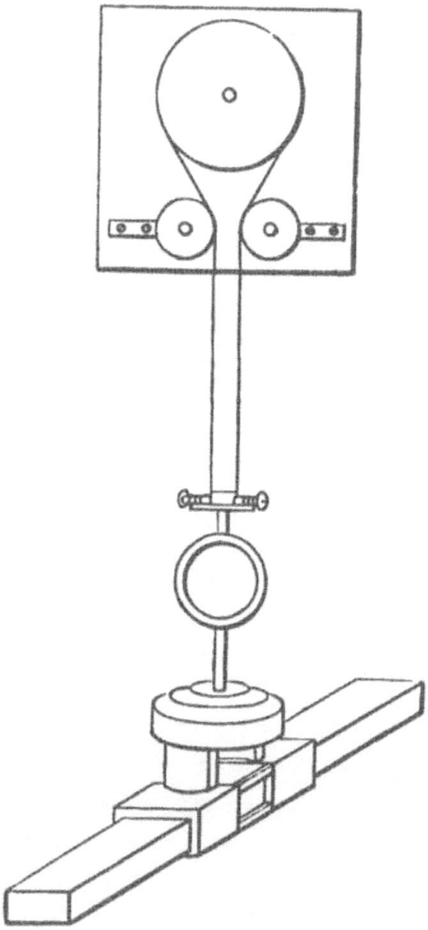

Fig. 16.

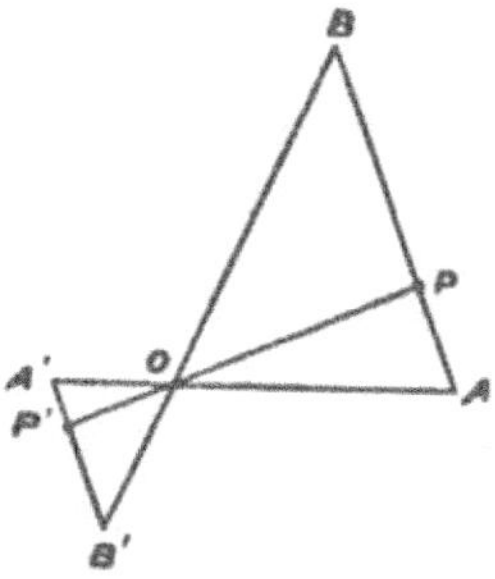

Fig. 17.

If m is the mass of the suspended apparatus, and g the intensity of gravity,

$$W+W' = mg. \tag{5}$$

If we also write

$$W-W' = nmg, \tag{6}$$

we find

$$L = \frac{1}{4}(1-n^2)mg\frac{ab}{h}\sin(\alpha-\beta). \tag{7}$$

The value of L is therefore a maximum with respect to n when n

is zero, that is, when the weight of the suspended mass is equally borne by the two wires.

We may adjust the tensions of the wires to equality by observing the time of vibration, and making it a minimum, or we may obtain a self-acting adjustment by attaching the ends of the wires, as in Fig. 16, to a pulley, which turns on its axis till the tensions are equal.

The distance of the upper ends of the suspension wires is regulated by means of two other pullies. The distance between the lower ends of the wires is also capable of adjustment.

By this adjustment of the tension, the couple arising from the tensions of the wires becomes

$$L = \frac{1}{4}\frac{ab}{h} mg \sin(\alpha-\beta).$$

The moment of the couple arising from the torsion of the wires is of the form

$$\tau(\gamma-\beta),$$

where τ is the sum of the coefficients of torsion of the wires.

The wires ought to be without torsion when $\alpha = \beta$, we may then make $\gamma = \alpha$.

The moment of the couple arising from the horizontal magnetic force is of the form

$$MH \sin(\delta-\theta),$$

where δ is the magnetic declination, and θ is the azimuth of the axis of the magnet. We shall avoid the introduction of unnecessary symbols without sacrificing generality if we assume that the axis of the magnet is parallel to BB', or that $\beta = \theta$.

The equation of motion then becomes

$$A\frac{d^2\theta}{dt^2} = MH\sin(\delta-\theta) + \frac{1}{4}\frac{ab}{h} mg\sin(\alpha-\theta) + \tau(\alpha-\theta). \qquad (8)$$

There are three principal positions of this apparatus.

(1) When α is nearly equal to δ. If T_1 is the time of a complete oscillation in this position, then

$$\frac{4\pi^2 A}{T_1^2} = \frac{1}{4}\frac{ab}{h} mg + \tau + MH. \qquad (9)$$

(2) When α is nearly equal to $\delta+\pi$. If T_2 is the time of a complete oscillation in this position, the north end of the magnet being now turned towards the south,

$$\frac{4\pi^2 A}{T_2^2} = \frac{1}{4}\frac{ab}{h} mg + \tau - MH. \qquad (10)$$

The quantity on the right-hand of this equation may be made

as small as we please by diminishing a or b, but it must not be made negative, or the equilibrium of the magnet will become unstable. The magnet in this position forms an instrument by which small variations in the *direction* of the magnetic force may be rendered sensible.

For when $\delta-\theta$ is nearly equal to π, $\sin(\delta-\theta)$ is nearly equal to $\theta-\delta$, and we find

$$\theta = a - \frac{MH}{\frac{1}{4}\frac{ab}{h}mg+\tau-MH}(\delta-a). \qquad (11)$$

By diminishing the denominator of the fraction in the last term we may make the variation of θ very large compared with that of δ. We should notice that the coefficient of δ in this expression is negative, so that when the direction of the magnetic force turns in one direction the magnet turns in the opposite direction.

(3) In the third position the upper part of the suspension-apparatus is turned round till the axis of the magnet is nearly perpendicular to the magnetic meridian.

If we make

$$\theta-\delta = \frac{\pi}{2}+\theta', \quad \text{and} \quad a-\theta = \beta-\theta', \qquad (12)$$

the equation of motion may be written

$$A\frac{d^2\theta'}{dt^2} = MH\cos\theta' + \frac{1}{4}\frac{ab}{h}mg\sin(\beta-\theta')+\tau(\beta-\theta'). \qquad (13)$$

If there is equilibrium when $H=H_0$ and $\theta'=0$,

$$MH_0 + \frac{1}{4}\frac{ab}{h}mg\sin\beta+\beta\tau = 0, \qquad (14)$$

and if H is the value of the horizontal force corresponding to a small angle θ',

$$H = H_0\left(1 - \frac{\frac{1}{4}\frac{ab}{h}mg\cos\beta+\tau}{\frac{1}{4}\frac{ab}{h}mg\sin\beta+\tau\beta}\theta'\right). \qquad (15)$$

In order that the magnet may be in stable equilibrium it is necessary that the numerator of the fraction in the second member should be positive, but the more nearly it approaches zero, the more sensitive will be the instrument in indicating changes in the value of the intensity of the horizontal component of terrestrial magnetism.

The statical method of estimating the intensity of the force depends upon the action of an instrument which of itself assumes different positions of equilibrium for different values of the force. Hence, by means of a mirror attached to the magnet and throwing a spot of light upon a photographic surface moved by clockwork, a curve may be traced, from which the intensity of the force at any instant may be determined according to a scale, which we may for the present consider an arbitrary one.

460.] In an observatory, where a continuous system of registration of declination and intensity is kept up either by eye observation or by the automatic photographic method, the absolute values of the declination and of the intensity, as well as the position and moment of the magnetic axis of a magnet, may be determined to a greater degree of accuracy.

For the declinometer gives the declination at every instant affected by a constant error, and the bifilar magnetometer gives the intensity at every instant multiplied by a constant coefficient. In the experiments we substitute for δ, $\delta+\delta_0$ where δ' is the reading of the declinometer at the given instant, and δ_0 is the unknown but constant error, so that $\delta'+\delta_0$ is the true declination at that instant.

In like manner for H, we substitute CH' where H' is the reading of the magnetometer on its arbitrary scale, and C is an unknown but constant multiplier which converts these readings into absolute measure, so that CH' is the horizontal force at a given instant.

The experiments to determine the absolute values of the quantities must be conducted at a sufficient distance from the declinometer and magnetometer, so that the different magnets may not sensibly disturb each other. The time of every observation must be noted and the corresponding values of δ' and H' inserted. The equations are then to be treated so as to find δ_0, the constant error of the declinometer, and C the coefficient to be applied to the readings of the magnetometer. When these are found the readings of both instruments may be expressed in absolute measure. The absolute measurements, however, must be frequently repeated in order to take account of changes which may occur in the magnetic axis and magnetic moment of the magnets.

461.] The methods of determining the vertical component of the terrestrial magnetic force have not been brought to the same degree of precision. The vertical force must act on a magnet which turns about a horizontal axis. Now a body which turns about a horizontal axis cannot be made so sensitive to the action of small forces as a body which is suspended by a fibre and turns about a vertical axis. Besides this, the weight of a magnet is so large compared with the magnetic force exerted upon it that a small displacement of the centre of inertia by unequal dilatation, &c. produces a greater effect on the position of the magnet than a considerable change of the magnetic force.

Hence the measurement of the vertical force, or the comparison of the vertical and the horizontal forces, is the least perfect part of the system of magnetic measurements.

The vertical part of the magnetic force is generally deduced from the horizontal force by determining the direction of the total force.

If i be the angle which the total force makes with its horizontal component, i is called the magnetic Dip or Inclination, and if H is the horizontal force already found, then the vertical force is $H \tan i$, and the total force is $H \sec i$.

The magnetic dip is found by means of the Dip Needle.

The theoretical dip-needle is a magnet with an axis which passes through its centre of inertia perpendicular to the magnetic axis of the needle. The ends of this axis are made in the form of cylinders of small radius, the axes of which are coincident with the line passing through the centre of inertia. These cylindrical ends rest on two horizontal planes and are free to roll on them.

When the axis is placed magnetic east and west, the needle is free to rotate in the plane of the magnetic meridian, and if the instrument is in perfect adjustment, the magnetic axis will set itself in the direction of the total magnetic force.

It is, however, practically impossible to adjust a dip-needle so that its weight does not influence its position of equilibrium, because its centre of inertia, even if originally in the line joining the centres of the rolling sections of the cylindrical ends, will cease to be in this line when the needle is imperceptibly bent or unequally expanded. Besides, the determination of the true centre of inertia of a magnet is a very difficult operation, owing to the interference of the magnetic force with that of gravity.

Let us suppose one end of the needle and one end of the pivot to be marked. Let a line, real or imaginary, be drawn on the needle, which we shall call the Line of Collimation. The position of this line is read off on a vertical circle. Let θ be the angle which this line makes with the radius to zero, which we shall suppose to be horizontal. Let λ be the angle which the magnetic axis makes with the line of collimation, so that when the needle is in this position the line of collimation is inclined $\theta + \lambda$ to the horizontal.

Let p be the perpendicular from the centre of inertia on the plane on which the axis rolls, then p will be a function of θ, whatever be the shape of the rolling surfaces. If both the rolling sections of the ends of the axis are circular,

$$p = c - a \sin(\theta + \alpha) \qquad (1)$$

where a is the distance of the centre of inertia from the line joining the centres of the rolling sections, and α is the angle which this line makes with the line of collimation.

If M is the magnetic moment, m the mass of the magnet, and g the force of gravity, I the total magnetic force, and i the dip, then, by the conservation of energy, when there is stable equilibrium,

$$MI \cos(\theta + \lambda - i) - mgp \qquad (2)$$

must be a maximum with respect to θ, or

$$MI \sin(\theta + \lambda - i) = -mg\frac{dp}{d\theta}, \qquad (3)$$

$$= -mga \cos(\theta + \alpha),$$

if the ends of the axis are cylindrical.

Also, if T be the time of vibration about the position of equilibrium,

$$MI + mga \sin(\theta + \alpha) = \frac{4\pi^2 A}{T^2} \qquad (4)$$

where A is the moment of inertia of the needle about its axis of rotation.

In determining the dip a reading is taken with the dip circle in the magnetic meridian and with the graduation towards the west. Let θ_1 be this reading, then we have

$$MI \sin(\theta_1 + \lambda - i) = -mga \cos(\theta_1 + \alpha). \qquad (5)$$

The instrument is now turned about a vertical axis through 180°, so that the graduation is to the east, and if θ_2 is the new reading,

$$MI \sin(\theta_2 + \lambda - \pi + i) = -mga \cos(\theta_2 + \alpha). \qquad (6)$$

Taking (6) from (5), and remembering that θ_1 is nearly equal to i, and θ_2 nearly equal to $\pi - i$, and that λ is a small angle, such that $mga\lambda$ may be neglected in comparison with MI,

$$MI(\theta_1 - \theta_2 + \pi - 2i) = -2\, mga \cos i \cos \alpha. \qquad (7)$$

Now take the magnet from its bearings and place it in the deflexion apparatus, Art. 453, so as to indicate its own magnetic moment by the deflexion of a suspended magnet, then

$$M = \tfrac{1}{2} r^3 HD \qquad (8)$$

where D is the tangent of the deflexion.

Next, reverse the magnetism of the needle and determine its new magnetic moment M', by observing a new deflexion, the tangent of which is D',

$$M' = \tfrac{1}{2} r^3 H D', \tag{9}$$

whence

$$MD' = M'D. \tag{10}$$

Then place it on its bearings and take two readings, θ_3 and θ_4, in which θ_3 is nearly $\pi + i$, and θ_4 nearly $-i$,

$$M'I' \sin(\theta_3 + \lambda' - \pi - i) = m g a \cos(\theta_3 + a), \tag{11}$$

$$M'I' \sin(\theta_4 + \lambda' + i) \quad = m g a \cos(\theta_4 + a), \tag{12}$$

whence, as before,

$$M'I(\theta_3 - \theta_4 - \pi - 2i) = 2 m g a \cos i \cos a, \tag{13}$$

adding (8),

$$MI(\theta_1 - \theta_2 + \pi - 2i) + M'I(\theta_3 - \theta_4 - \pi - 2i) = 0, \tag{14}$$

or

$$D(\theta_1 - \theta_2 + \pi - 2i) + D'(\theta_3 - \theta_4 - \pi - 2i) = 0, \tag{15}$$

whence we find the dip

$$i = \frac{D(\theta_1 - \theta_2 + \pi) + D'(\theta_3 - \theta_4 - \pi)}{2D + 2D'}, \tag{16}$$

where D and D' are the tangents of the deflexions produced by the needle in its first and second magnetizations respectively.

In taking observations with the dip circle the vertical axis is carefully adjusted so that the plane bearings upon which the axis of the magnet rests are horizontal in every azimuth. The magnet being magnetized so that the end A dips, is placed with its axis on the plane bearings, and observations are taken with the plane of the circle in the magnetic meridian, and with the graduated side of the circle east. Each end of the magnet is observed by means of reading microscopes carried on an arm which moves concentric with the dip circle. The cross wires of the microscope are made to coincide with the image of a mark on the magnet, and the position of the arm is then read off on the dip circle by means of a vernier.

We thus obtain an observation of the end A and another of the end B when the graduations are east. It is necessary to observe both ends in order to eliminate any error arising from the axle of the magnet not being concentric with the dip circle.

The graduated side is then turned west, and two more observations are made.

The magnet is then turned round so that the ends of the axle are reversed, and four more observations are made looking at the other side of the magnet.

The magnetization of the magnet is then reversed so that the end B dips, the magnetic moment is ascertained, and eight observations are taken in this state, and the sixteen observations combined to determine the true dip.

462.] It is found that in spite of the utmost care the dip, as thus deduced from observations made with one dip circle, differs perceptibly from that deduced from observations with another dip circle at the same place. Mr. Broun has pointed out the effect due to ellipticity of the bearings of the axle, and how to correct it by taking observations with the magnet magnetized to different strengths.

The principle of this method may be stated thus. We shall suppose that the error of any one observation is a small quantity not exceeding a degree. We shall also suppose that some unknown but regular force acts upon the magnet, disturbing it from its true position.

If L is the moment of this force, θ_0 the true dip, and θ the observed dip, then

$$L = MI\sin(\theta-\theta_0), \tag{17}$$

$$= MI(\theta-\theta_0), \tag{18}$$

since $\theta-\theta_0$ is small.

It is evident that the greater M becomes the nearer does the needle approach its proper position. Now let the operation of taking the dip be performed twice, first with the magnetization equal to M_1, the greatest that the needle is capable of, and next with the magnetization equal to M_2, a much smaller value but sufficient to make the readings distinct and the error still moderate. Let θ_1 and θ_2 be the dips deduced from these two sets of observations, and let L be the mean value of the unknown disturbing force for the eight positions of each determination, which we shall suppose the same for both determinations. Then

$$L = M_1 I(\theta_1-\theta_0) = M_2 I(\theta_2-\theta_0). \tag{19}$$

Hence $$\theta_0 = \frac{M_1\theta_1 - M_2\theta_2}{M_1-M_2}, \qquad L = M_1 M_2 I\frac{\theta_1-\theta_2}{M_2-M_1}. \tag{20}$$

If we find that several experiments give nearly equal values for L, then we may consider that θ_0 must be very nearly the true value of the dip.

463.] Dr. Joule has recently constructed a new dip-circle, in which the axis of the needle, instead of rolling on horizontal agate planes, is slung on two filaments of silk or spider's thread, the ends of the filaments being attached to the arms of a delicate balance. The axis of the needle thus rolls on two loops of silk fibre, and Dr. Joule finds that its freedom of motion is much greater than when it rolls on agate planes.

In Fig. 18, *NS* is the needle, *CC′* is its axis, consisting of a straight cylindrical wire, and *PCQ*, *P′C′Q′* are the filaments on which the axis rolls. *POQ* is the balance, consisting of a double bent lever supported by a wire, *OO*, stretched horizontally between the prongs of a forked piece, and having a counterpoise *R* which can be screwed up or down, so that the balance is in neutral equilibrium about *OO*.

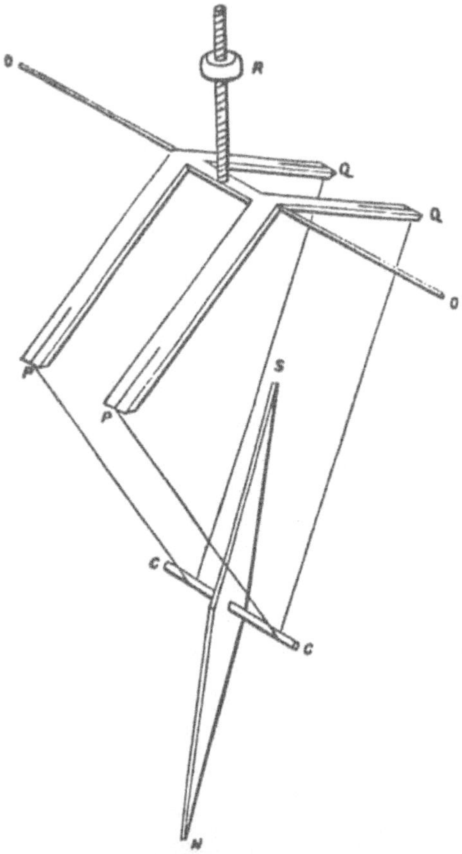

Fig. 18.

In order that the needle may be in neutral equilibrium as the needle rolls on the filaments the centre of gravity must neither rise nor fall. Hence the distance *OC* must remain constant as the needle rolls. This condition will be fulfilled if the arms of the balance *OP* and *OQ* are equal, and if the filaments are at right angles to the arms.

Dr. Joule finds that the needle should not be more than five inches long. When it is eight inches long, the bending of the needle tends to diminish the apparent dip by a fraction of a minute. The axis of the needle was originally of steel wire, straightened by being brought to a red heat while stretched by a weight, but Dr. Joule found that with the new suspension it is not necessary to use steel wire, for platinum and even standard gold are hard enough.

The balance is attached to a wire *OO* about a foot long stretched horizontally between the prongs of a fork. This fork is turned round in azimuth by means of a circle at the top of a tripod which supports the whole. Six complete observations of the dip can be obtained in one hour, and the average error of a single observation is a fraction of a minute of arc.

It is proposed that the dip needle in the Cambridge Physical Laboratory shall be observed by means of a double image instrument, consisting of two totally reflecting prisms placed as in Fig. 19 and mounted on a vertical graduated circle, so that the plane of reflexion may be turned round a horizontal axis nearly coinciding with the prolongation of the axis of the suspended dip-needle. The needle is viewed by means of a telescope placed behind the prisms, and the two ends of the needle are seen together as in Fig. 20. By turning the prisms about the axis of the vertical circle, the images of two lines drawn on the needle may be made to coincide. The inclination of the needle is thus determined from the reading of the vertical circle.

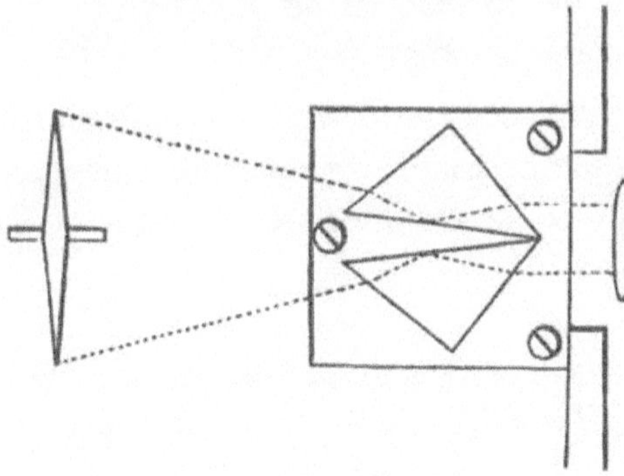

Fig. 19.

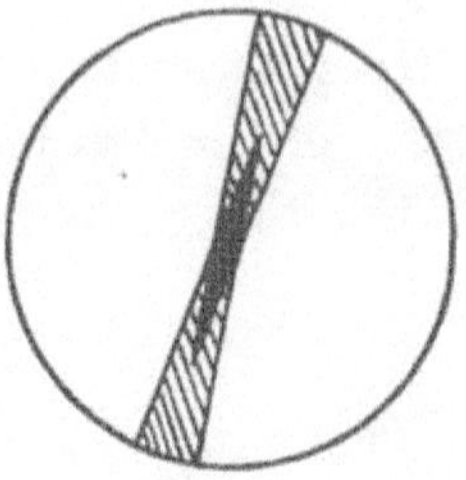

Fig. 20.

The total intensity I of the magnetic force in the line of dip may be deduced as follows from the times of vibration T_1, T_2, T_3, T_4 in the four positions already described,

$$I = \frac{4\pi^2 A}{2M+2M'}\left\{\frac{1}{T_1^2}+\frac{1}{T_2^2}+\frac{1}{T_3^2}+\frac{1}{T_4^2}\right\}.$$

The values of M and M' must be found by the method of deflexion and vibration formerly described, and A is the moment of inertia of the magnet about its axle.

The observations with a magnet suspended by a fibre are so much more accurate that it is usual to deduce the total force from the horizontal force from the equation

$$I = H \sec\theta,$$

where I is the total force, H the horizontal force, and θ the dip.

464.] The process of determining the dip being a tedious one, is not suitable for determining the continuous variation of the magnetic force. The most convenient instrument for continuous observations is the vertical force magnetometer, which is simply a magnet balanced on knife edges so as to be in stable equilibrium with its magnetic axis nearly horizontal.

If Z is the vertical component of the magnetic force, M the magnetic moment, and θ the small angle which the magnetic axis makes with the horizon

$$MZ = m g a \cos(\alpha - \theta),$$

where m is the mass of the magnet, g the force of gravity, a the distance of the centre of gravity from the axis of suspension, and α the angle which the plane through the axis and the centre of gravity makes with the magnetic axis.

Hence, for the small variation of vertical force δZ, there will be a variation of the angular position of the magnet $\delta\theta$ such that

$$M \delta Z = m g a \sin(\alpha - \theta)\, \delta\theta.$$

In practice this instrument is not used to determine the absolute value of the vertical force, but only to register its small variations.

For this purpose it is sufficient to know the absolute value of Z when $\theta = 0$, and the value of $\frac{dZ}{d\theta}$.

The value of Z, when the horizontal force and the dip are known, is found from the equation $Z = H \tan\theta_0$, where θ_0 is the dip and H the horizontal force.

To find the deflexion due to a given variation of Z, take a magnet and place it with its axis east and west, and with its centre at a known distance r_1 east or west from the declinometer, as in experiments on deflexion, and let the tangent of deflexion be D_1.

Then place it with its axis vertical and with its centre at a distance r_2 above or below the centre of the vertical force magnetometer, and let the tangent of the deflexion produced in the magnetometer be D_2. Then, if the moment of the deflecting magnet is M,

$$M = H r_1^3 D_1 = \frac{dZ}{d\theta} r_2^3 D_2.$$

Hence $$\frac{dZ}{d\theta} = H \frac{r_1^3}{r_2^3} \frac{D_1}{D_2}.$$

The actual value of the vertical force at any instant is

$$Z = Z_0 + \theta \frac{dZ}{d\theta},$$

where Z_0 is the value of Z when $\theta = 0$.

For continuous observations of the variations of magnetic force at a fixed observatory the Unifilar Declinometer, the Bifilar Horizontal Force Magnetometer, and the Balance Vertical Force Magnetometer are the most convenient instruments.

At several observatories photographic traces are now produced on prepared paper moved by clock work, so that a continuous record of the indications of the three instruments at every instant is formed. These traces indicate the variation of the three rectangular components of the force from their standard values. The declinometer gives the force towards mean magnetic west, the bifilar magnetometer gives the variation of the force towards magnetic north, and the balance magnetometer gives the variation of the vertical force. The standard values of these forces, or their values when these instruments indicate their several zeros, are deduced by frequent observations of the absolute declination, horizontal force, and dip.

CHAPTER VIII.

ON TERRESTRIAL MAGNETISM.

465.] OUR knowledge of Terrestrial Magnetism is derived from the study of the distribution of magnetic force on the earth's surface at any one time, and of the changes in that distribution at different times.

The magnetic force at any one place and time is known when its three coordinates are known. These coordinates may be given in the form of the declination or azimuth of the force, the dip or inclination to the horizon, and the total intensity.

The most convenient method, however, for investigating the general distribution of magnetic force on the earth's surface is to consider the magnitudes of the three components of the force,

$$\left.\begin{aligned} X &= H\cos\delta, \text{ directed due north,} \\ Y &= H\sin\delta, \text{ directed due west,} \\ Z &= H\tan\theta, \text{ directed vertically downwards,} \end{aligned}\right\} \quad (1)$$

where H denotes the horizontal force, δ the declination, and θ the dip.

If V is the magnetic potential at the earth's surface, and if we consider the earth a sphere of radius a, then

$$X = \frac{1}{a}\frac{dV}{dl}, \quad Y = \frac{1}{a\cos l}\frac{dV}{d\lambda}, \quad Z = \frac{dV}{dr}, \quad (2)$$

where l is the latitude, and λ the longitude, and r the distance from the centre of the earth.

A knowledge of V over the surface of the earth may be obtained from the observations of horizontal force alone as follows.

Let V_0 be the value of V at the true north pole, then, taking the line-integral along any meridian, we find,

$$V = a\int_{\frac{\pi}{2}}^{l} X\,dl + V_0, \quad (3)$$

for the value of the potential on that meridian at latitude l.

Thus the potential may be found for any point on the earth's surface provided we know the value of X, the northerly component at every point, and V_0, the value of V at the pole.

Since the forces depend not on the absolute value of V but on its derivatives, it is not necessary to fix any particular value for V_0.

The value of V at any point may be ascertained if we know the value of X along any given meridian, and also that of Y over the whole surface.

Let
$$V_l = a\int_{\frac{\pi}{2}}^{l} X\,dl + V_0, \tag{4}$$
where the integration is performed along the given meridian from the pole to the parallel l, then
$$V = V_l + a\int_{\lambda_0}^{\lambda} Y\cos l\,d\lambda, \tag{5}$$
where the integration is performed along the parallel l from the given meridian to the required point.

These methods imply that a complete magnetic survey of the earth's surface has been made, so that the values of X or of Y or of both are known for every point of the earth's surface at a given epoch. What we actually know are the magnetic components at a certain number of stations. In the civilized parts of the earth these stations are comparatively numerous; in other places there are large tracts of the earth's surface about which we have no data.

Magnetic Survey.

466.] Let us suppose that in a country of moderate size, whose greatest dimensions are a few hundred miles, observations of the declination and the horizontal force have been taken at a considerable number of stations distributed fairly over the country.

Within this district we may suppose the value of V to be represented with sufficient accuracy by the formula
$$V = V_0 + a(A_1 l + A_2\lambda + \tfrac{1}{2}B_1 l^2 + B_2 l\lambda + \tfrac{1}{2}B_3\lambda^2 + \&c.), \tag{6}$$
whence
$$X = A_1 + B_1 l + B_2\lambda, \tag{7}$$
$$Y\cos l = A_2 + B_2 l + B_3\lambda. \tag{8}$$

Let there be n stations whose latitudes are $l_1, l_2, \ldots$ &c. and longitudes λ_1, λ_2, &c., and let X and Y be found for each station.

Let
$$l_0 = \frac{1}{n}\Sigma(l), \quad \text{and} \quad \lambda_0 = \frac{1}{n}\Sigma(\lambda), \tag{9}$$

l_0 and λ_0 may be called the latitude and longitude of the central station. Let

$$X_0 = \frac{1}{n}\Sigma(X), \quad \text{and} \quad Y_0 \cos l_0 = \frac{1}{n}\Sigma(Y\cos l), \tag{10}$$

then X_0 and Y_0 are the values of X and Y at the imaginary central station, then

$$X = X_0 + B_1(l-l_0) + B_2(\lambda-\lambda_0), \tag{11}$$

$$Y\cos l = Y_0 \cos l_0 + B_2(l-l_0) + B_3(\lambda-\lambda_0). \tag{12}$$

We have n equations of the form of (11) and n of the form (12). If we denote the probable error in the determination of X by ξ, and that of $Y\cos l$ by η, then we may calculate ξ and η on the supposition that they arise from errors of observation of H and δ.

Let the probable error of H be h, and that of δ, d, then since

$$dX = \cos\delta\,.\,dH - H\sin\delta\,.\,d\delta,$$

$$\xi^2 = h^2\cos^2\delta + d^2 H^2 \sin^2\delta.$$

Similarly $$\eta^2 = h^2 \sin^2\delta + d^2 H^2 \cos^2\delta.$$

If the variations of X and Y from their values as given by equations of the form (11) and (12) considerably exceed the probable errors of observation, we may conclude that they are due to local attractions, and then we have no reason to give the ratio of ξ to η any other value than unity.

According to the method of least squares we multiply the equations of the form (11) by η, and those of the form (12) by ξ to make their probable error the same. We then multiply each equation by the coefficient of one of the unknown quantities B_1, B_2, or B_3 and add the results, thus obtaining three equations from which to find B_1, B_2, and B_3.

$$P_1 = B_1 b_1 + B_2 b_2,$$

$$(\eta^2 P_2 + \xi^2 Q_1) = B_1 \eta^2 b_2 + B_2(\xi^2 b_1 + \eta^2 b_3) + B_3 \xi^2 b_2,$$

$$Q_2 = \qquad B_2 b_2 \qquad + B_3 b_3;$$

in which we write for conciseness,

$$b_1 = \Sigma(l^2) - n\,l_0^2, \quad b_2 = \Sigma(l\lambda) - n\,l_0\lambda_0, \quad b_3 = \Sigma(\lambda^2) - n\lambda_0^2,$$

$$P_1 = \Sigma(lX) - n\,l_0 X_0, \qquad Q_1 = \Sigma(lY\cos l) - n\,l_0 Y_0 \cos l_0,$$

$$P_2 = \Sigma(\lambda X) - n\lambda_0 X_0, \qquad Q_2 = \Sigma(\lambda Y\cos l) - n\lambda_0 Y_0 \cos l_0.$$

By calculating B_1, B_2, and B_3, and substituting in equations (11) and (12), we can obtain the values of X and Y at any point within the limits of the survey free from the local disturbances which are found to exist where the rock near the station is magnetic, as most igneous rocks are.

Surveys of this kind can be made only in countries where magnetic instruments can be carried about and set up in a great many stations. For other parts of the world we must be content to find the distribution of the magnetic elements by interpolation between their values at a few stations at great distances from each other.

467.] Let us now suppose that by processes of this kind, or by the equivalent graphical process of constructing charts of the lines of equal values of the magnetic elements, the values of X and Y, and thence of the potential V, are known over the whole surface of the globe. The next step is to expand V in the form of a series of spherical surface harmonics.

If the earth were magnetized uniformly and in the same direction throughout its interior, V would be an harmonic of the first degree, the magnetic meridians would be great circles passing through two magnetic poles diametrically opposite, the magnetic equator would be a great circle, the horizontal force would be equal at all points of the magnetic equator, and if H_0 is this constant value, the value at any other point would be $H = H_0 \cos l'$, where l' is the magnetic latitude. The vertical force at any point would be $Z = 2 H_0 \sin l'$, and if θ is the dip, $\tan\theta = 2\tan l'$.

In the case of the earth, the magnetic equator is defined to be the line of no dip. It is not a great circle of the sphere.

The magnetic poles are defined to be the points where there is no horizontal force or where the dip is 90°. There are two such points, one in the northern and one in the southern regions, but they are not diametrically opposite, and the line joining them is not parallel to the magnetic axis of the earth.

468.] The magnetic poles are the points where the value of V on the surface of the earth is a maximum or minimum, or is stationary.

At any point where the potential is a minimum the north end of the dip-needle points vertically downwards, and if a compass-needle be placed anywhere near such a point, the north end will point towards that point.

At points where the potential is a maximum the south end of the dip-needle points downwards, and the south end of the compass-needle points towards the point.

If there are p minima of V on the earth's surface there must be $p-1$ other points, where the north end of the dip-needle points downwards, but where the compass-needle, when carried in a circle round the point, instead of revolving so that its north end points constantly to the centre, revolves in the opposite direction, so as to turn sometimes its north end and sometimes its south end towards the point.

If we call the points where the potential is a minimum true north poles, then these other points may be called false north poles, because the compass-needle is not true to them. If there are p true north poles, there must be $p-1$ false north poles, and in like manner, if there are q true south poles, there must be $q-1$ false south poles. The number of poles of the same name must be odd, so that the opinion at one time prevalent, that there are two north poles and two south poles, is erroneous. According to Gauss there is in fact only one true north pole and one true south pole on the earth's surface, and therefore there are no false poles. The line joining these poles is not a diameter of the earth, and it is not parallel to the earth's magnetic axis.

469.] Most of the early investigators into the nature of the earth's magnetism endeavoured to express it as the result of the action of one or more bar magnets, the position of the poles of which were to be determined. Gauss was the first to express the distribution of the earth's magnetism in a perfectly general way by expanding its potential in a series of solid harmonics, the coefficients of which he determined for the first four degrees. These coefficients are 24 in number, 3 for the first degree, 5 for the second, 7 for the third, and 9 for the fourth. All these terms are found necessary in order to give a tolerably accurate representation of the actual state of the earth's magnetism.

To find what Part of the Observed Magnetic Force is due to External and what to Internal Causes.

470.] Let us now suppose that we have obtained an expansion of the magnetic potential of the earth in spherical harmonics, consistent with the actual direction and magnitude of the horizontal force at every point on the earth's surface, then Gauss has shewn how to determine, from the observed vertical force, whether the magnetic forces are due to causes, such as magnetization or electric currents, within the earth's surface, or whether any part is directly due to causes exterior to the earth's surface.

Let V be the actual potential expanded in a double series of spherical harmonics,

$$V = A_1 \frac{r}{a} + \&c. + A_i \left(\frac{r}{a}\right)^i,$$
$$+ B_1 \left(\frac{r}{a}\right)^{-2} + \&c. + B_i \left(\frac{r}{a}\right)^{-(i+1)}.$$

The first series represents the part of the potential due to causes exterior to the earth, and the second series represents the part due to causes within the earth.

The observations of horizontal force give us the sum of these series when $r = a$, the radius of the earth. The term of the order i is

$$V_i = A_i + B_i.$$

The observations of vertical force give us

$$Z = \frac{dV}{dr},$$

and the term of the order i in aZ is

$$a Z_i = i A_i - (i+1) B_i.$$

Hence the part due to external causes is

$$A_i = \frac{(i+1) V_i + a Z_i}{2i+1},$$

and the part due to causes within the earth is

$$B_i = \frac{i V_i - a Z_i}{2i+1}.$$

The expansion of V has hitherto been calculated only for the mean value of V at or near certain epochs. No appreciable part of this mean value appears to be due to causes external to the earth.

471.] We do not yet know enough of the form of the expansion of the solar and lunar parts of the variations of V to determine *by this method* whether any part of these variations arises from magnetic force acting from without. It is certain, however, as the calculations of MM. Stoney and Chambers have shewn, that the principal part of these variations cannot arise from any direct magnetic action of the sun or moon, supposing these bodies to be magnetic *.

472.] The principal changes in the magnetic force to which attention has been directed are as follows.

* Professor Hornstein of Prague has discovered a periodic change in the magnetic elements, the period of which is 26.33 days, almost exactly equal to that of the synodic revolution of the sun, as deduced from the observation of sun-spots near his equator. This method of discovering the time of rotation of the unseen solid body of the sun by its effects on the magnetic needle is the first instalment of the repayment by Magnetism of its debt to Astronomy. *Akad.*, Wien, June 15, 1871. See *Proc. R. S.*, Nov. 16, 1871.

I. *The more Regular Variations.*

(1) The Solar variations, depending on the hour of the day and the time of the year.

(2) The Lunar variations, depending on the moon's hour angle and on her other elements of position.

(3) These variations do not repeat themselves in different years, but seem to be subject to a variation of longer period of about eleven years.

(4) Besides this, there is a secular alteration in the state of the earth's magnetism, which has been going on ever since magnetic observations have been made, and is producing changes of the magnetic elements of far greater magnitude than any of the variations of small period.

II. *The Disturbances.*

473.] Besides the more regular changes, the magnetic elements are subject to sudden disturbances of greater or less amount. It is found that these disturbances are more powerful and frequent at one time than at another, and that at times of great disturbance the laws of the regular variations are masked, though they are very distinct at times of small disturbance. Hence great attention has been paid to these disturbances, and it has been found that disturbances of a particular kind are more likely to occur at certain times of the day, and at certain seasons and intervals of time, though each individual disturbance appears quite irregular. Besides these more ordinary disturbances, there are occasionally times of excessive disturbance, in which the magnetism is strongly disturbed for a day or two. These are called Magnetic Storms. Individual disturbances have been sometimes observed at the same instant in stations widely distant.

Mr. Airy has found that a large proportion of the disturbances at Greenwich correspond with the electric currents collected by electrodes placed in the earth in the neighbourhood, and are such as would be directly produced in the magnet if the earth-current, retaining its actual direction, were conducted through a wire placed *underneath* the magnet.

It has been found that there is an epoch of maximum disturbance every eleven years, and that this appears to coincide with the epoch of maximum number of spots in the sun.

474.] The field of investigation into which we are introduced by the study of terrestrial magnetism is as profound as it is extensive.

We know that the sun and moon act on the earth's magnetism. It has been proved that this action cannot be explained by supposing these bodies magnets. The action is therefore indirect. In the case of the sun part of it may be thermal action, but in the case of the moon we cannot attribute it to this cause. Is it possible that the attraction of these bodies, by causing strains in the interior of the earth, produces (Art. 447) changes in the magnetism already existing in the earth, and so by a kind of tidal action causes the semidiurnal variations?

But the amount of all these changes is very small compared with the great secular changes of the earth's magnetism.

What cause, whether exterior to the earth or in its inner depths, produces such enormous changes in the earth's magnetism, that its magnetic poles move slowly from one part of the globe to another? When we consider that the intensity of the magnetization of the great globe of the earth is quite comparable with that which we produce with much difficulty in our steel magnets, these immense changes in so large a body force us to conclude that we are not yet acquainted with one of the most powerful agents in nature, the scene of whose activity lies in those inner depths of the earth, to the knowledge of which we have so few means of access.

PART IV.

ELECTROMAGNETISM.

CHAPTER I.

ELECTROMAGNETIC FORCE.

475.] It had been noticed by many different observers that in certain cases magnetism is produced or destroyed in needles by electric discharges through them or near them, and conjectures of various kinds had been made as to the relation between magnetism and electricity, but the laws of these phenomena, and the form of these relations, remained entirely unknown till Hans Christian Örsted *, at a private lecture to a few advanced students at Copenhagen, observed that a wire connecting the ends of a voltaic battery affected a magnet in its vicinity. This discovery he published in a tract entitled *Experimenta circa effectum Conflictûs Electrici in Acum Magneticam,* dated July 21, 1820.

Experiments on the relation of the magnet to bodies charged with electricity had been tried without any result till Örsted endeavoured to ascertain the effect of a wire *heated* by an electric current. He discovered, however, that the current itself, and not the heat of the wire, was the cause of the action, and that the 'electric conflict acts in a revolving manner,' that is, that a magnet placed near a wire transmitting an electric current tends to set itself perpendicular to the wire, and with the same end always pointing forwards as the magnet is moved round the wire.

476.] It appears therefore that in the space surrounding a wire

* See another account of Örsted's discovery in a letter from Professor Hansteen in the *Life of Faraday* by Dr. Bence Jones, vol. ii. p. 395.

transmitting an electric current a magnet is acted on by forces depending on the position of the wire and on the strength of the current. The space in which these forces act may therefore be considered as a magnetic field, and we may study it in the same way as we have already studied the field in the neighbourhood of ordinary magnets, by tracing the course of the lines of magnetic force, and measuring the intensity of the force at every point.

477.] Let us begin with the case of an indefinitely long straight wire carrying an electric current. If a man were to place himself in imagination in the position of the wire, so that the current should flow from his head to his feet, then a magnet suspended freely before him would set itself so that the end which points north would, under the action of the current, point to his right hand.

The lines of magnetic force are everywhere at right angles to planes drawn through the wire, and are therefore circles each in a plane perpendicular to the wire, which passes through its centre. The pole of a magnet which points north, if carried round one of these circles from left to right, would experience a force acting always in the direction of its motion. The other pole of the same magnet would experience a force in the opposite direction.

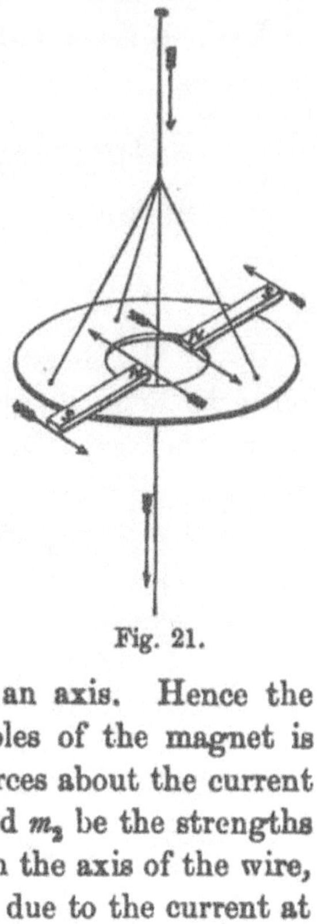

Fig. 21.

478.] To compare these forces let the wire be supposed vertical, and the current a descending one, and let a magnet be placed on an apparatus which is free to rotate about a vertical axis coinciding with the wire. It is found that under these circumstances the current has no effect in causing the rotation of the apparatus as a whole about itself as an axis. Hence the action of the vertical current on the two poles of the magnet is such that the statical moments of the two forces about the current as an axis are equal and opposite. Let m_1 and m_2 be the strengths of the two poles, r_1 and r_2 their distances from the axis of the wire, T_1 and T_2 the intensities of the magnetic force due to the current at the two poles respectively, then the force on m_1 is $m_1 T_1$, and since it is at right angles to the axis its moment is $m_1 T_1 r_1$. Similarly that of the force on the other pole is $m_2 T_2 r_2$, and since there is no motion observed,

$$m_1 T_1 r_1 + m_2 T_2 r_2 = 0.$$

But we know that in all magnets

$$m_1 + m_2 = 0.$$

Hence $$T_1 r_1 = T_2 r_2,$$

or the electromagnetic force due to a straight current of infinite length is perpendicular to the current, and varies inversely as the distance from it.

479.] Since the product Tr depends on the strength of the current it may be employed as a measure of the current. This method of measurement is different from that founded upon electrostatic phenomena, and as it depends on the magnetic phenomena produced by electric currents it is called the Electromagnetic system of measurement. In the electromagnetic system if i is the current,

$$Tr = 2i.$$

480.] If the wire be taken for the axis of z, then the rectangular components of T are

$$X = -2i\frac{y}{r^2}, \qquad Y = 2i\frac{x}{r^2}, \qquad Z = 0.$$

Here $X dx + Y dy + Z dz$ is a complete differential, being that of

$$2i \tan^{-1}\frac{y}{x} + C.$$

Hence the magnetic force in the field can be deduced from a potential function, as in several former instances, but the potential is in this case a function having an infinite series of values whose common difference is $4\pi i$. The differential coefficients of the potential with respect to the coordinates have, however, definite and single values at every point.

The existence of a potential function in the field near an electric current is not a self-evident result of the principle of the conservation of energy, for in all actual currents there is a continual expenditure of the electric energy of the battery in overcoming the resistance of the wire, so that unless the amount of this expenditure were accurately known, it might be suspected that part of the energy of the battery may be employed in causing work to be done on a magnet moving in a cycle. In fact, if a magnetic pole, m, moves round a closed curve which embraces the wire, work is actually done to the amount of $4\pi m i$. It is only for closed paths which do not embrace the wire that the line-integral of the force vanishes. We must therefore for the present consider the law of force and the existence of a potential as resting on the evidence of the experiment already described.

481.] If we consider the space surrounding an infinite straight line we shall see that it is a cyclic space, because it returns into itself. If we now conceive a plane, or any other surface, commencing at the straight line and extending on one side of it to infinity, this surface may be regarded as a diaphragm which reduces the cyclic space to an acyclic one. If from any fixed point lines be drawn to any other point without cutting the diaphragm, and the potential be defined as the line-integral of the force taken along one of these lines, the potential at any point will then have a single definite value.

The magnetic field is now identical in all respects with that due to a magnetic shell coinciding with this surface, the strength of the shell being i. This shell is bounded on one edge by the infinite straight line. The other parts of its boundary are at an infinite distance from the part of the field under consideration.

482.] In all actual experiments the current forms a closed circuit of finite dimensions. We shall therefore compare the magnetic action of a finite circuit with that of a magnetic shell of which the circuit is the bounding edge.

It has been shewn by numerous experiments, of which the earliest are those of Ampère, and the most accurate those of Weber, that the magnetic action of a small plane circuit at distances which are great compared with the dimensions of the circuit is the same as that of a magnet whose axis is normal to the plane of the circuit, and whose magnetic moment is equal to the area of the circuit multiplied by the strength of the current.

If the circuit be supposed to be filled up by a surface bounded by the circuit and thus forming a diaphragm, and if a magnetic shell of strength i coinciding with this surface be substituted for the electric current, then the magnetic action of the shell on all distant points will be identical with that of the current.

483.] Hitherto we have supposed the dimensions of the circuit to be small compared with the distance of any part of it from the part of the field examined. We shall now suppose the circuit to be of any form and size whatever, and examine its action at any point P not in the conducting wire itself. The following method, which has important geometrical applications, was introduced by Ampère for this purpose.

Conceive any surface S bounded by the circuit and not passing through the point P. On this surface draw two series of lines crossing each other so as to divide it into elementary portions, the dimensions of which are small compared with their distance from P, and with the radii of curvature of the surface.

Round each of these elements conceive a current of strength i to flow, the direction of circulation being the same in all the elements as it is in the original circuit.

Along every line forming the division between two contiguous elements two equal currents of strength i flow in opposite directions.

The effect of two equal and opposite currents in the same place is absolutely zero, in whatever aspect we consider the currents. Hence their magnetic effect is zero. The only portions of the elementary circuits which are not neutralized in this way are those which coincide with the original circuit. The total effect of the elementary circuits is therefore equivalent to that of the original circuit.

484.] Now since each of the elementary circuits may be considered as a small plane circuit whose distance from P is great compared with its dimensions, we may substitute for it an elementary magnetic shell of strength i whose bounding edge coincides with the elementary circuit. The magnetic effect of the elementary shell on P is equivalent to that of the elementary circuit. The whole of the elementary shells constitute a magnetic shell of strength i, coinciding with the surface S and bounded by the original circuit, and the magnetic action of the whole shell on P is equivalent to that of the circuit.

It is manifest that the action of the circuit is independent of the form of the surface S, which was drawn in a perfectly arbitrary manner so as to fill it up. We see from this that the action of a magnetic shell depends only on the form of its edge and not on the form of the shell itself. This result we obtained before, at Art. 410, but it is instructive to see how it may be deduced from electromagnetic considerations.

The magnetic force due to the circuit at any point is therefore identical in magnitude and direction with that due to a magnetic shell bounded by the circuit and not passing through the point, the strength of the shell being numerically equal to that of the current. The direction of the current in the circuit is related to the direction of magnetization of the shell, so that if a man were to stand with his feet on that side of the shell which we call the positive side, and which tends to point to the north, the current in front of him would be from right to left.

485.] The magnetic potential of the circuit, however, differs from that of the magnetic shell for those points which are in the substance of the magnetic shell.

If ω is the solid angle subtended at the point P by the magnetic shell, reckoned positive when the positive or austral side of the shell is next to P, then the magnetic potential at any point not in the shell itself is $\omega\phi$, where ϕ is the strength of the shell. At any point in the substance of the shell itself we may suppose the shell divided into two parts whose strengths are ϕ_1 and ϕ_2, where $\phi_1+\phi_2=\phi$, such that the point is on the positive side of ϕ_1 and on the negative side of ϕ_2. The potential at this point is

$$\omega(\phi_1+\phi_2)-4\pi\phi_2.$$

On the negative side of the shell the potential becomes $\phi(\omega-4\pi)$. In this case therefore the potential is continuous, and at every point has a single determinate value. In the case of the electric circuit, on the other hand, the magnetic potential at every point not in the conducting wire itself is equal to $i\omega$, where i is the strength of the current, and ω is the solid angle subtended by the circuit at the point, and is reckoned positive when the current, as seen from P, circulates in the direction opposite to that of the hands of a watch.

The quantity $i\omega$ is a function having an infinite series of values whose common difference is $4\pi i$. The differential coefficients of $i\omega$ with respect to the coordinates have, however, single and determinate values for every point of space.

486.] If a long thin flexible solenoidal magnet were placed in the neighbourhood of an electric circuit, the north and south ends of the solenoid would tend to move in opposite directions round the wire, and if they were free to obey the magnetic force the magnet would finally become wound round the wire in a close coil. If it were possible to obtain a magnet having only one pole, or poles of unequal strength, such a magnet would be moved round and round the wire continually in one direction, but since the poles of every magnet are equal and opposite, this result can never occur. Faraday, however, has shewn how to produce the continuous rotation of one pole of a magnet round an electric current by making it possible for one pole to go round and round the current while the other pole does not. That this process may be repeated indefinitely, the body of the magnet must be transferred from one side of the current to the other once in each revolution. To do this without interrupting the flow of electricity, the current is split into two branches, so that when one branch is opened to let the magnet pass the current continues to flow through the other. Faraday used for this purpose a circular trough of mercury, as shewn in Fig. 23, Art. 491. The current enters the trough through the wire *AB*, it is divided at *B*, and after flowing through the arcs *BQP* and *BRP* it unites at *P*, and leaves the trough through the wire *PO*, the cup of mercury *O*, and a vertical wire beneath *O*, down which the current flows.

The magnet (not shewn in the figure) is mounted so as to be capable of revolving about a vertical axis through *O*, and the wire *OP* revolves with it. The body of the magnet passes through the aperture of the trough, one pole, say the north pole, being beneath the plane of the trough, and the other above it. As the magnet and the wire *OP* revolve about the vertical axis, the current is gradually transferred from the branch of the trough which lies in front of the magnet to that which lies behind it, so that in every complete revolution the magnet passes from one side of the current to the other. The north pole of the magnet revolves about the descending current in the direction N.E.S.W. and if ω, ω' are the solid angles (irrespective of sign) subtended by the circular trough at the two poles, the work done by the electromagnetic force in a complete revolution is

$$mi\,(4\pi-\omega-\omega'),$$

where m is the strength of either pole, and i the strength of the current.

487.] Let us now endeavour to form a notion of the state of the magnetic field near a linear electric circuit.

Let the value of ω, the solid angle subtended by the circuit, be found for every point of space, and let the surfaces for which ω is constant be described. These surfaces will be the equipotential surfaces. Each of these surfaces will be bounded by the circuit, and any two surfaces, ω_1 and ω_2, will meet in the circuit at an angle $\frac{1}{2}(\omega_1-\omega_2)$.

Figure XVIII, at the end of this volume, represents a section of the equipotential surfaces due to a circular current. The small circle represents a section of the conducting wire, and the horizontal line at the bottom of the figure is the perpendicular to the plane of the circular current through its centre. The equipotential surfaces, 24 of which are drawn corresponding to a series of values of ω differing by $\frac{\pi}{6}$, are surfaces of revolution, having this line for their common axis. They are evidently oblate figures, being flattened in the direction of the axis. They meet each other in the line of the circuit at angles of 15°.

The force acting on a magnetic pole placed at any point of an equipotential surface is perpendicular to this surface, and varies inversely as the distance between consecutive surfaces. The closed curves surrounding the section of the wire in Fig. XVIII are the lines of force. They are copied from Sir W. Thomson's Paper on 'Vortex Motion*.' See also Art. 702.

Action of an Electric Circuit on any Magnetic System.

488.] We are now able to deduce the action of an electric circuit on any magnetic system in its neighbourhood from the theory of magnetic shells. For if we construct a magnetic shell, whose strength is numerically equal to the strength of the current, and whose edge coincides in position with the circuit, while the shell itself does not pass through any part of the magnetic system, the action of the shell on the magnetic system will be identical with that of the electric circuit.

Reaction of the Magnetic System on the Electric Circuit.

489.] From this, applying the principle that action and reaction are equal and opposite, we conclude that the mechanical action of the magnetic system on the electric circuit is identical with its action on a magnetic shell having the circuit for its edge.

The potential energy of a magnetic shell of strength ϕ placed in a field of magnetic force of which the potential is V, is, by Art. 410,

$$M = \phi \iint \left(l \frac{dV}{dx} + m \frac{dV}{dy} + n \frac{dV}{dz}\right) dS,$$

where l, m, n are the direction-cosines of the normal drawn from the positive side of the element dS of the shell, and the integration is extended over the surface of the shell.

Now the surface-integral

$$N = \iint (la + mb + nc)\, dS,$$

where a, b, c are the components of the magnetic induction, represents the quantity of magnetic induction through the shell, or,

* *Trans. R. S. Edin.*, vol. xxv. p. 217, (1869).

in the language of Faraday, the *number* of lines of magnetic induction, reckoned algebraically, which pass through the shell from the negative to the positive side, lines which pass through the shell in the opposite direction being reckoned negative.

Remembering that the shell does not belong to the magnetic system to which the potential V is due, and that the magnetic force is therefore equal to the magnetic induction, we have

$$a = -\frac{dV}{dx}, \quad b = -\frac{dV}{dy}, \quad c = -\frac{dV}{dz},$$

and we may write the value of M,

$$M = -\phi N.$$

If δx_1 represents any displacement of the shell, and X_1 the force acting on the shell so as to aid the displacement, then by the principle of conservation of energy,

$$X_1 \delta x_1 + \delta M = 0,$$

or

$$X = \phi \frac{\delta N}{\delta x}.$$

We have now determined the nature of the force which corresponds to any given displacement of the shell. It aids or resists that displacement accordingly as the displacement increases or diminishes N, the number of lines of induction which pass through the shell.

The same is true of the equivalent electric circuit. Any displacement of the circuit will be aided or resisted accordingly as it increases or diminishes the number of lines of induction which pass through the circuit in the positive direction.

We must remember that the positive direction of a line of magnetic induction is the direction in which the pole of a magnet which points north tends to move along the line, and that a line of induction passes through the circuit in the positive direction, when the direction of the line of induction is related to the direction of the current of vitreous electricity in the circuit as the longitudinal to the rotational motion of a right-handed screw. See Art. 23.

490.] It is manifest that the force corresponding to any displacement of the circuit as a whole may be deduced at once from the theory of the magnetic shell. But this is not all. If a portion of the circuit is flexible, so that it may be displaced independently of the rest, we may make the edge of the shell capable of the same kind of displacement by cutting up the surface of the shell into a sufficient number of portions connected by flexible joints. Hence we conclude that if by the displacement of any portion of the circuit in a given direction the number of lines of induction which pass through the circuit can be increased, this displacement will be aided by the electromagnetic force acting on the circuit.

Every portion of the circuit therefore is acted on by a force urging it across the lines of magnetic induction so as to include a greater number of these lines within the embrace of the circuit, and the work done by the force during this displacement is numerically equal to the number of the additional lines of induction multiplied by the strength of the current.

Let the element ds of a circuit, in which a current of strength i is flowing, be moved parallel to itself through a space δx, it will sweep out an area in the form of a parallelogram whose sides are parallel and equal to ds and δx respectively.

If the magnetic induction is denoted by $\mathfrak{B}$, and if its direction makes an angle ϵ with the normal to the parallelogram, the value of the increment of N corresponding to the displacement is found by multiplying the area of the parallelogram by $\mathfrak{B}\cos\epsilon$. The result of this operation is represented geometrically by the volume of a parallelepiped whose edges represent in magnitude and direction δx, ds, and $\mathfrak{B}$, and it is to be reckoned positive if when we point in these three directions in the order here given the pointer moves round the diagonal of the parallelepiped in the direction of the hands of a watch. The volume of this parallelepiped is equal to $X\delta x$.

If θ is the angle between ds and $\mathfrak{B}$, the area of the parallelogram is $ds\,.\,\mathfrak{B}\sin\theta$, and if η is the angle which the displacement δx makes with the normal to this parallelogram, the volume of the parallelepiped is

$$ds\,.\,\mathfrak{B}\sin\theta\,.\,\delta x\cos\eta = \delta N.$$

Now $$X\delta x = i\delta N = i\,ds\,.\,\mathfrak{B}\sin\theta\,\delta x\cos\eta,$$

and $$X = i\,ds\,.\,\mathfrak{B}\sin\theta\cos\eta$$

is the force which urges ds, resolved in the direction δx.

The direction of this force is therefore perpendicular to the parallelogram, and is equal to $i\,.\,ds\,.\,\mathfrak{B}\sin\theta$.

This is the area of a parallelogram whose sides represent in magnitude and direction $i\,ds$ and $\mathfrak{B}$. The force acting on ds is therefore represented in magnitude by the area of this parallelogram, and in direction by a normal to its plane drawn in the direction of the longitudinal motion of a right-handed screw, the handle of which is turned from the direction of the current $i\,ds$ to that of the magnetic induction $\mathfrak{B}$.

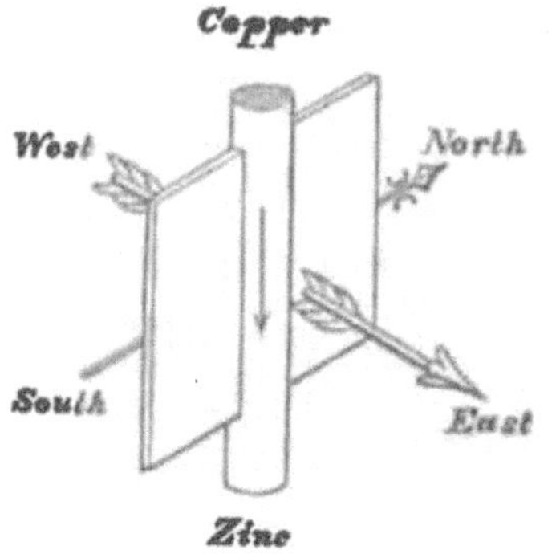

Fig. 22.

We may express in the language of Quaternions, both the direction and the magnitude of this force by saying that it is the vector part of the result of multiplying the vector $i\,ds$, the element of the current, by the vector $\mathfrak{B}$, the magnetic induction.

491.] We have thus completely determined the force which acts on any portion of an electric circuit placed in a magnetic field. If the circuit is moved in any way so that, after assuming various forms and positions, it returns to its original place, the strength of the current remaining constant during the motion, the whole amount of work done by the electromagnetic forces will be zero. Since this is true of any cycle of motions of the circuit, it follows that it is impossible to maintain by electromagnetic forces a motion of continuous rotation in any part of a linear circuit of constant strength against the resistance of friction, &c.

It is possible, however, to produce continuous rotation provided that at some part of the course of the electric current it passes from one conductor to another which slides or glides over it.

When in a circuit there is sliding contact of a conductor over the surface of a smooth solid or a fluid, the circuit can no longer be considered as a single linear circuit of constant strength, but must be regarded as a system of two or of some greater number of circuits of variable strength, the current being so distributed among them that those for which N is increasing have currents in the positive direction, while those for which N is diminishing have currents in the negative direction.

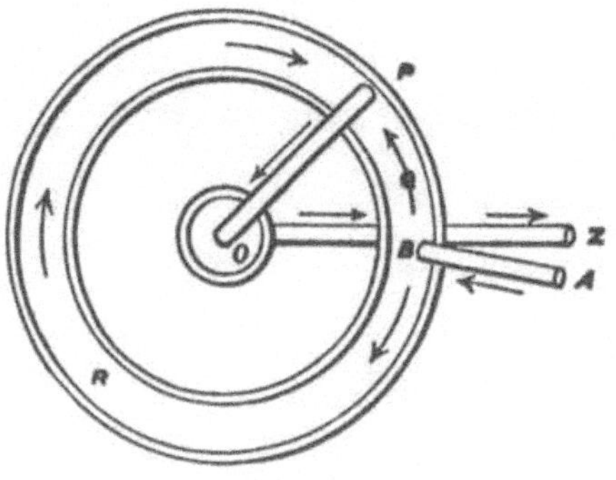

Fig. 23.

Thus, in the apparatus represented in Fig. 23, OP is a moveable conductor, one end of which rests in a cup of mercury O, while the other dips into a circular trough of mercury concentric with O.

The current i enters along AB, and divides in the circular trough into two parts, one of which, x, flows along the arc BQP, while the other, y, flows along BRP. These currents, uniting at P, flow along the moveable conductor PO and the electrode OZ to the zinc end of the battery. The strength of the current along OP and OZ is $x+y$ or i.

Here we have two circuits, $ABQPOZ$, the strength of the current in which is x, flowing in the positive direction, and $ABRPOZ$, the strength of the current in which is y, flowing in the negative direction.

Let $\mathfrak{B}$ be the magnetic induction, and let it be in an upward direction, normal to the plane of the circle.

While OP moves through an angle θ in the direction opposite to that of the hands of a watch, the area of the first circuit increases by $\frac{1}{2}OP^2.\theta$, and that of the second diminishes by the same quantity. Since the strength of the current in the first circuit is x, the work done by it is $\frac{1}{2}x.OP^2.\theta.\mathfrak{B}$, and since the strength of the second is $-y$, the work done by it is $\frac{1}{2}y.OP^2.\theta\mathfrak{B}$. The whole work done is therefore

$$\frac{1}{2}(x+y)OP^2.\theta\mathfrak{B} \quad \text{or} \quad \frac{1}{2}i.OP^2.\theta\mathfrak{B},$$

depending only on the strength of the current in PO. Hence, if i is maintained constant, the arm OP will be carried round and round the circle with a uniform force whose moment is $\frac{1}{2}i.OP^2\mathfrak{B}$. If, as in northern latitudes, $\mathfrak{B}$ acts downwards, and if the current is inwards, the rotation will be in the negative direction, that is, in the direction $PQBR$.

492.] We are now able to pass from the mutual action of magnets and currents to the action of one current on another. For we know that the magnetic properties of an electric circuit C_1, with respect to any magnetic system M_2, are identical with those of a magnetic shell S_1, whose edge coincides with the circuit, and whose strength is numerically equal to that of the electric current. Let the magnetic system M_2 be a magnetic shell S_2, then the mutual action between S_1 and S_2 is identical with that between S_1 and a circuit C_2, coinciding with the edge of S_2 and equal in numerical strength, and this latter action is identical with that between C_1 and C_2.

Hence the mutual action between two circuits, C_1 and C_2, is identical with that between the corresponding magnetic shells S_1 and S_2.

We have already investigated, in Art. 423, the mutual action of two magnetic shells whose edges are the closed curves s_1 and s_2. If we make

$$M = \int_0^{s_2}\int_0^{s_1} \frac{\cos\epsilon}{r}\, ds_1\, ds_2,$$

where ϵ is the angle between the directions of the elements ds_1 and ds_2, and r is the distance between them, the integration being extended once round s_2 and once round s_1, and if we call M the potential of the two closed curves s_1 and s_2, then the potential energy due to the mutual action of two magnetic shells whose strengths are i_1 and i_2 bounded by the two circuits is

$$-i_1 i_2 M,$$

and the force X, which aids any displacement δx, is

$$i_1 i_2 \frac{\delta M}{\delta x}.$$

The whole theory of the force acting on any portion of an electric circuit due to the action of another electric circuit may be deduced from this result.

493.] The method which we have followed in this chapter is that of Faraday. Instead of beginning, as we shall do, following Ampère, in the next chapter, with the direct action of a portion of one circuit on a portion of another, we shew, first, that a circuit produces the same effect on a magnet as a magnetic shell, or, in other words, we determine the nature of the magnetic field due to the circuit. We shew, secondly, that a circuit when placed in any magnetic field experiences the same force as a magnetic shell. We thus determine the force acting on the circuit placed in any magnetic field. Lastly, by supposing the magnetic field to be due to a second electric circuit we determine the action of one circuit on the whole or any portion of the other.

494.] Let us apply this method to the case of a straight current of infinite length acting on a portion of a parallel straight conductor.

Let us suppose that a current i in the first conductor is flowing vertically downwards. In this case the end of a magnet which points north will point to the right-hand of a man looking at it from the axis of the current.

The lines of magnetic induction are therefore horizontal circles, having their centres in the axis of the current, and their positive direction is north, east, south, west.

Let another descending vertical current be placed due west of the first. The lines of magnetic induction due to the first current are here directed towards the north. The direction of the force acting on the second current is to be determined by turning the handle of a right-handed screw from the nadir, the direction of the current, to the north, the direction of the magnetic induction. The screw will then move towards the east, that is, the force acting on the second current is directed towards the first current, or, in general, since the phenomenon depends only on the relative position of the currents, two parallel currents in the same direction attract each other.

In the same way we may shew that two parallel currents in opposite directions repel one another.

495.] The intensity of the magnetic induction at a distance r from a straight current of strength i is, as we have shewn in Art. 479,

$$2\frac{i}{r}.$$

Hence, a portion of a second conductor parallel to the first, and carrying a current i' in the same direction, will be attracted towards the first with a force

$$F = 2ii'\frac{a}{r},$$

where a is the length of the portion considered, and r is its distance from the first conductor.

Since the ratio of a to r is a numerical quantity independent of the absolute value of either of these lines, the product of two currents measured in the electromagnetic system must be of the dimensions of a force, hence the dimensions of the unit current are

$$[i] = [F^{\frac{1}{2}}] = [M^{\frac{1}{2}} L^{\frac{1}{2}} T^{-1}].$$

496.] Another method of determining the direction of the force which acts on a current is to consider the relation of the magnetic action of the current to that of other currents and magnets.

If on one side of the wire which carries the current the magnetic action due to the current is in the same or nearly the same direction as that due to other currents, then, on the other side of the wire, these forces will be in opposite or nearly opposite directions, and the force acting on the wire will be from the side on which the forces strengthen each other to the side on which they oppose each other.

Thus, if a descending current is placed in a field of magnetic force directed towards the north, its magnetic action will be to the north on the west side, and to the south on the east side. Hence the forces strengthen each other on the west side and oppose each other on the east side, and the current will therefore be acted on by a force from west to east. See Fig. 22, p. 138.

In Fig. XVII at the end of this volume the small circle represents a section of the wire carrying a descending current, and placed in a uniform field of magnetic force acting towards the left-hand of the figure. The magnetic force is greater below the wire than above it. It will therefore be urged from the bottom towards the top of the figure.

497.] If two currents are in the same plane but not parallel, we may apply this principle. Let one of the conductors be an infinite straight wire in the plane of the paper, supposed horizontal. On the right side of the current the magnetic force acts downward, and on the left side it acts upwards. The same is true of the magnetic force due to any short portion of a second current in the same plane. If the second current is on the right side of the first, the magnetic forces will strengthen each other on its right side and oppose each other on its left side. Hence the second current will be acted on by a force urging it from its right side to its left side. The magnitude of this force depends only on the position of the second current and not on its direction. If the second current is on the left side of the first it will be urged from left to right.

Hence, if the second current is in the same direction as the first it is attracted, if in the opposite direction it is repelled, if it flows at right angles to the first and away from it, it is urged in the direction of the first current, and if it flows toward the first current, it is urged in the direction opposite to that in which the first current flows.

In considering the mutual action of two currents it is not necessary to bear in mind the relations between electricity and magnetism which we have endeavoured to illustrate by means of a right-handed screw. Even if we have forgotten these relations we shall arrive at correct results, provided we adhere consistently to one of the two possible forms of the relation.

498.] Let us now bring together the magnetic phenomena of the electric circuit so far as we have investigated them.

We may conceive the electric circuit to consist of a voltaic battery, and a wire connecting its extremities, or of a thermoelectric arrangement, or of a charged Leyden jar with a wire connecting its positive and negative coatings, or of any other arrangement for producing an electric current along a definite path.

The current produces magnetic phenomena in its neighbourhood.

If any closed curve be drawn, and the line-integral of the magnetic force taken completely round it, then, if the closed curve is not linked with the circuit, the line-integral is zero, but if it is linked with the circuit, so that the current i flows through the closed curve, the line-integral is $4\pi i$, and is positive if the direction of integration round the closed curve would coincide with that of the hands of a watch as seen by a person passing through it in the direction in which the electric current flows. To a person moving along the closed curve in the direction of integration, and passing through the electric circuit, the direction of the current would appear to be that of the hands of a watch. We may express this in another way by saying that the relation between the directions of the two closed curves may be expressed by describing a right-handed screw round the electric circuit and a right-handed screw round the closed curve. If the direction of rotation of the thread of either, as we pass along it, coincides with the positive direction in the other, then the line-integral will be positive, and in the opposite case it will be negative.

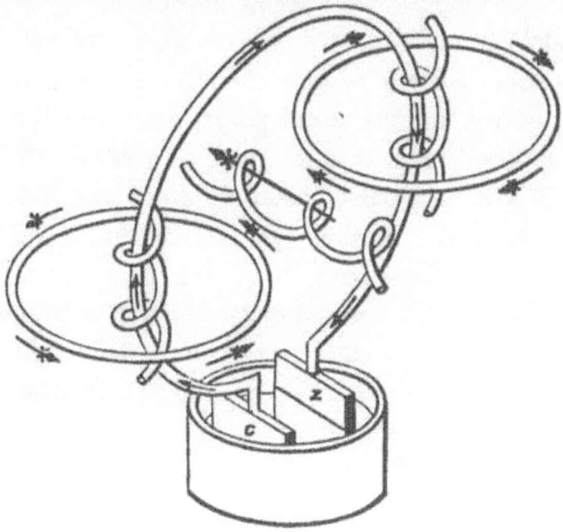

Fig. 24.

Relation between the electric current and the lines of magnetic induction indicated by a right-handed screw.

499.] *Note.*—The line-integral $4\pi i$ depends solely on the quantity of the current, and not on any other thing whatever. It does not depend on the nature of the conductor through which the current is passing, as, for instance, whether it be a metal or an electrolyte, or an imperfect conductor. We have reason for believing that even when there is no proper conduction, but merely a variation of electric displacement, as in the glass of a Leyden jar during charge or discharge, the magnetic effect of the electric movement is precisely the same.

Again, the value of the line-integral $4\pi i$ does not depend on the nature of the medium in which the closed curve is drawn. It is the same whether the closed curve is drawn entirely through air, or passes through a magnet, or soft iron, or any other substance, whether paramagnetic or diamagnetic.

500.] When a circuit is placed in a magnetic field the mutual action between the current and the other constituents of the field depends on the surface-integral of the magnetic induction through any surface bounded by that circuit. If by any given motion of the circuit, or of part of it, this surface-integral can be *increased*, there will be a mechanical force tending to move the conductor or the portion of the conductor in the given manner.

The kind of motion of the conductor which increases the surface-integral is motion of the conductor perpendicular to the direction of the current and across the lines of induction.

If a parallelogram be drawn, whose sides are parallel and proportional to the strength of the current at any point, and to the magnetic induction at the same point, then the force on unit of length of the conductor is numerically equal to the area of this parallelogram, and is perpendicular to its plane, and acts in the direction in which the motion of turning the handle of a right-handed screw from the direction of the current to the direction of the magnetic induction would cause the screw to move.

Hence we have a new electromagnetic definition of a line of magnetic induction. It is that line to which the force on the conductor is always perpendicular.

It may also be defined as a line along which, if an electric current be transmitted, the conductor carrying it will experience no force.

501.] It must be carefully remembered, that the mechanical force which urges a conductor carrying a current across the lines of magnetic force, acts, not on the electric current, but on the conductor which carries it. If the conductor be a rotating disk or a fluid it will move in obedience to this force, and this motion may or may not be accompanied with a change of position of the electric current which it carries. But if the current itself be free to choose any path through a fixed solid conductor or a network of wires, then, when a constant magnetic force is made to act on the system, the path of the current through the conductors is not permanently altered, but after certain transient phenomena, called induction currents, have subsided, the distribution of the current will be found to be the same as if no magnetic force were in action.

The only force which acts on electric currents is electromotive force, which must be distinguished from the mechanical force which is the subject of this chapter.

Fig. 25.

Relations between the positive directions of motion and of rotation indicated by three right-handed screws.

CHAPTER II.

AMPÈRE'S INVESTIGATION OF THE MUTUAL ACTION OF ELECTRIC CURRENTS.

502.] We have considered in the last chapter the nature of the magnetic field produced by an electric current, and the mechanical action on a conductor carrying an electric current placed in a magnetic field. From this we went on to consider the action of one electric circuit upon another, by determining the action on the first due to the magnetic field produced by the second. But the action of one circuit upon another was originally investigated in a direct manner by Ampère almost immediately after the publication of Örsted's discovery. We shall therefore give an outline of Ampère's method, resuming the method of this treatise in the next chapter.

The ideas which guided Ampère belong to the system which admits direct action at a distance, and we shall find that a remarkable course of speculation and investigation founded on these ideas has been carried on by Gauss, Weber, J. Neumann, Riemann, Betti, C. Neumann, Lorenz, and others, with very remarkable results both in the discovery of new facts and in the formation of a theory of electricity. See Arts. 846–866.

The ideas which I have attempted to follow out are those of action through a medium from one portion to the contiguous portion. These ideas were much employed by Faraday, and the development of them in a mathematical form, and the comparison of the results with known facts, have been my aim in several published papers. The comparison, from a philosophical point of view, of the results of two methods so completely opposed in their first principles must lead to valuable data for the study of the conditions of scientific speculation.

503.] Ampère's theory of the mutual action of electric currents is founded on four experimental facts and one assumption.

Ampère's fundamental experiments are all of them examples of what has been called the null method of comparing forces. See Art. 214. Instead of measuring the force by the dynamical effect of communicating motion to a body, or the statical method of placing it in equilibrium with the weight of a body or the elasticity of a fibre, in the null method two forces, due to the same source, are made to act simultaneously on a body already in equilibrium, and no effect is produced, which shews that these forces are themselves in equilibrium. This method is peculiarly valuable for comparing the effects of the electric current when it passes through circuits of different forms. By connecting all the conductors in one continuous series, we ensure that the strength of the current is the same at every point of its course, and since the current begins everywhere throughout its course almost at the same instant, we may prove that the forces due to its action on a suspended body are in equilibrium by observing that the body is not at all affected by the starting or the stopping of the current.

504.] Ampère's balance consists of a light frame capable of revolving about a vertical axis, and carrying a wire which forms two circuits of equal area, in the same plane or in parallel planes, in which the current flows in opposite directions. The object of this arrangement is to get rid of the effects of terrestrial magnetism on the conducting wire. When an electric circuit is free to move it tends to place itself so as to embrace the largest possible number of the lines of induction. If these lines are due to terrestrial magnetism, this position, for a circuit in a vertical plane, will be when the plane of the circuit is east and west, and when the direction of the current is opposed to the apparent course of the sun.

By rigidly connecting two circuits of equal area in parallel planes, in which equal currents run in opposite directions, a combination is formed which is unaffected by terrestrial magnetism, and is therefore called an Astatic Combination, see Fig. 26. It is acted on, however, by forces arising from currents or magnets which are so near it that they act differently on the two circuits.

505.] Ampère's first experiment is on the effect of two equal currents close together in opposite directions. A wire covered with insulating material is doubled on itself, and placed near one of the circuits of the astatic balance. When a current is made to pass through the wire and the balance, the equilibrium of the balance remains undisturbed, shewing that two equal currents close together in opposite directions neutralize each other. If, instead of two wires side by side, a wire be insulated in the middle of a metal

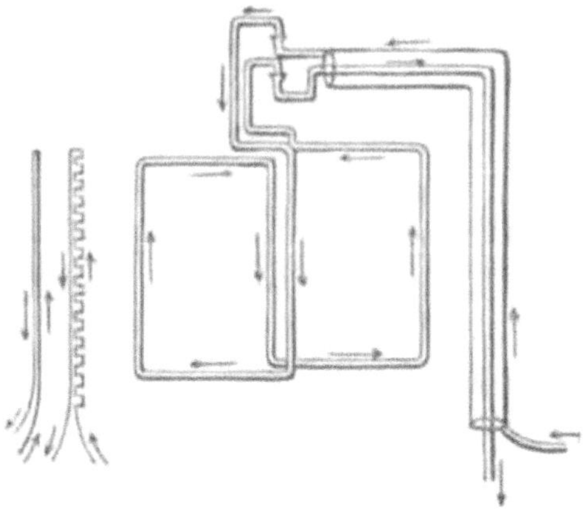

Fig. 26.

tube, and if the current pass through the wire and back by the tube, the action outside the tube is not only approximately but accurately null. This principle is of great importance in the construction of electric apparatus, as it affords the means of conveying the current to and from any galvanometer or other instrument in such a way that no electromagnetic effect is produced by the current on its passage to and from the instrument. In practice it is generally sufficient to bind the wires together, care being taken that they are kept perfectly insulated from each other, but where they must pass near any sensitive part of the apparatus it is better to make one of the conductors a tube and the other a wire inside it. See Art. 683.

506.] In Ampère's second experiment one of the wires is bent and crooked with a number of small sinuosities, but so that in every part of its course it remains very near the straight wire. A current, flowing through the crooked wire and back again through the straight wire, is found to be without influence on the astatic balance. This proves that the effect of the current running through any crooked part of the wire is equivalent to the same current running in the straight line joining its extremities, provided the crooked line is in no part of its course far from the straight one. Hence any small element of a circuit is equivalent to two or more component elements, the relation between the component elements and the resultant element being the same as that between component and resultant displacements or velocities.

507.] In the third experiment a conductor capable of moving

only in the direction of its length is substituted for the astatic balance, the current enters the conductor and leaves it at fixed points of space, and it is found that no closed circuit placed in the neighbourhood is able to move the conductor.

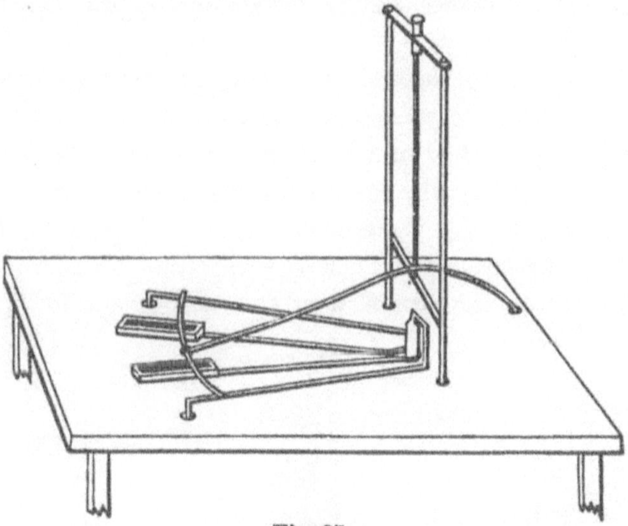

Fig. 27.

The conductor in this experiment is a wire in the form of a circular arc suspended on a frame which is capable of rotation about a vertical axis. The circular arc is horizontal, and its centre coincides with the vertical axis. Two small troughs are filled with mercury till the convex surface of the mercury rises above the level of the troughs. The troughs are placed under the circular arc and adjusted till the mercury touches the wire, which is of copper well amalgamated. The current is made to enter one of these troughs, to traverse the part of the circular arc between the troughs, and to escape by the other trough. Thus part of the circular arc is traversed by the current, and the arc is at the same time capable of moving with considerable freedom in the direction of its length. Any closed currents or magnets may now be made to approach the moveable conductor without producing the slightest tendency to move it in the direction of its length.

508.] In the fourth experiment with the astatic balance two circuits are employed, each similar to one of those in the balance, but one of them, C, having dimensions n times greater, and the other, A, n times less. These are placed on opposite sides of the circuit of the balance, which we shall call B, so that they are similarly placed with respect to it, the distance of C from B being n times greater than the distance of B from A. The direction and strength of the current is the same in A and C. Its direction in B may be the same or opposite. Under these circumstances it is found that B is in equilibrium under the action of A and C, whatever be the forms and distances of the three circuits, provided they have the relations given above.

Since the actions between the complete circuits may be considered to be due to actions between the elements of the circuits, we may use the following method of determining the law of these actions.

Let A_1, B_1, C_1, Fig. 28, be corresponding elements of the three circuits, and let A_2, B_2, C_2 be also corresponding elements in another part of the circuits. Then the situation of B_1 with respect to A_2 is similar to the situation of C_1 with respect to B_2, but the

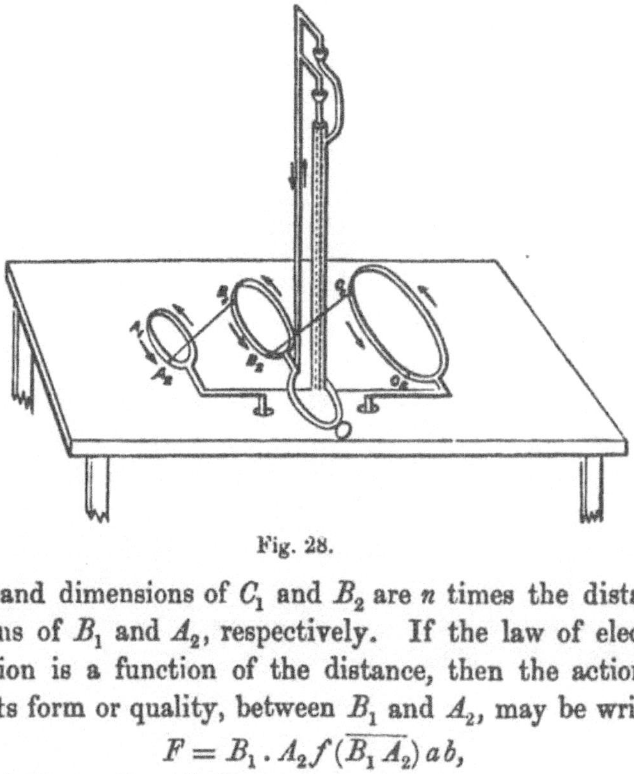

Fig. 28.

distance and dimensions of C_1 and B_2 are n times the distance and dimensions of B_1 and A_2, respectively. If the law of electromagnetic action is a function of the distance, then the action, whatever be its form or quality, between B_1 and A_2, may be written

$$F = B_1 . A_2 f(\overline{B_1 A_2})\, ab,$$

and that between C_1 and B_2

$$F' = C_1 . B_2 f(\overline{C_1 B_2})\, bc,$$

where a, b, c are the strengths of the currents in A, B, C. But $nB_1 = C_1$, $nA_2 = B_2$, $n\overline{B_1 A_2} = \overline{C_1 B_2}$, and $a = c$. Hence

$$F' = n^2 B_1 . A_2 f(n\overline{B_1 A_2})\, ab,$$

and this is equal to F by experiment, so that we have

$$n^2 f(n\overline{A_2 B_1}) = f(\overline{A_2 B_1});$$

or, *the force varies inversely as the square of the distance.*

509.] It may be observed with reference to these experiments that every electric current forms a closed circuit. The currents used by Ampère, being produced by the voltaic battery, were of course in closed circuits. It might be supposed that in the case of the current of discharge of a conductor by a spark we might have a current forming an open finite line, but according to the views of this book even this case is that of a closed circuit. No experiments on the mutual action of unclosed currents have been made. Hence no statement about the mutual action of two elements of circuits can be said to rest on purely experimental grounds. It is true we may render a portion of a circuit moveable, so as to ascertain the action of the other currents upon it, but these currents, together with that in the moveable portion, necessarily form closed circuits, so that the ultimate result of the experiment is the action of one or more closed currents upon the whole or a part of a closed current.

510.] In the analysis of the phenomena, however, we may regard the action of a closed circuit on an element of itself or of another circuit as the resultant of a number of separate forces, depending on the separate parts into which the first circuit may be conceived, for mathematical purposes, to be divided.

This is a merely mathematical analysis of the action, and is therefore perfectly legitimate, whether these forces can really act separately or not.

511.] We shall begin by considering the purely geometrical relations between two lines in space representing the circuits, and between elementary portions of these lines.

Let there be two curves in space in each of which a fixed point is taken from which the arcs are measured in a defined direction along the curve. Let A, A' be these points. Let PQ and $P'Q'$ be elements of the two curves.

Let
$$\left.\begin{aligned} AP &= s, & A'P' &= s', \\ PQ &= ds, & P'Q' &= ds', \end{aligned}\right\} \quad (1)$$
and let the distance PP' be denoted by r. Let the angle $P'PQ$ be denoted by θ, and $PP'Q'$ by θ', and let the angle between the planes of these angles be denoted by η.

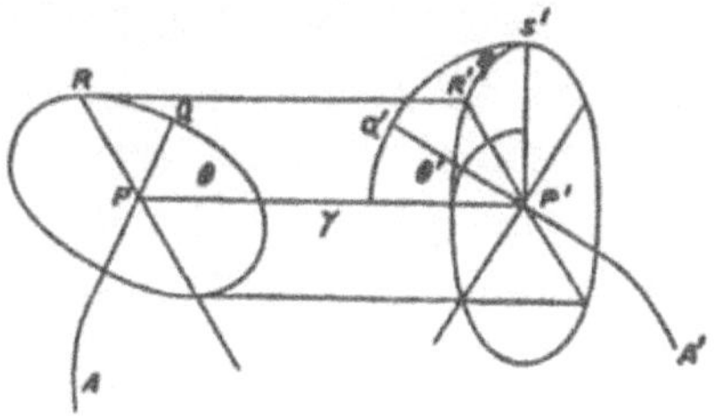

Fig. 29.

The relative position of the two elements is sufficiently defined by their distance r and the three angles θ, θ', and η, for if these be given their relative position is as completely determined as if they formed part of the same rigid body.

512.] If we use rectangular coordinates and make x, y, z the coordinates of P, and x', y', z' those of P', and if we denote by l, m, n and by l', m', n' the direction-cosines of PQ, and of $P'Q'$ respectively, then

$$\left.\begin{aligned} \frac{dx}{ds} &= l, & \frac{dy}{ds} &= m, & \frac{dz}{ds} &= n, \\ \frac{dx'}{ds'} &= l', & \frac{dy'}{ds'} &= m', & \frac{dz'}{ds'} &= n', \end{aligned}\right\} \tag{2}$$

and

$$\left.\begin{aligned} l\,(x'-x)+m\,(y'-y)+n\,(z'-z) &= r\cos\theta, \\ l'(x'-x)+m'(y'-y)+n'(z'-z) &= -r\cos\theta', \\ ll'+mm'+nn' &= \cos\epsilon, \end{aligned}\right\} \tag{3}$$

where ϵ is the angle between the directions of the elements themselves, and

$$\cos\epsilon = -\cos\theta\cos\theta' + \sin\theta\sin\theta'\cos\eta. \tag{4}$$

Again

$$r^2 = (x'-x)^2 + (y'-y)^2 + (z'-z)^2, \tag{5}$$

whence

$$\left.\begin{aligned} r\frac{dr}{ds} &= -(x'-x)\frac{dx}{ds} - (y'-y)\frac{dy}{ds} - (z'-z)\frac{dz}{ds}, \\ &= -r\cos\theta. \end{aligned}\right.$$

Similarly

$$\left.\begin{aligned} r\frac{dr}{ds'} &= (x'-x)\frac{dx'}{ds'} + (y'-y)\frac{dy'}{ds'} + (z'-z)\frac{dz'}{ds'}, \\ &= -r\cos\theta'; \end{aligned}\right\} \tag{6}$$

and differentiating $r\frac{dr}{ds}$ with respect to s',

$$\left.\begin{aligned} r\frac{d^2r}{ds\,ds'} + \frac{dr}{ds}\frac{dr}{ds'} &= -\frac{dx}{ds}\frac{dx'}{ds'} - \frac{dy}{ds}\frac{dy'}{ds'} - \frac{dz}{ds}\frac{dz'}{ds'}, \\ &= -(ll'+mm'+nn') \\ &= -\cos\epsilon. \end{aligned}\right\} \tag{7}$$

We can therefore express the three angles θ, θ', and η, and the auxiliary angle ϵ in terms of the differential coefficients of r with respect to s and s' as follows,

$$\left.\begin{aligned} \cos\theta &= -\frac{dr}{ds}, \\ \cos\theta' &= -\frac{dr}{ds'}, \\ \cos\epsilon &= -r\frac{d^2r}{ds\,ds'} - \frac{dr}{ds}\frac{dr}{ds'}, \\ \sin\theta\sin\theta'\cos\eta &= -r\frac{d^2r}{ds\,ds'}. \end{aligned}\right\} \tag{8}$$

513.] We shall next consider in what way it is mathematically conceivable that the elements PQ and $P'Q'$ might act on each other, and in doing so we shall not at first assume that their mutual action is necessarily in the line joining them.

We have seen that we may suppose each element resolved into other elements, provided that these components, when combined according to the rule of addition of vectors, produce the original element as their resultant.

We shall therefore consider ds as resolved into $\cos\theta\, ds = \alpha$ in the direction of r, and $\sin\theta\, ds = \beta$ in a direction perpendicular to r in the plane $P'PQ$.

Fig. 30.

We shall also consider ds' as resolved into $\cos\theta'\, ds' = \alpha'$ in the direction of r reversed, $\sin\theta' \cos\eta\, ds' = \beta$ in a direction parallel to that in which β was measured, and $\sin\theta' \sin\eta\, ds' = \gamma'$ in a direction perpendicular to α' and β'.

Let us consider the action between the components α and β on the one hand, and α', β', γ' on the other.

(1) α and α' are in the same straight line. The force between them must therefore be in this line. We shall suppose it to be an attraction $= A\alpha\alpha' ii'$,
where A is a function of r, and i, i' are the intensities of the currents in ds and ds' respectively. This expression satisfies the condition of changing sign with i and with i'.

(2) β and β' are parallel to each other and perpendicular to the line joining them. The action between them may be written

$$B\beta\beta' ii'.$$

This force is evidently in the line joining β and β', for it must be in the plane in which they both lie, and if we were to measure β and β' in the reversed direction, the value of this expression would remain the same, which shews that, if it represents a force, that force has no component in the direction of β, and must therefore be directed along r. Let us assume that this expression, when positive, represents an attraction.

(3) β and γ' are perpendicular to each other and to the line joining them. The only action possible between elements so related is a couple whose axis is parallel to r. We are at present engaged with forces, so we shall leave this out of account.

(4) The action of α and β', if they act on each other, must be expressed by $C\alpha\beta' ii'$.

The sign of this expression is reversed if we reverse the direction in which we measure β'. It must therefore represent either a force in the direction of β', or a couple in the plane of α and β'. As we are not investigating couples, we shall take it as a force acting on α in the direction of β'.

There is of course an equal force acting on β' in the opposite direction.

We have for the same reason a force

$$C\alpha\gamma' ii'$$

acting on α in the direction of γ', and a force

$$C\beta\alpha' ii'$$

acting on β in the opposite direction.

514.] Collecting our results, we find that the action on ds is compounded of the following forces,

$$\left.\begin{aligned} X &= (A\alpha\alpha' + B\beta\beta')ii' \text{ in the direction of } r, \\ Y &= C(\alpha\beta' - \alpha'\beta)ii' \text{ in the direction of } \beta, \\ \text{and} \quad Z &= C\alpha\gamma' ii' \text{ in the direction of } \gamma'. \end{aligned}\right\} \qquad (9)$$

Let us suppose that this action on ds is the resultant of three forces, $Rii'dsds'$ acting in the direction of r, $Sii'dsds'$ acting in the direction of ds, and $S'ii'dsds'$ acting in the direction of ds', then in terms of θ, θ', and η,

$$\left.\begin{aligned} R &= \quad A\cos\theta\cos\theta' + B\sin\theta\sin\theta'\cos\eta, \\ S &= -C\cos\theta', \qquad S' = C\cos\theta. \end{aligned}\right\} \qquad (10)$$

In terms of the differential coefficients of r

$$\left.\begin{aligned} R &= \quad A\frac{dr}{ds}\frac{dr}{ds'} - Br\frac{d^2 r}{ds\,ds'}, \\ S &= +C\frac{dr}{ds'}, \qquad S' = -C\frac{dr}{ds}, \end{aligned}\right\} \qquad (11)$$

In terms of l, m, n, and l', m', n',

$$\left.\begin{aligned} R &= -(A+B)\frac{1}{r^2}(l\xi + m\eta + n\zeta)(l'\xi + m'\eta + n'\zeta) + B(ll' + mm' + nn'), \\ S &= C\frac{1}{r}(l'\xi + m'\eta + n'\zeta), \qquad S' = C\frac{1}{r}(l\xi + m\eta + n\zeta), \end{aligned}\right\} \qquad (12)$$

where ξ, η, ζ are written for $x'-x$, $y'-y$, and $z'-z$ respectively.

515.] We have next to calculate the force with which the finite current s' acts on the finite current s. The current s extends from A, where $s = 0$, to P, where it has the value s. The current s' extends from A', where $s' = 0$, to P', where it has the value s'.

The coordinates of points on either current are functions of s or of s'.

If F is any function of the position of a point, then we shall use the subscript $_{(s,0)}$ to denote the excess of its value at P over that at A, thus

$$F_{(s,0)} = F_P - F_A.$$

Such functions necessarily disappear when the circuit is closed.

Let the components of the total force with which $A'P'$ acts on AA be $ii'X$, $ii'Y$, and $ii'Z$. Then the component parallel to X of the force with which ds' acts on ds will be $ii'\dfrac{d^2X}{ds\,ds'}ds\,ds'$.

Hence
$$\frac{d^2X}{ds\,ds'} = R\frac{\xi}{r} + Sl + S'l'. \tag{13}$$

Substituting the values of R, S, and S' from (12), remembering that
$$l'\xi + m'\eta + n'\zeta = r\frac{dr}{ds'}, \tag{14}$$

and arranging the terms with respect to l, m, n, we find

$$\begin{aligned}\frac{d^2X}{ds\,ds'} = {} & l\left\{-(A+B)\frac{1}{r^2}\frac{dr}{ds'}\xi^2 + C\frac{dr}{ds'} + (B+C)\frac{l'\xi}{r}\right\},\\ & +m\left\{-(A+B)\frac{1}{r^2}\frac{dr}{ds'}\xi\eta + C\frac{l'\eta}{r} + B\frac{m'\xi}{r}\right\},\\ & +n\left\{-(A+B)\frac{1}{r^2}\frac{dr}{ds'}\xi\zeta + C\frac{l'\zeta}{r} + B\frac{n'\xi}{r}\right\}.\end{aligned} \tag{15}$$

Since A, B, and C are functions of r, we may write

$$P = \int_r^\infty (A+B)\frac{1}{r^2}\,dr, \qquad Q = \int_r^\infty C\,dr, \tag{16}$$

the integration being taken between r and ∞ because A, B, C vanish when $r = \infty$.

Hence
$$(A+B)\frac{1}{r^2} = -\frac{dP}{dr}, \quad \text{and} \quad C = -\frac{dQ}{dr}. \tag{17}$$

516.] Now we know, by Ampère's third case of equilibrium, that when s' is a closed circuit, the force acting on ds is perpendicular to the direction of ds, or, in other words, the component of the force in the direction of ds itself is zero. Let us therefore assume the direction of the axis of x so as to be parallel to ds by making $l = 1$, $m = 0$, $n = 0$. Equation (15) then becomes

$$\frac{d^2X}{ds\,ds'} = \frac{dP}{ds'}\xi^2 - \frac{dQ}{ds'} + (B+C)\frac{l'\xi}{r}. \tag{18}$$

To find $\dfrac{dX}{ds}$, the force on ds referred to unit of length, we must integrate this expression with respect to s'. Integrating the first term by parts, we find

$$\frac{dX}{ds} = (P\xi^2 - Q)_{(s',0)} - \int_0^{s'} (2Pr - B - C)\frac{l'\xi}{r}\,ds'. \tag{19}$$

When s' is a closed circuit this expression must be zero. The first term will disappear of itself. The second term, however, will not in general disappear in the case of a closed circuit unless the quantity under the sign of integration is always zero. Hence, to satisfy Ampère's condition,

$$P = \frac{1}{2r}(B+C). \tag{20}$$

517.] We can now eliminate P, and find the general value of $\frac{dX}{ds}$,

$$\frac{dX}{ds} = \left\{\frac{B+C}{2}\frac{\xi}{r}(l\xi+m\eta+n\zeta)+Q\right\}_{(s',0)}$$
$$+m\int_0^{s'}\frac{B-C}{2}\frac{m'\xi-l'\eta}{r}\,ds' - n\int_0^{s'}\frac{B-C}{2}\frac{l'\zeta-n'\xi}{r}\,ds'. \tag{21}$$

When s' is a closed circuit the first term of this expression vanishes, and if we make

$$\left.\begin{aligned} \alpha' &= \int_0^{s'}\frac{B-C}{2}\frac{n'\eta-m'\zeta}{r}\,ds', \\ \beta' &= \int_0^{s'}\frac{B-C}{2}\frac{l'\zeta-n'\xi}{r}\,ds', \\ \gamma' &= \int_0^{s'}\frac{B-C}{2}\frac{m'\xi-l'\eta}{r}\,ds', \end{aligned}\right\} \tag{22}$$

where the integration is extended round the closed circuit s', we may write

$$\left.\begin{aligned} \frac{dX}{ds} &= m\gamma' - n\beta'. \\ \text{Similarly}\qquad \frac{dY}{ds} &= n\alpha' - l\gamma', \\ \frac{dZ}{ds} &= l\beta' - m\alpha'. \end{aligned}\right\} \tag{23}$$

The quantities α', β', γ' are sometimes called the determinants of the circuit s' referred to the point P. Their resultant is called by Ampère the directrix of the electrodynamic action.

It is evident from the equation, that the force whose components are $\frac{dX}{ds}$, $\frac{dY}{ds}$, and $\frac{dZ}{ds}$ is perpendicular both to ds and to this directrix, and is represented numerically by the area of the parallelogram whose sides are ds and the directrix.

In the language of quaternions, the resultant force on ds is the vector part of the product of the directrix multiplied by ds.

Since we already know that the directrix is the same thing as the magnetic force due to a unit current in the circuit s', we shall henceforth speak of the directrix as the magnetic force due to the circuit.

518.] We shall now complete the calculation of the components of the force acting between two finite currents, whether closed or open.

Let ρ be a new function of r, such that

$$\rho = \tfrac{1}{2}\int_r^\infty (B-C)\,dr, \tag{24}$$

then by (17) and (20)

$$A+B = r\frac{d^2}{dr^2}(Q+\rho) - \frac{d}{dr}(Q+\rho), \tag{25}$$

and equations (11) become

$$\left.\begin{aligned} R &= -\frac{d\rho}{dr}\cos\epsilon + r\frac{d^2}{ds\,ds'}(Q+\rho), \\ S &= -\frac{dQ}{ds'}, \qquad S' = \frac{dQ}{ds}. \end{aligned}\right\} \tag{26}$$

With these values of the component forces, equation (13) becomes

$$\begin{aligned} \frac{d^2X}{ds\,ds'} &= -\cos\epsilon\frac{d\rho}{dr}\frac{\xi}{r} + \xi\frac{d^2}{ds\,ds'}(Q+\rho) - l\frac{dQ}{ds'} + l'\frac{dQ}{ds}, \\ &= \cos\epsilon\frac{d\rho}{dx} + \frac{d^2(Q+\rho)\xi}{ds\,ds'} + l\frac{d\rho}{ds'} - l'\frac{d\rho}{ds}. \end{aligned} \tag{27}$$

519.] Let

$$F = \int_0^s l\rho\,ds, \qquad G = \int_0^s m\rho\,ds, \qquad H = \int_0^s n\rho\,ds, \tag{28}$$

$$F' = \int_0^{s'} l'\rho\,ds', \qquad G' = \int_0^{s'} m'\rho\,ds', \qquad H' = \int_0^{s''} n'\rho\,ds'. \tag{29}$$

These quantities have definite values for any given point of space. When the circuits are closed, they correspond to the components of the vector-potentials of the circuits.

Let L be a new function of r, such that

$$L = \int_0^r r(Q+\rho)\,dr, \tag{30}$$

and let M be the double integral

$$M = \int_0^{s'}\int_0^s \rho\cos\epsilon\,ds\,ds', \tag{31}$$

which, when the circuits are closed, becomes their mutual potential, then (27) may be written

$$\frac{d^2X}{ds\,ds'} = \frac{d^2}{ds\,ds'}\left\{\frac{dM}{dx} - \frac{dL}{dx} + F' - F\right\}. \tag{32}$$

520.] Integrating, with respect to s and s', between the given limits, we find

$$X = \frac{dM}{dx} - \frac{d}{dx}(L_{PP'} - L_{AP'} - L_{A'P} + L_{AA'}),$$
$$+ F'_P - F'_A - F_{P'} + F_{A'}, \tag{33}$$

where the subscripts of L indicate the distance, r, of which the quantity L is a function, and the subscripts of F and F' indicate the points at which their values are to be taken.

The expressions for Y and Z may be written down from this. Multiplying the three components by dx, dy, and dz respectively, we obtain

$$X\,dx + Y\,dy + Z\,dz = DM - D(L_{PP'} - L_{AP'} - L_{A'P} + L_{AA'}),$$
$$+ (F'dx + G'dy + H'dz)_{(P-A)},$$
$$-(F\,dx + G\,dy + H\,dz)_{(P'-A)'}, \tag{34}$$

where D is the symbol of a complete differential.

Since $F\,dx + G\,dy + H\,dz$ is not in general a complete differential of a function of x, y, z, $X\,dx + Y\,dy + Z\,dz$ is not a complete differential for currents either of which is not closed.

521.] If, however, both currents are closed, the terms in L, F, G, H, F', G', H' disappear, and

$$X\,dx + Y\,dy + Z\,dz = DM, \tag{35}$$

where M is the mutual potential of two closed circuits carrying unit currents. The quantity M expresses the work done by the electromagnetic forces on either conducting circuit when it is moved parallel to itself from an infinite distance to its actual position. Any alteration of its position, by which M is *increased*, will be *assisted* by the electromagnetic forces.

It may be shewn, as in Arts. 490, 596, that when the motion of the circuit is not parallel to itself the forces acting on it are still determined by the variation of M, the potential of the one circuit on the other.

522.] The only experimental fact which we have made use of in this investigation is the fact established by Ampère that the action of a closed current on any portion of another current is perpendicular to the direction of the latter. Every other part of the investigation depends on purely mathematical considerations depending on the properties of lines in space. The reasoning therefore may be presented in a much more condensed and appropriate form by the use of the ideas and language of the mathematical method specially adapted to the expression of such geometrical relations—the *Quaternions* of Hamilton.

This has been done by Professor Tait in the *Quarterly Mathematical Journal,* 1866, and in his treatise on *Quaternions*, § 399, for Ampère's original investigation, and the student can easily adapt the same method to the somewhat more general investigation given here.

523.] Hitherto we have made no assumption with respect to the quantities A, B, C, except that they are functions of r, the distance between the elements. We have next to ascertain the form of these functions, and for this purpose we make use of Ampère's fourth case of equilibrium, Art. 508, in which it is shewn that if all the linear dimensions and distances of a system of two circuits be altered in the same proportion, the currents remaining the same, the force between the two circuits will remain the same.

Now the force between the circuits for unit currents is $\dfrac{dM}{dx}$, and since this is independent of the dimensions of the system, it must be a numerical quantity. Hence M itself, the coefficient of the mutual potential of the circuits, must be a quantity of the dimensions of a line. It follows, from equation (31), that ρ must be the reciprocal of a line, and therefore by (24), $B-C$ must be the inverse square of a line. But since B and C are both functions of r, $B-C$ must be the inverse square of r or some numerical multiple of it.

524.] The multiple we adopt depends on our system of measurement. If we adopt the electromagnetic system, so called because it agrees with the system already established for magnetic measurements, the value of M ought to coincide with that of the potential of two magnetic shells of strength unity whose boundaries are the two circuits respectively. The value of M in that case is, by Art. 423,

$$M = \iint \frac{\cos\epsilon}{r}\, ds\, ds', \tag{36}$$

the integration being performed round both circuits in the positive direction. Adopting this as the numerical value of M, and comparing with (31), we find

$$\rho = \frac{1}{r}, \quad \text{and} \quad B-C = \frac{2}{r^2}. \tag{37}$$

525.] We may now express the components of the force on ds arising from the action of ds' in the most general form consistent with experimental facts.

The force on ds is compounded of an attraction

$$R = \frac{1}{r^2}\left(\frac{dr}{ds}\frac{dr}{ds'} - 2r\frac{d^2 r}{ds\,ds'}\right)ii'\,ds\,ds' + r\frac{d^2 Q}{ds\,ds'}ii'\,ds\,ds'$$

in the direction of r,

$$S = -\frac{dQ}{ds'}ii'\,ds\,ds' \text{ in the direction of } ds, \tag{38}$$

and $$S' = \frac{dQ}{ds}ii'\,ds\,ds' \text{ in the direction of } ds',$$

where $Q = \int_r^\infty C\,dr$, and since C is an unknown function of r, we know only that Q is some function of r.

526.] The quantity Q cannot be determined, without assumptions of some kind, from experiments in which the active current forms a closed circuit. If we suppose with Ampère that the action between the elements ds and ds' is in the line joining them, then S and S' must disappear, and Q must be constant, or zero. The force is then reduced to an attraction whose value is

$$R = \frac{1}{r^2}\left(\frac{dr}{ds}\frac{dr}{ds'} - 2r\frac{d^2 r}{ds\,ds'}\right)ii'\,ds\,ds'. \tag{39}$$

Ampère, who made this investigation long before the magnetic system of units had been established, uses a formula having a numerical value half of this, namely

$$R = \frac{1}{r^2}\left(\frac{1}{2}\frac{dr}{ds}\frac{dr}{ds'} - r\frac{dr}{ds\,ds'}\right)jj'\,ds\,ds'. \tag{40}$$

Here the strength of the current is measured in what is called electrodynamic measure. If i, i' are the strength of the currents in electromagnetic measure, and j, j' the same in electrodynamic measure, then it is plain that

$$jj' = 2ii', \quad \text{or} \quad j = \sqrt{2}\,i. \tag{41}$$

Hence the unit current adopted in electromagnetic measure is greater than that adopted in electrodynamic measure in the ratio of $\sqrt{2}$ to 1.

The only title of the electrodynamic unit to consideration is that it was originally adopted by Ampère, the discoverer of the law of action between currents. The continual recurrence of $\sqrt{2}$ in calculations founded on it is inconvenient, and the electromagnetic system has the great advantage of coinciding numerically with all our magnetic formulae. As it is difficult for the student to bear in mind whether he is to multiply or to divide by $\sqrt{2}$, we shall henceforth use only the electromagnetic system, as adopted by Weber and most other writers.

Since the form and value of Q have no effect on any of the experiments hitherto made, in which the active current at least is always a closed one, we may, if we please, adopt any value of Q which appears to us to simplify the formulae.

Thus Ampère assumes that the force between two elements is in the line joining them. This gives $Q = 0$,

$$R = \frac{1}{r^2}\left(\frac{dr}{ds}\frac{dr}{ds'} - 2r\frac{d^2r}{ds\,ds'}\right) ii\,ds\,ds', \quad S = 0, \quad S' = 0. \qquad (42)$$

Grassmann* assumes that two elements in the same straight line have no mutual action. This gives

$$Q = -\frac{1}{2r}, \quad R = -\frac{3}{2r}\frac{d^2r}{ds\,ds'}, \quad S = -\frac{1}{2r^2}\frac{dr}{ds'}, \quad S' = \frac{1}{2r^2}\frac{dr}{ds}. \qquad (43)$$

We might, if we pleased, assume that the attraction between two elements at a given distance is proportional to the cosine of the angle between them. In this case

$$Q = -\frac{1}{r}, \quad R = \frac{1}{r^2}\cos\epsilon, \quad S = -\frac{1}{r^2}\frac{dr}{ds'}, \quad S' = \frac{1}{r^2}\frac{dr}{ds}. \qquad (44)$$

Finally, we might assume that the attraction and the oblique forces depend only on the angles which the elements make with the line joining them, and then we should have

$$Q = -\frac{2}{r}, \quad R = -3\frac{1}{r^2}\frac{dr}{ds}\frac{dr}{ds'}, \quad S = -\frac{2}{r^2}\frac{dr}{ds'}, \quad S' = \frac{2}{r^2}\frac{dr}{ds}. \qquad (45)$$

527.] Of these four different assumptions that of Ampère is undoubtedly the best, since it is the only one which makes the forces on the two elements not only equal and opposite but in the straight line which joins them.

* Pogg., *Ann.* lxiv. p. 1 (1845).

CHAPTER III.

ON THE INDUCTION OF ELECTRIC CURRENTS.

528.] The discovery by Örsted of the magnetic action of an electric current led by a direct process of reasoning to that of magnetization by electric currents, and of the mechanical action between electric currents. It was not, however, till 1831 that Faraday, who had been for some time endeavouring to produce electric currents by magnetic or electric action, discovered the conditions of magneto-electric induction. The method which Faraday employed in his researches consisted in a constant appeal to experiment as a means of testing the truth of his ideas, and a constant cultivation of ideas under the direct influence of experiment. In his published researches we find these ideas expressed in language which is all the better fitted for a nascent science, because it is somewhat alien from the style of physicists who have been accustomed to established mathematical forms of thought.

The experimental investigation by which Ampère established the laws of the mechanical action between electric currents is one of the most brilliant achievements in science.

The whole, theory and experiment, seems as if it had leaped, full grown and full armed, from the brain of the 'Newton of electricity.' It is perfect in form, and unassailable in accuracy, and it is summed up in a formula from which all the phenomena may be deduced, and which must always remain the cardinal formula of electro-dynamics.

The method of Ampère, however, though cast into an inductive form, does not allow us to trace the formation of the ideas which guided it. We can scarcely believe that Ampère really discovered the law of action by means of the experiments which he describes. We are led to suspect, what, indeed, he tells us himself*, that he

* *Théorie des Phenomènes Electrodynamiques*, p. 9.

discovered the law by some process which he has not shewn us, and that when he had afterwards built up a perfect demonstration he removed all traces of the scaffolding by which he had raised it.

Faraday, on the other hand, shews us his unsuccessful as well as his successful experiments, and his crude ideas as well as his developed ones, and the reader, however inferior to him in inductive power, feels sympathy even more than admiration, and is tempted to believe that, if he had the opportunity, he too would be a discoverer. Every student therefore should read Ampère's research as a splendid example of scientific style in the statement of a discovery, but he should also study Faraday for the cultivation of a scientific spirit, by means of the action and reaction which will take place between newly discovered facts and nascent ideas in his own mind.

It was perhaps for the advantage of science that Faraday, though thoroughly conscious of the fundamental forms of space, time, and force, was not a professed mathematician. He was not tempted to enter into the many interesting researches in pure mathematics which his discoveries would have suggested if they had been exhibited in a mathematical form, and he did not feel called upon either to force his results into a shape acceptable to the mathematical taste of the time, or to express them in a form which mathematicians might attack. He was thus left at leisure to do his proper work, to coordinate his ideas with his facts, and to express them in natural, untechnical language.

It is mainly with the hope of making these ideas the basis of a mathematical method that I have undertaken this treatise.

529.] We are accustomed to consider the universe as made up of parts, and mathematicians usually begin by considering a single particle, and then conceiving its relation to another particle, and so on. This has generally been supposed the most natural method. To conceive of a particle, however, requires a process of abstraction, since all our perceptions are related to extended bodies, so that the idea of the *all* that is in our consciousness at a given instant is perhaps as primitive an idea as that of any individual thing. Hence there may be a mathematical method in which we proceed from the whole to the parts instead of from the parts to the whole. For example, Euclid, in his first book, conceives a line as traced out by a point, a surface as swept out by a line, and a solid as generated by a surface. But he also defines a surface as the boundary of a solid, a line as the edge of a surface, and a point as the extremity of a line.

In like manner we may conceive the potential of a material system as a function found by a certain process of integration with respect to the masses of the bodies in the field, or we may suppose these masses themselves to have no other mathematical meaning than the volume-integrals of $\frac{1}{4\pi}\nabla^2\Psi$, where Ψ is the potential.

In electrical investigations we may use formulae in which the quantities involved are the distances of certain bodies, and the electrifications or currents in these bodies, or we may use formulae which involve other quantities, each of which is continuous through all space.

The mathematical process employed in the first method is integration along lines, over surfaces, and throughout finite spaces, those employed in the second method are partial differential equations and integrations throughout all space.

The method of Faraday seems to be intimately related to the second of these modes of treatment. He never considers bodies as existing with nothing between them but their distance, and acting on one another according to some function of that distance. He conceives all space as a field of force, the lines of force being in general curved, and those due to any body extending from it on all sides, their directions being modified by the presence of other bodies. He even speaks * of the lines of force belonging to a body as in some sense part of itself, so that in its action on distant bodies it cannot be said to act where it is not. This, however, is not a dominant idea with Faraday. I think he would rather have said that the field of space is full of lines of force, whose arrangement depends on that of the bodies in the field, and that the mechanical and electrical action on each body is determined by the lines which abut on it.

PHENOMENA OF MAGNETO-ELECTRIC INDUCTION †.

530.] 1. *Induction by Variation of the Primary Current.*

Let there be two conducting circuits, the Primary and the Secondary circuit. The primary circuit is connected with a voltaic

* *Exp. Res.*, ii. p. 293; iii. p. 447.

† Read Faraday's *Experimental Researches*, series i and ii.

battery by which the primary current may be produced, maintained, stopped, or reversed. The secondary circuit includes a galvanometer to indicate any currents which may be formed in it. This galvanometer is placed at such a distance from all parts of the primary circuit that the primary current has no sensible direct influence on its indications.

Let part of the primary circuit consist of a straight wire, and part of the secondary circuit of a straight wire near, and parallel to the first, the other parts of the circuits being at a greater distance from each other.

It is found that at the instant of sending a current through the straight wire of the primary circuit the galvanometer of the secondary circuit indicates a current in the secondary straight wire in the *opposite* direction. This is called the induced current. If the primary current is maintained constant, the induced current soon disappears, and the primary current appears to produce no effect on the secondary circuit. If now the primary current is stopped, a secondary current is observed, which is in the *same* direction as the primary current. Every variation of the primary current produces electromotive force in the secondary circuit. When the primary current increases, the electromotive force is in the opposite direction to the current. When it diminishes, the electromotive force is in the same direction as the current. When the primary current is constant, there is no electromotive force.

These effects of induction are increased by bringing the two wires nearer together. They are also increased by forming them into two circular or spiral coils placed close together, and still more by placing an iron rod or a bundle of iron wires inside the coils.

2. *Induction by Motion of the Primary Circuit.*

We have seen that when the primary current is maintained constant and at rest the secondary current rapidly disappears.

Now let the primary current be maintained constant, but let the primary straight wire be made to approach the secondary straight wire. During the approach there will be a secondary current in the *opposite* direction from the primary.

If the primary circuit be moved away from the secondary, there will be a secondary current in the *same* direction as the primary.

3. *Induction by Motion of the Secondary Circuit.*

If the secondary circuit be moved, the secondary current is

opposite to the primary when the secondary wire is approaching the primary wire, and in the same direction when it is receding from it.

In all cases the direction of the secondary current is such that the mechanical action between the two conductors is opposite to the direction of motion, being a repulsion when the wires are approaching, and an attraction when they are receding. This very important fact was established by Lenz *.

4. *Induction by the Relative Motion of a Magnet and the Secondary Circuit.*

If we substitute for the primary circuit a magnetic shell, whose edge coincides with the circuit, whose strength is numerically equal to that of the current in the circuit, and whose austral face corresponds to the positive face of the circuit, then the phenomena produced by the relative motion of this shell and the secondary circuit are the same as those observed in the case of the primary circuit.

531.] The whole of these phenomena may be summed up in one law. When the number of lines of magnetic induction which pass through the secondary circuit in the positive direction is altered, an electromotive force acts round the circuit, which is measured by the rate of decrease of the magnetic induction through the circuit.

532.] For instance, let the rails of a railway be insulated from the earth, but connected at one terminus through a galvanometer, and let the circuit be completed by the wheels and axle of a railway carriage at a distance x from the terminus. Neglecting the height of the axle above the level of the rails, the induction through the secondary circuit is due to the vertical component of the earth's magnetic force, which in northern latitudes is directed downwards. Hence, if b is the gauge of the railway, the horizontal area of the circuit is bx, and the surface-integral of the magnetic induction through it is Zbx, where Z is the vertical component of the magnetic force of the earth. Since Z is downwards, the lower face of the circuit is to be reckoned positive, and the positive direction of the circuit itself is north, east, south, west, that is, in the direction of the sun's apparent diurnal course.

Now let the carriage be set in motion, then x will vary, and

* Pogg., *Ann*. xxi. 483 (1834.)

there will be an electromotive force in the circuit whose value is $-Zb\frac{dx}{dt}$.

If x is increasing, that is, if the carriage is moving away from the terminus, this electromotive force is in the negative direction, or north, west, south, east. Hence the direction of this force through the axle is from right to left. If x were diminishing, the absolute direction of the force would be reversed, but since the direction of the motion of the carriage is also reversed, the electromotive force on the axle is still from right to left, the observer in the carriage being always supposed to move face forwards. In southern latitudes, where the south end of the needle dips, the electromotive force on a moving body is from left to right.

Hence we have the following rule for determining the electromotive force on a wire moving through a field of magnetic force. Place, in imagination, your head and feet in the position occupied by the ends of a compass needle which point north and south respectively; turn your face in the forward direction of motion, then the electromotive force due to the motion will be from left to right.

533.] As these directional relations are important, let us take another illustration. Suppose a metal girdle laid round the earth at the equator, and a metal wire laid along the meridian of Greenwich from the equator to the north pole.

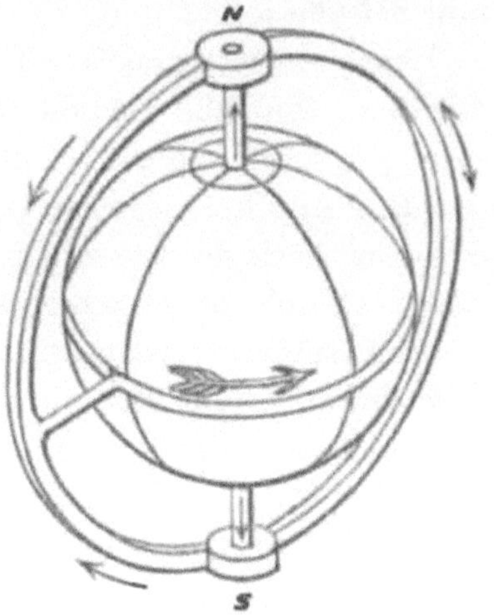

Fig. 31.

Let a great quadrantal arch of metal be constructed, of which one extremity is pivoted on the north pole, while the other is carried round the equator, sliding on the great girdle of the earth, and following the sun in his daily course. There will then be an electromotive force along the moving quadrant, acting from the pole towards the equator.

The electromotive force will be the same whether we suppose the earth at rest and the quadrant moved from east to west, or whether we suppose the quadrant at rest and the earth turned from west to east. If we suppose the earth to rotate, the electromotive force will be the same whatever be the form of the part of the circuit fixed in space of which one end touches one of the poles and the other the equator. The current in this part of the circuit is from the pole to the equator.

The other part of the circuit, which is fixed with respect to the earth, may also be of any form, and either within or without the earth. In this part the current is from the equator to either pole.

534.] The intensity of the electromotive force of magneto-electric induction is entirely independent of the nature of the substance of the conductor in which it acts, and also of the nature of the conductor which carries the inducing current.

To shew this, Faraday * made a conductor of two wires of different metals insulated from one another by a silk covering, but twisted together, and soldered together at one end. The other ends of the wires were connected with a galvanometer. In this way the wires were similarly situated with respect to the primary circuit, but if the electromotive force were stronger in the one wire than in the other it would produce a current which would be indicated by the galvanometer. He found, however, that such a combination may be exposed to the most powerful electromotive forces due to induction without the galvanometer being affected. He also found that whether the two branches of the compound conductor consisted of two metals, or of a metal and an electrolyte, the galvanometer was not affected †.

Hence the electromotive force on any conductor depends only on the form and the motion of that conductor, together with the strength, form, and motion of the electric currents in the field.

535.] Another negative property of electromotive force is that it has of itself no tendency to cause the mechanical motion of any body, but only to cause a current of electricity within it.

If it actually produces a current in the body, there will be mechanical action due to that current, but if we prevent the current from being formed, there will be no mechanical action on the body itself. If the body is electrified, however, the electromotive force will move the body, as we have described in Electrostatics.

536.] The experimental investigation of the laws of the induction of electric currents in fixed circuits may be conducted with considerable accuracy by methods in which the electromotive force, and therefore the current, in the galvanometer circuit is rendered zero.

For instance, if we wish to shew that the induction of the coil

* *Exp. Res.*, 195. † Ib., 200.

A on the coil *X* is equal to that of *B* upon *Y*, we place the first pair of coils *A* and *X* at a sufficient distance from the second pair

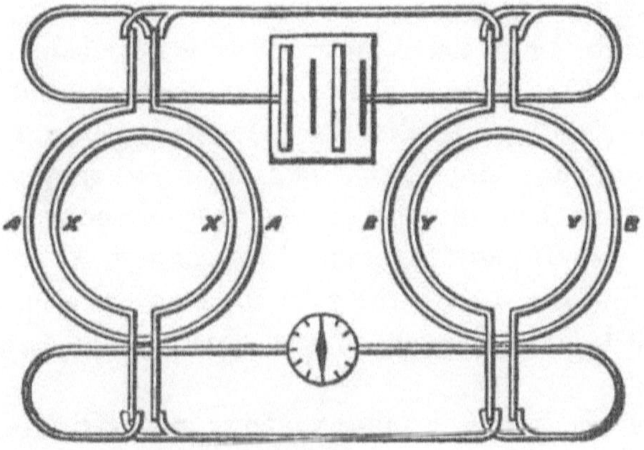

Fig. 32.

B and *Y*. We then connect *A* and *B* with a voltaic battery, so that we can make the same primary current flow through *A* in the positive direction and then through *B* in the negative direction. We also connect *X* and *Y* with a galvanometer, so that the secondary current, if it exists, shall flow in the same direction through *X* and *Y* in series.

Then, if the induction of *A* on *X* is equal to that of *B* on *Y*, the galvanometer will indicate no induction current when the battery circuit is closed or broken.

The accuracy of this method increases with the strength of the primary current and the sensitiveness of the galvanometer to instantaneous currents, and the experiments are much more easily performed than those relating to electromagnetic attractions, where the conductor itself has to be delicately suspended.

A very instructive series of well devised experiments of this kind is described by Professor Felici of Pisa *.

I shall only indicate briefly some of the laws which may be proved in this way.

(1) The electromotive force of the induction of one circuit on another is independent of the area of the section of the conductors and of the material of which they are made.

For we can exchange any one of the circuits in the experiment for another of a different section and material, but of the same form, without altering the result.

* *Annales de Chimie*, xxxiv. p. 66 (1852), and *Nuovo Cimento*, ix. p. 345 (1859).

(2) The induction of the circuit A on the circuit X is equal to that of X upon A.

For if we put A in the galvanometer circuit, and X in the battery circuit, the equilibrium of electromotive force is not disturbed.

(3) The induction is proportional to the inducing current.

For if we have ascertained that the induction of A on X is equal to that of B on Y, and also to that of C on Z, we may make the battery current first flow through A, and then divide itself in any proportion between B and C. Then if we connect X reversed, Y and Z direct, all in series, with the galvanometer, the electromotive force in X will balance the sum of the electromotive forces in Y and Z.

(4) In pairs of circuits forming systems geometrically similar the induction is proportional to their linear dimensions.

For if the three pairs of circuits above mentioned are all similar, but if the linear dimension of the first pair is the sum of the corresponding linear dimensions of the second and third pairs, then, if A, B, and C are connected in series with the battery, and X reversed, Y and Z also in series with the galvanometer, there will be equilibrium.

(5) The electromotive force produced in a coil of n windings by a current in a coil of m windings is proportional to the product mn.

537.] For experiments of the kind we have been considering the galvanometer should be as sensitive as possible, and its needle as light as possible, so as to give a sensible indication of a very small transient current. The experiments on induction due to motion require the needle to have a somewhat longer period of vibration, so that there may be time to effect certain motions of the conductors while the needle is not far from its position of equilibrium. In the former experiments, the electromotive forces in the galvanometer circuit were in equilibrium during the whole time, so that no current passed through the galvanometer coil. In those now to be described, the electromotive forces act first in one direction and then in the other, so as to produce in succession two currents in opposite directions through the galvanometer, and we have to shew that the impulses on the galvanometer needle due to these successive currents are in certain cases equal and opposite.

The theory of the application of the galvanometer to the measurement of transient currents will be considered more at length in Art. 748. At present it is sufficient for our purpose to observe that as long as the galvanometer needle is near its position of equilibrium the deflecting force of the current is proportional to the current itself, and if the whole time of action of the current is small compared with the period of vibration of the needle, the final velocity of the magnet will be proportional to the total quantity of electricity in the current. Hence, if two currents pass in rapid succession, conveying equal quantities of electricity in opposite directions, the needle will be left without any final velocity.

Thus, to shew that the induction-currents in the secondary circuit, due to the closing and the breaking of the primary circuit, are equal in total quantity but opposite in direction, we may arrange the primary circuit in connexion with the battery, so that by touching a key the current may be sent through the primary circuit, or by removing the finger the contact may be broken at pleasure. If the key is pressed down for some time, the galvanometer in the secondary circuit indicates, at the time of making contact, a transient current in the opposite direction to the primary current. If contact be maintained, the induction current simply passes and disappears. If we now break contact, another transient current passes in the opposite direction through the secondary circuit, and the galvanometer needle receives an impulse in the opposite direction.

But if we make contact only for an instant, and then break contact, the two induced currents pass through the galvanometer in such rapid succession that the needle, when acted on by the first current, has not time to move a sensible distance from its position of equilibrium before it is stopped by the second, and, on account of the exact equality between the quantities of these transient currents, the needle is stopped dead.

If the needle is watched carefully, it appears to be jerked suddenly from one position of rest to another position of rest very near the first.

In this way we prove that the quantity of electricity in the induction current, when contact is broken, is exactly equal and opposite to that in the induction current when contact is made.

538.] Another application of this method is the following, which is given by Felici in the second series of his *Researches*.

It is always possible to find many different positions of the secondary coil B, such that the making or the breaking of contact in the primary coil A produces no induction current in B. The positions of the two coils are in such cases said to be *conjugate* to each other.

Let B_1 and B_2 be two of these positions. If the coil B be suddenly moved from the position B_1 to the position B_2, the algebraical sum of the transient currents in the coil B is exactly zero, so that the galvanometer needle is left at rest when the motion of B is completed.

This is true in whatever way the coil B is moved from B_1 to B_2, and also whether the current in the primary coil A be continued constant, or made to vary during the motion.

Again, let B' be any other position of B not conjugate to A, so that the making or breaking of contact in A produces an induction current when B is in the position B'.

Let the contact be made when B is in the conjugate position B_1, there will be no induction current. Move B to B', there will be an induction current due to the motion, but if B is moved rapidly to B', and the primary contact then broken, the induction current due to breaking contact will exactly annul the effect of that due to the motion, so that the galvanometer needle will be left at rest. Hence the current due to the motion from a conjugate position to any other position is equal and opposite to the current due to breaking contact in the latter position.

Since the effect of making contact is equal and opposite to that of breaking it, it follows that the effect of making contact when the coil B is in any position B' is equal to that of bringing the coil from any conjugate position B_1 to B' while the current is flowing through A.

If the change of the relative position of the coils is made by moving the primary circuit instead of the secondary, the result is found to be the same.

539.] It follows from these experiments that the total induction current in B during the simultaneous motion of A from A_1 to A_2, and of B from B_1 to B_2, while the current in A changes from γ_1 to γ_2, depends only on the initial state A_1, B_1, γ_1, and the final state A_2, B_2, γ_2, and not at all on the nature of the intermediate states through which the system may pass.

Hence the value of the total induction current must be of the form

$$F(A_2, B_2, \gamma_2) - F(A_1, B_1, \gamma_1),$$

where F is a function of A, B, and γ.

With respect to the form of this function, we know, by Art. 536, that when there is no motion, and therefore $A_1 = A_2$ and $B_1 = B_2$, the induction current is proportional to the primary current. Hence γ enters simply as a factor, the other factor being a function of the form and position of the circuits A and B.

We also know that the value of this function depends on the relative and not on the absolute positions of A and B, so that it must be capable of being expressed as a function of the distances of the different elements of which the circuits are composed, and of the angles which these elements make with each other.

Let M be this function, then the total induction current may be written

$$C\{M_1\gamma_1 - M_2\gamma_2\},$$

where C is the conductivity of the secondary circuit, and M_1, γ_1 are the original, and M_2, γ_2 the final values of M and γ.

These experiments, therefore, shew that the total current of induction depends on the change which takes place in a certain quantity, $M\gamma$, and that this change may arise either from variation of the primary current γ, or from any motion of the primary or secondary circuit which alters M.

540.] The conception of such a quantity, on the changes of which, and not on its absolute magnitude, the induction current depends, occurred to Faraday at an early stage of his researches *. He observed that the secondary circuit, when at rest in an electromagnetic field which remains of constant intensity, does not shew any electrical effect, whereas, if the same state of the field had been suddenly produced, there would have been a current. Again, if the primary circuit is removed from the field, or the magnetic forces abolished, there is a current of the opposite kind. He therefore recognised in the secondary circuit, when in the electromagnetic field, a 'peculiar electrical condition of matter,' to which he gave the name of the Electrotonic State. He afterwards found that he could dispense with this idea by means of considerations founded on the lines of magnetic force †, but even in his latest researches ‡, he says, 'Again and again the idea of an *electrotonic* state § has been forced upon my mind.'

The whole history of this idea in the mind of Faraday, as shewn in his published researches, is well worthy of study. By a course of experiments, guided by intense application of thought, but without the aid of mathematical calculations, he was led to recognise the existence of something which we now know to be a mathematical quantity, and which may even be called the fundamental

* *Exp. Res.*, series i. 60.

† Ib., series ii. (242).

‡ Ib., 3269.

§ Ib., 60, 1114, 1661, 1729, 1733.

quantity in the theory of electromagnetism. But as he was led up to this conception by a purely experimental path, he ascribed to it a physical existence, and supposed it to be a peculiar condition of matter, though he was ready to abandon this theory as soon as he could explain the phenomena by any more familiar forms of thought.

Other investigators were long afterwards led up to the same idea by a purely mathematical path, but, so far as I know, none of them recognised, in the refined mathematical idea of the potential of two circuits, Faraday's bold hypothesis of an electrotonic state. Those, therefore, who have approached this subject in the way pointed out by those eminent investigators who first reduced its laws to a mathematical form, have sometimes found it difficult to appreciate the scientific accuracy of the statements of laws which Faraday, in the first two series of his *Researches*, has given with such wonderful completeness.

The scientific value of Faraday's conception of an electrotonic state consists in its directing the mind to lay hold of a certain quantity, on the changes of which the actual phenomena depend. Without a much greater degree of development than Faraday gave it, this conception does not easily lend itself to the explanation of the phenomena. We shall return to this subject again in Art. 584.

541.] A method which, in Faraday's hands, was far more powerful is that in which he makes use of those lines of magnetic force which were always in his mind's eye when contemplating his magnets or electric currents, and the delineation of which by means of iron filings he rightly regarded * as a most valuable aid to the experimentalist.

Faraday looked on these lines as expressing, not only by their direction that of the magnetic force, but by their number and concentration the intensity of that force, and in his later researches † he shews how to conceive of unit lines of force. I have explained in various parts of this treatise the relation between the properties which Faraday recognised in the lines of force and the mathematical conditions of electric and magnetic forces, and how Faraday's notion of unit lines and of the number of lines within certain limits may be made mathematically precise. See Arts. 82, 404, 490.

In the first series of his *Researches* ‡ he shews clearly how the direction of the current in a conducting circuit, part of which is

* *Exp. Res.*, 3234. † Ib., 3122. ‡ Ib., 114.

moveable, depends on the mode in which the moving part cuts through the lines of magnetic force.

In the second series* he shews how the phenomena produced by variation of the strength of a current or a magnet may be explained, by supposing the system of lines of force to expand from or contract towards the wire or magnet as its power rises or falls.

I am not certain with what degree of clearness he then held the doctrine afterwards so distinctly laid down by him †, that the moving conductor, as it cuts the lines of force, sums up the action due to an area or section of the lines of force. This, however, appears no new view of the case after the investigations of the second series ‡ have been taken into account.

The conception which Faraday had of the continuity of the lines of force precludes the possibility of their suddenly starting into existence in a place where there were none before. If, therefore, the number of lines which pass through a conducting circuit is made to vary, it can only be by the circuit moving across the lines of force, or else by the lines of force moving across the circuit. In either case a current is generated in the circuit.

The number of the lines of force which at any instant pass through the circuit is mathematically equivalent to Faraday's earlier conception of the electrotonic state of that circuit, and it is represented by the quantity $M\gamma$.

It is only since the definitions of electromotive force, Arts. 69, 274, and its measurement have been made more precise, that we can enunciate completely the true law of magneto-electric induction in the following terms:—

The total electromotive force acting round a circuit at any instant is measured by the rate of decrease of the number of lines of magnetic force which pass through it.

When integrated with respect to the time this statement becomes:—

The time-integral of the total electromotive force acting round any circuit, together with the number of lines of magnetic force which pass through the circuit, is a constant quantity.

Instead of speaking of the number of lines of magnetic force, we may speak of the magnetic induction through the circuit, or the surface-integral of magnetic induction extended over any surface bounded by the circuit.

* *Exp. Res.*, 238.

† Ib., 3082, 3087, 3113.

‡ Ib., 217, &c.

We shall return again to this method of Faraday. In the mean time we must enumerate the theories of induction which are founded on other considerations.

Lenz's Law.

542.] In 1834, Lenz* enunciated the following remarkable relation between the phenomena of the mechanical action of electric currents, as defined by Ampère's formula, and the induction of electric currents by the relative motion of conductors. An earlier attempt at a statement of such a relation was given by Ritchie in the *Philosophical Magazine* for January of the same year, but the direction of the induced current was in every case stated wrongly. Lenz's law is as follows.—

If a constant current flows in the primary circuit A, and if, by the motion of A, or of the secondary circuit B, a current is induced in B, the direction of this induced current will be such that, by its electromagnetic action on A, it tends to oppose the relative motion of the circuits.

On this law J. Neumann † founded his mathematical theory of induction, in which he established the mathematical laws of the induced currents due to the motion of the primary or secondary conductor. He shewed that the quantity M, which we have called the potential of the one circuit on the other, is the same as the electromagnetic potential of the one circuit on the other, which we have already investigated in connexion with Ampère's formula. We may regard J. Neumann, therefore, as having completed for the induction of currents the mathematical treatment which Ampère had applied to their mechanical action.

543.] A step of still greater scientific importance was soon after made by Helmholtz in his *Essay on the Conservation of Force* ‡, and by Sir W. Thomson §, working somewhat later, but independently of Helmholtz. They shewed that the induction of electric currents discovered by Faraday could be mathematically deduced from the electromagnetic actions discovered by Örsted and Ampère by the application of the principle of the Conservation of Energy.

Helmholtz takes the case of a conducting circuit of resistance R, in which an electromotive force A, arising from a voltaic or thermo-

* Pogg., *Ann.* xxxi. 483 (1834).

† *Berlin Acad.*, 1845 and 1847.

‡ Read before the Physical Society of Berlin, July 23, 1847. Translated in Taylor's 'Scientific Memoirs,' part ii. p. 114.

§ *Trans. Brit. Ass.*, 1848, and *Phil. Mag.*, Dec. 1851. See also his paper on 'Transient Electric Currents,' *Phil. Mag.*, 1853.

electric arrangement, acts. The current in the circuit at any instant is I. He supposes that a magnet is in motion in the neighbourhood of the circuit, and that its potential with respect to the conductor is V, so that, during any small interval of time dt, the energy communicated to the magnet by the electromagnetic action is $I\frac{dV}{dt}dt$.

The work done in generating heat in the circuit is, by Joule's law, Art. 242, $I^2 R dt$, and the work spent by the electromotive force A, in maintaining the current I during the time dt, is $A I dt$. Hence, since the total work done must be equal to the work spent,

$$A I dt = I^2 R dt + I\frac{dV}{dt}dt,$$

whence we find the intensity of the current

$$I = \frac{A - \frac{dV}{dt}}{R}.$$

Now the value of A may be what we please. Let, therefore, $A = 0$, and then

$$I = -\frac{1}{R}\frac{dV}{dt},$$

or, there will be a current due to the motion of the magnet, equal to that due to an electromotive force $-\frac{dV}{dt}$.

The whole induced current during the motion of the magnet from a place where its potential is V_1 to a place where its potential is V_2, is

$$\int I dt = -\frac{1}{R}\int \frac{dV}{dt} dt = \frac{1}{R}(V_1 - V_2),$$

and therefore the total current is independent of the velocity or the path of the magnet, and depends only on its initial and final positions.

In Helmholtz's original investigation he adopted a system of units founded on the measurement of the heat generated in the conductor by the current. Considering the unit of current as arbitrary, the unit of resistance is that of a conductor in which this unit current generates unit of heat in unit of time. The unit of electromotive force in this system is that required to produce the unit of current in the conductor of unit resistance. The adoption of this system of units necessitates the introduction into the equations of a quantity a, which is the mechanical equivalent of the unit of heat. As we invariably adopt either the electrostatic or the electromagnetic system of units, this factor does not occur in the equations here given.

544.] Helmholtz also deduces the current of induction when a conducting circuit and a circuit carrying a constant current are made to move relatively to one another.

Let R_1, R_2 be the resistances, I_1, I_2 the currents, A_1, A_2 the external electromotive forces, and V the potential of the one circuit on the other due to unit current in each, then we have, as before,

$$A_1 I_1 + A_2 I_2 = I_1^2 R_1 + I_2^2 R_2 + I_1 I_2 \frac{dV}{dt}.$$

If we suppose I_1 to be the primary current, and I_2 so much less than I_1, that it does not by its induction produce any sensible alteration in I_1, so that we may put $I_1 = \frac{A_1}{R_1}$, then

$$I_2 = \frac{A_2 - I_1 \frac{dV}{dt}}{R_2},$$

a result which may be interpreted exactly as in the case of the magnet.

If we suppose I_2 to be the primary current, and I_1 to be very much smaller than I_2, we get for I_1,

$$I_1 = \frac{A_1 - I_2 \frac{dV}{dt}}{R_1}.$$

This shews that for equal currents the electromotive force of the first circuit on the second is equal to that of the second on the first, whatever be the forms of the circuits.

Helmholtz does not in this memoir discuss the case of induction due to the strengthening or weakening of the primary current, or the induction of a current on itself. Thomson * applied the same principle to the determination of the mechanical value of a current, and pointed out that when work is done by the mutual action of two constant currents, their mechanical value is *increased* by the same amount, so that the battery has to supply *double* that amount of work, in addition to that required to maintain the currents against the resistance of the circuits †.

545.] The introduction, by W. Weber, of a system of absolute

* Mechanical Theory of Electrolysis, *Phil. Mag.*, Dec., 1851.

† Nichol's *Cyclopaedia of Physical Science*, ed. 1860, Article 'Magnetism, Dynamical Relations of,' and *Reprint*, § 571.

units for the measurement of electrical quantities is one of the most important steps in the progress of the science. Having already, in conjunction with Gauss, placed the measurement of magnetic quantities in the first rank of methods of precision, Weber proceeded in his *Electrodynamic Measurements* not only to lay down sound principles for fixing the units to be employed, but to make determinations of particular electrical quantities in terms of these units, with a degree of accuracy previously unattempted. Both the electromagnetic and the electrostatic systems of units owe their development and practical application to these researches.

Weber has also formed a general theory of electric action from which he deduces both electrostatic and electromagnetic force, and also the induction of electric currents. We shall consider this theory, with some of its more recent developments, in a separate chapter. See Art. 846.

N 2

N 2

CHAPTER IV.

ON THE INDUCTION OF A CURRENT ON ITSELF.

546.] FARADAY has devoted the ninth series of his *Researches* to the investigation of a class of phenomena exhibited by the current in a wire which forms the coil of an electromagnet.

Mr. Jenkin had observed that, although it is impossible to produce a sensible shock by the direct action of a voltaic system consisting of only one pair of plates, yet, if the current is made to pass through the coil of an electromagnet, and if contact is then broken between the extremities of two wires held one in each hand, a smart shock will be felt. No such shock is felt on making contact.

Faraday shewed that this and other phenomena, which he describes, are due to the same inductive action which he had already observed the current to exert on neighbouring conductors. In this case, however, the inductive action is exerted on the same conductor which carries the current, and it is so much the more powerful as the wire itself is nearer to the different elements of the current than any other wire can be.

547.] He observes, however *, that 'the first thought that arises in the mind is that the electricity circulates with something like momentum or inertia in the wire.' Indeed, when we consider one particular wire only, the phenomena are exactly analogous to those of a pipe full of water flowing in a continued stream. If while the stream is flowing we suddenly close the end of the tube, the momentum of the water produces a sudden pressure, which is much greater than that due to the head of water, and may be sufficient to burst the pipe.

If the water has the means of escaping through a narrow jet

* *Exp. Res.*, 1077.

when the principal aperture is closed, it will be projected with a velocity much greater than that due to the head of water, and if it can escape through a valve into a chamber, it will do so, even when the pressure in the chamber is greater than that due to the head of water.

It is on this principle that the hydraulic ram is constructed, by which a small quantity of water may be raised to a great height by means of a large quantity flowing down from a much lower level.

548.] These effects of the inertia of the fluid in the tube depend solely on the quantity of fluid running through the tube, on its length, and on its section in different parts of its length. They do not depend on anything outside the tube, nor on the form into which the tube may be bent, provided its length remains the same.

In the case of the wire conveying a current this is not the case, for if a long wire is doubled on itself the effect is very small, if the two parts are separated from each other it is greater, if it is coiled up into a helix it is still greater, and greatest of all if, when so coiled, a piece of soft iron is placed inside the coil.

Again, if a second wire is coiled up with the first, but insulated from it, then, if the second wire does not form a closed circuit, the phenomena are as before, but if the second wire forms a closed circuit, an induction current is formed in the second wire, and the effects of self-induction in the first wire are retarded.

549.] These results shew clearly that, if the phenomena are due to momentum, the momentum is certainly not that of the electricity in the wire, because the same wire, conveying the same current, exhibits effects which differ according to its form; and even when its form remains the same, the presence of other bodies, such as a piece of iron or a closed metallic circuit, affects the result.

550.] It is difficult, however, for the mind which has once recognised the analogy between the phenomena of self-induction and those of the motion of material bodies, to abandon altogether the help of this analogy, or to admit that it is entirely superficial and misleading. The fundamental dynamical idea of matter, as capable by its motion of becoming the recipient of momentum and of energy, is so interwoven with our forms of thought that, whenever we catch a glimpse of it in any part of nature, we feel that a path is before us leading, sooner or later, to the complete understanding of the subject.

551.] In the case of the electric current, we find that, when the electromotive force begins to act, it does not at once produce the full current, but that the current rises gradually. What is the electromotive force doing during the time that the opposing resistance is not able to balance it? It is increasing the electric current.

Now an ordinary force, acting on a body in the direction of its motion, increases its momentum, and communicates to it kinetic energy, or the power of doing work on account of its motion.

In like manner the unresisted part of the electromotive force has been employed in increasing the electric current. Has the electric current, when thus produced, either momentum or kinetic energy?

We have already shewn that it has something very like momentum, that it resists being suddenly stopped, and that it can exert, for a short time, a great electromotive force.

But a conducting circuit in which a current has been set up has the power of doing work in virtue of this current, and this power cannot be said to be something very like energy, for it is really and truly energy.

Thus, if the current be left to itself, it will continue to circulate till it is stopped by the resistance of the circuit. Before it is stopped, however, it will have generated a certain quantity of heat, and the amount of this heat in dynamical measure is equal to the energy originally existing in the current.

Again, when the current is left to itself, it may be made to do mechanical work by moving magnets, and the inductive effect of these motions will, by Lenz's law, stop the current sooner than the resistance of the circuit alone would have stopped it. In this way part of the energy of the current may be transformed into mechanical work instead of heat.

552.] It appears, therefore, that a system containing an electric current is a seat of energy of some kind; and since we can form no conception of an electric current except as a kinetic phenomenon *, its energy must be kinetic energy, that is to say, the energy which a moving body has in virtue of its motion.

We have already shewn that the electricity in the wire cannot be considered as the moving body in which we are to find this energy, for the energy of a moving body does not depend on anything external to itself, whereas the presence of other bodies near the current alters its energy.

* Faraday, *Exp. Res.* (283.)

We are therefore led to enquire whether there may not be some motion going on in the space outside the wire, which is not occupied by the electric current, but in which the electromagnetic effects of the current are manifested.

I shall not at present enter on the reasons for looking in one place rather than another for such motions, or for regarding these motions as of one kind rather than another.

What I propose now to do is to examine the consequences of the assumption that the phenomena of the electric current are those of a moving system, the motion being communicated from one part of the system to another by forces, the nature and laws of which we do not yet even attempt to define, because we can eliminate these forces from the equations of motion by the method given by Lagrange for any connected system.

In the next five chapters of this treatise I propose to deduce the main structure of the theory of electricity from a dynamical hypothesis of this kind, instead of following the path which has led Weber and other investigators to many remarkable discoveries and experiments, and to conceptions, some of which are as beautiful as they are bold. I have chosen this method because I wish to shew that there are other ways of viewing the phenomena which appear to me more satisfactory, and at the same time are more consistent with the methods followed in the preceding parts of this book than those which proceed on the hypothesis of direct action at a distance.

CHAPTER V.

ON THE EQUATIONS OF MOTION OF A CONNECTED SYSTEM.

553.] IN the fourth section of the second part of his *Mécanique Analytique*, Lagrange has given a method of reducing the ordinary dynamical equations of the motion of the parts of a connected system to a number equal to that of the degrees of freedom of the system.

The equations of motion of a connected system have been given in a different form by Hamilton, and have led to a great extension of the higher part of pure dynamics *.

As we shall find it necessary, in our endeavours to bring electrical phenomena within the province of dynamics, to have our dynamical ideas in a state fit for direct application to physical questions, we shall devote this chapter to an exposition of these dynamical ideas from a physical point of view.

554.] The aim of Lagrange was to bring dynamics under the power of the calculus. He began by expressing the elementary dynamical relations in terms of the corresponding relations of pure algebraical quantities, and from the equations thus obtained he deduced his final equations by a purely algebraical process. Certain quantities (expressing the reactions between the parts of the system called into play by its physical connexions) appear in the equations of motion of the component parts of the system, and Lagrange's investigation, as seen from a mathematical point of view, is a method of eliminating these quantities from the final equations.

In following the steps of this elimination the mind is exercised in calculation, and should therefore be kept free from the intrusion of dynamical ideas. Our aim, on the other hand, is to cultivate

* See Professor Cayley's 'Report on Theoretical Dynamics,' *British Association*, 1857; and Thomson and Tait's *Natural Philosophy*.

our dynamical ideas. We therefore avail ourselves of the labours of the mathematicians, and retranslate their results from the language of the calculus into the language of dynamics, so that our words may call up the mental image, not of some algebraical process, but of some property of moving bodies.

The language of dynamics has been considerably extended by those who have expounded in popular terms the doctrine of the Conservation of Energy, and it will be seen that much of the following statement is suggested by the investigation in Thomson and Tait's *Natural Philosophy*, especially the method of beginning with the theory of impulsive forces.

I have applied this method so as to avoid the explicit consideration of the motion of any part of the system except the coordinates or variables, on which the motion of the whole depends. It is doubtless important that the student should be able to trace the connexion of the motion of each part of the system with that of the variables, but it is by no means necessary to do this in the process of obtaining the final equations, which are independent of the particular form of these connexions.

The Variables.

555.] The number of degrees of freedom of a system is the number of data which must be given in order completely to determine its position. Different forms may be given to these data, but their number depends on the nature of the system itself, and cannot be altered.

To fix our ideas we may conceive the system connected by means of suitable mechanism with a number of moveable pieces, each capable of motion along a straight line, and of no other kind of motion. The imaginary mechanism which connects each of these pieces with the system must be conceived to be free from friction, destitute of inertia, and incapable of being strained by the action of the applied forces. The use of this mechanism is merely to assist the imagination in ascribing position, velocity, and momentum to what appear, in Lagrange's investigation, as pure algebraical quantities.

Let q denote the position of one of the moveable pieces as defined by its distance from a fixed point in its line of motion. We shall distinguish the values of q corresponding to the different pieces by the suffixes $_1$, $_2$, &c. When we are dealing with a set of quantities belonging to one piece only we may omit the suffix.

When the values of all the variables (q) are given, the position of each of the moveable pieces is known, and, in virtue of the imaginary mechanism, the configuration of the entire system is determined.

The Velocities.

556.] During the motion of the system the configuration changes in some definite manner, and since the configuration at each instant is fully defined by the values of the variables (q), the velocity of every part of the system, as well as its configuration, will be completely defined if we know the values of the variables (q), together with their velocities $\left(\frac{dq}{dt}\text{, or, according to Newton's notation, } \dot{q}\right)$.

The Forces.

557.] By a proper regulation of the motion of the variables, any motion of the system, consistent with the nature of the connexions, may be produced. In order to produce this motion by moving the variable pieces, forces must be applied to these pieces.

We shall denote the force which must be applied to any variable q_r by F_r. The system of forces (F) is mechanically equivalent (in virtue of the connexions of the system) to the system of forces, whatever it may be, which really produces the motion.

The Momenta.

558.] When a body moves in such a way that its configuration, with respect to the force which acts on it, remains always the same, (as, for instance, in the case of a force acting on a single particle in the line of its motion,) the moving force is measured by the rate of increase of the momentum. If F is the moving force, and p the momentum,

$$F = \frac{dp}{dt},$$

whence

$$p = \int F\,dt.$$

The time-integral of a force is called the Impulse of the force; so that we may assert that the momentum is the impulse of the force which would bring the body from a state of rest into the given state of motion.

In the case of a connected system in motion, the configuration is continually changing at a rate depending on the velocities $(\dot{q})$, so that we can no longer assume that the momentum is the time-integral of the force which acts on it.

But the increment δq of any variable cannot be greater than $\dot{q}'\delta t$, where δt is the time during which the increment takes place, and $\dot{q}'$ is the greatest value of the velocity during that time. In the case of a system moving from rest under the action of forces always in the same direction, this is evidently the final velocity.

If the final velocity and configuration of the system are given, we may conceive the velocity to be communicated to the system in a very small time δt, the original configuration differing from the final configuration by quantities δq_1, δq_2, &c., which are less than $\dot{q}_1 \delta t$, $\dot{q}_2 \delta t$, &c., respectively.

The smaller we suppose the increment of time δt, the greater must be the impressed forces, but the time-integral, or impulse, of each force will remain finite. The limiting value of the impulse, when the time is diminished and ultimately vanishes, is defined as the *instantaneous* impulse, and the momentum p, corresponding to any variable q, is defined as the impulse corresponding to that variable, when the system is brought instantaneously from a state of rest into the given state of motion.

This conception, that the momenta are capable of being produced by instantaneous impulses on the system at rest, is introduced only as a method of defining the magnitude of the momenta, for the momenta of the system depend only on the instantaneous state of motion of the system, and not on the process by which that state was produced.

In a connected system the momentum corresponding to any variable is in general a linear function of the velocities of all the variables, instead of being, as in the dynamics of a particle, simply proportional to the velocity.

The impulses required to change the velocities of the system suddenly from $\dot{q}_1$, $\dot{q}_2$, &c. to $\dot{q}_1'$, $\dot{q}_2'$, &c. are evidently equal to $p_1' - p_1$, $p_2' - p_2$, the changes of momentum of the several variables.

Work done by a Small Impulse.

559.] The work done by the force F_1 during the impulse is the space-integral of the force, or

$$W = \int F_1 \, dq_1,$$

$$= \int F_1 \dot{q}_1 \, dt.$$

If $\dot{q}_1'$ is the greatest and $\dot{q}_1''$ the least value of the velocity $\dot{q}_1$ during the action of the force, W must be less than

$$\dot{q}_1' \int F\,dt \quad \text{or} \quad \dot{q}_1'(p_1' - p_1),$$

and greater than $\dot{q}_1'' \int F\,dt$ or $\dot{q}_1''(p_1' - p_1)$.

If we now suppose the impulse $\int F\,dt$ to be diminished without limit, the values of $\dot{q}_1'$ and q_1'' will approach and ultimately coincide with that of $\dot{q}_1$, and we may write $p_1' - p_1 = \delta p_1$, so that the work done is ultimately

$$\delta W_1 = \dot{q}_1 \delta p_1,$$

or, *the work done by a very small impulse is ultimately the product of the impulse and the velocity.*

Increment of the Kinetic Energy.

560.] When work is done in setting a conservative system in motion, energy is communicated to it, and the system becomes capable of doing an equal amount of work against resistances before it is reduced to rest.

The energy which a system possesses in virtue of its motion is called its Kinetic Energy, and is communicated to it in the form of the work done by the forces which set it in motion.

If T be the kinetic energy of the system, and if it becomes $T + \delta T$, on account of the action of an infinitesimal impulse whose components are δp_1, δp_2, &c., the increment δT must be the sum of the quantities of work done by the components of the impulse, or in symbols,

$$\delta T = \dot{q}_1 \delta p_1 + \dot{q}_2 \delta p_2 + \&c.,$$

$$= \Sigma (\dot{q}\, \delta p). \qquad (1)$$

The instantaneous state of the system is completely defined if the variables and the momenta are given. Hence the kinetic energy, which depends on the instantaneous state of the system, can be expressed in terms of the variables (q), and the momenta (p). This is the mode of expressing T introduced by Hamilton. When T is expressed in this way we shall distinguish it by the suffix $_p$, thus, T_p.

The complete variation of T_p is

$$\delta T_p = \Sigma\left(\frac{dT_p}{dp}\delta p\right) + \Sigma\left(\frac{dT_p}{dq}\delta q\right). \qquad (2)$$

The last term may be written

$$\Sigma\left(\frac{dT_p}{dq}\dot{q}\,\delta t\right),$$

which diminishes with δt, and ultimately vanishes with it when the impulse becomes instantaneous.

Hence, equating the coefficients of δp in equations (1) and (2), we obtain

$$\dot{q} = \frac{dT_p}{dp}, \tag{3}$$

or, *the velocity corresponding to the variable q is the differential coefficient of T_p with respect to the corresponding momentum p.*

We have arrived at this result by the consideration of impulsive forces. By this method we have avoided the consideration of the change of configuration during the action of the forces. But the instantaneous state of the system is in all respects the same, whether the system was brought from a state of rest to the given state of motion by the transient application of impulsive forces, or whether it arrived at that state in any manner, however gradual.

In other words, the variables, and the corresponding velocities and momenta, depend on the actual state of motion of the system at the given instant, and not on its previous history.

Hence, the equation (3) is equally valid, whether the state of motion of the system is supposed due to impulsive forces, or to forces acting in any manner whatever.

We may now therefore dismiss the consideration of impulsive forces, together with the limitations imposed on their time of action, and on the changes of configuration during their action.

Hamilton's Equations of Motion.

561.] We have already shewn that

$$\frac{dT_p}{dp} = \dot{q}. \tag{4}$$

Let the system move in any arbitrary way, subject to the conditions imposed by its connexions, then the variations of p and q are

$$\delta p = \frac{dp}{dt}\delta t, \qquad \delta q = \dot{q}\,\delta t. \tag{5}$$

Hence

$$\frac{dT_p}{dp}\delta p = \frac{dp}{dt}\dot{q}\,\delta t,$$

$$= \frac{dp}{dt}\delta q, \tag{6}$$

and the complete variation of T_p is

$$\delta T_p = \Sigma \left(\frac{dT_p}{dp}\delta p + \frac{dT_p}{dq}\delta q\right),$$

$$= \Sigma \left(\left(\frac{dp}{dt} + \frac{dT_p}{dq}\right) dq\right). \qquad (7)$$

But the increment of the kinetic energy arises from the work done by the impressed forces, or

$$\delta T_p = \Sigma (F \delta q). \qquad (8)$$

In these two expressions the variations δq are all independent of each other, so that we are entitled to equate the coefficients of each of them in the two expressions (7) and (8). We thus obtain

$$F_r = \frac{dp_r}{dt} + \frac{dT_p}{dq_r}, \qquad (9)$$

where the momentum p_r and the force F_r belong to the variable q_r.

There are as many equations of this form as there are variables. These equations were given by Hamilton. They shew that the force corresponding to any variable is the sum of two parts. The first part is the rate of increase of the momentum of that variable with respect to the time. The second part is the rate of increase of the kinetic energy per unit of increment of the variable, the other variables and all the momenta being constant.

The Kinetic Energy expressed in Terms of the Momenta and Velocities.

562.] Let p_1, p_2, &c. be the momenta, and $\dot{q}_1$, $\dot{q}_2$, &c. the velocities at a given instant, and let p_1, p_2, &c., $\dot{\mathrm{q}}_1$, $\dot{\mathrm{q}}_2$, &c. be another system of momenta and velocities, such that

$$\mathrm{p}_1 = np_1, \qquad \dot{\mathrm{q}}_1 = n\dot{q}_1, \text{ \&c.} \qquad (10)$$

It is manifest that the systems p, $\dot{\mathrm{q}}$ will be consistent with each other if the systems p, $\dot{q}$ are so.

Now let n vary by δn. The work done by the force F_1 is

$$F_1 \delta q_1 = \dot{\mathrm{q}}_1 \delta \mathrm{p}_1 = \dot{q}_1 p_1 n \delta n. \qquad (11)$$

Let n increase from 0 to 1, then the system is brought from a state of rest into the state of motion $(\dot{q}p)$, and the whole work expended in producing this motion is

$$(\dot{q}_1 p_1 + \dot{q}_2 p_2 + \text{\&c.}) \int_0^1 n\, dn. \qquad (12)$$

But $$\int_0^1 n\, dn = \tfrac{1}{2},$$

and the work spent in producing the motion is equivalent to the kinetic energy. Hence

$$T_{p\dot{q}} = \tfrac{1}{2}(p_1\dot{q}_1 + p_2\dot{q}_2 + \&c.), \tag{13}$$

where $T_{p\dot{q}}$ denotes the kinetic energy expressed in terms of the momenta and velocities. The variables q_1, q_2, &c. do not enter into this expression.

The kinetic energy is therefore half the sum of the products of the momenta into their corresponding velocities.

When the kinetic energy is expressed in this way we shall denote it by the symbol $T_{p\dot{q}}$. It is a function of the momenta and velocities only, and does not involve the variables themselves.

563.] There is a third method of expressing the kinetic energy, which is generally, indeed, regarded as the fundamental one. By solving the equations (3) we may express the momenta in terms of the velocities, and then, introducing these values in (13), we shall have an expression for T involving only the velocities and the variables. When T is expressed in this form we shall indicate it by the symbol $T_{\dot{q}}$. This is the form in which the kinetic energy is expressed in the equations of Lagrange.

564.] It is manifest that, since T_p, $T_{\dot{q}}$, and $T_{p\dot{q}}$ are three different expressions for the same thing,

$$T_p + T_{\dot{q}} - 2\,T_{p\dot{q}} = 0,$$

or

$$T_p + T_{\dot{q}} - p_1\dot{q}_1 - p_2\dot{q}_2 - \&c. = 0. \tag{14}$$

Hence, if all the quantities p, q, and $\dot{q}$ vary,

$$\begin{aligned}&\left(\frac{dT_p}{dp_1} - \dot{q}_1\right)\delta p_1 + \left(\frac{dT_p}{dp_2} - \dot{q}_2\right)\delta p_2 + \&c.\\ &+\left(\frac{dT_{\dot{q}}}{d\dot{q}} - p_1\right)\delta\dot{q}_1 + \left(\frac{dT_{\dot{q}}}{d\dot{q}_2} - p_2\right)\delta\dot{q}_2 + \&c.\\ &+\left(\frac{dT_p}{dq_1} + \frac{dT_{\dot{q}}}{dq_1}\right)\delta q_1 + \left(\frac{dT_p}{dq_2} + \frac{dT_{\dot{q}}}{dq_2}\right)\delta q_2 + \&c. = 0.\end{aligned} \tag{15}$$

The variations δp are not independent of the variations δq and $\delta\dot{q}$, so that we cannot at once assert that the coefficient of each variation in this equation is zero. But we know, from equations (3), that

$$\frac{dT_p}{dp_1} - \dot{q}_1 = 0, \&c., \tag{16}$$

so that the terms involving the variations δp vanish of themselves.

The remaining variations $\delta\dot{q}$ and δq are now all independent, so that we find, by equating to zero the coefficients of $\delta\dot{q}_1$, &c ,

$$p_1 = \frac{dT_{\dot{q}}}{d\dot{q}_1}, \qquad p_2 = \frac{dT_{\dot{q}}}{d\dot{q}_2}, \&c.; \tag{17}$$

or, *the components of momentum are the differential coefficients of $T_{\dot q}$ with respect to the corresponding velocities.*

Again, by equating to zero the coefficients of δq_1, &c.,

$$\frac{dT_p}{dq_1}+\frac{dT_{\dot q}}{dq_1}=0; \tag{18}$$

or, *the differential coefficient of the kinetic energy with respect to any variable q_1 is equal in magnitude but opposite in sign when T is expressed as a function of the velocities instead of as a function of the momenta.*

In virtue of equation (18) we may write the equation of motion (9),

$$F_1=\frac{dp_1}{dt}-\frac{dT_{\dot q}}{dq_1}, \tag{19}$$

or

$$F_1=\frac{d}{dt}\frac{dT_{\dot q}}{d\dot q_1}-\frac{dT_{\dot q}}{dq_1}, \tag{20}$$

which is the form in which the equations of motion were given by Lagrange.

565.] In the preceding investigation we have avoided the consideration of the form of the function which expresses the kinetic energy in terms either of the velocities or of the momenta. The only explicit form which we have assigned to it is

$$T_{p\dot q}=\tfrac{1}{2}(p_1\dot q_1+p_2\dot q+\&c.), \tag{21}$$

in which it is expressed as half the sum of the products of the momenta each into its corresponding velocity.

We may express the velocities in terms of the differential coefficients of Tp with respect to the momenta, as in equation (3),

$$T_p=\tfrac{1}{2}\left(p_1\frac{dT_p}{dp_1}+p_2\frac{dT_p}{dp_2}+\&c.\right) \tag{22}$$

This shews that T_p is a homogeneous function of the second degree of the momenta p_1, p_2, &c.

We may also express the momenta in terms of $T_{\dot q}$, and we find

$$T_{\dot q}=\tfrac{1}{2}\left(\dot q_1\frac{dT_{\dot q}}{d\dot q_1}+\dot q_2\frac{dT_{\dot q}}{d\dot q_2}+\&c.\right) \tag{23}$$

which shews that $T_{\dot q}$ is a homogeneous function of the second degree with respect to the velocities $\dot q_1, \dot q_2$, &c.

If we write

$$P_{11} \text{ for } \frac{d^2T_{\dot q}}{d\dot q_1^{\,2}}, \qquad P_{12} \text{ for } \frac{d^2T_{\dot q}}{d\dot q_1\, d\dot q_2}, \ \&c.$$

and

$$Q_{11} \text{ for } \frac{d^2T_p}{dp_1^{\,2}}, \qquad Q_{12} \text{ for } \frac{d^2T_p}{dp_1\, dp_2}, \ \&c.;$$

then, since both $T_{\dot{q}}$ and T_p are functions of the second degree of $\dot{q}$ and of p respectively, both the P's and the Q's will be functions of the variables q only, and independent of the velocities and the momenta. We thus obtain the expressions for T,

$$2\,T_{\dot{q}} = P_{11}\,\dot{q}_1^{\,2} + 2\,P_{12}\,\dot{q}_1\,\dot{q}_2 + \&c., \tag{24}$$

$$2\,T_p = Q_{11}\,p_1^{\,2} + 2\,Q_{12}\,p_1\,p_2 + \&c. \tag{25}$$

The momenta are expressed in terms of the velocities by the linear equations

$$p_1 = P_{11}\,\dot{q}_1 + P_{12}\,\dot{q}_2 + \&c., \tag{26}$$

and the velocities are expressed in terms of the momenta by the linear equations

$$\dot{q}_1 = Q_{11}\,p_1 + Q_{12}\,p_2 + \&c. \tag{27}$$

In treatises on the dynamics of a rigid body, the coefficients corresponding to P_{11}, in which the suffixes are the same, are called Moments of Inertia, and those corresponding to P_{12}, in which the suffixes are different, are called Products of Inertia. We may extend these names to the more general problem which is now before us, in which these quantities are not, as in the case of a rigid body, absolute constants, but are functions of the variables q_1, q_2, &c.

In like manner we may call the coefficients of the form Q_{11} Moments of Mobility, and those of the form Q_{12}, Products of Mobility. It is not often, however, that we shall have occasion to speak of the coefficients of mobility.

566.] The kinetic energy of the system is a quantity essentially positive or zero. Hence, whether it be expressed in terms of the velocities, or in terms of the momenta, the coefficients must be such that no real values of the variables can make T negative.

We thus obtain a set of necessary conditions which the values of the coefficients P must satisfy.

The quantities P_{11}, P_{22}, &c., and all determinants of the symmetrical form

$$\begin{vmatrix} P_{11} & P_{12} & P_{13} & . \\ P_{12} & P_{22} & P_{23} & . \\ P_{13} & P_{23} & P_{33} & . \\ . & . & . & . \end{vmatrix}$$

which can be formed from the system of coefficients must be positive or zero. The number of such conditions for n variables is $2^n - 1$.

The coefficients Q are subject to conditions of the same kind.

567.] In this outline of the fundamental principles of the dynamics of a connected system, we have kept out of view the mechanism by which the parts of the system are connected. We have not even written down a set of equations to indicate how the motion of any part of the system depends on the variation of the variables. We have confined our attention to the variables, their velocities and momenta, and the forces which act on the pieces representing the variables. Our only assumptions are, that the connexions of the system are such that the time is not explicitly contained in the equations of condition, and that the principle of the conservation of energy is applicable to the system.

Such a description of the methods of pure dynamics is not unnecessary, because Lagrange and most of his followers, to whom we are indebted for these methods, have in general confined themselves to a demonstration of them, and, in order to devote their attention to the symbols before them, they have endeavoured to banish all ideas except those of pure quantity, so as not only to dispense with diagrams, but even to get rid of the ideas of velocity, momentum, and energy, after they have been once for all supplanted by symbols in the original equations. In order to be able to refer to the results of this analysis in ordinary dynamical language, we have endeavoured to retranslate the principal equations of the method into language which may be intelligible without the use of symbols.

As the development of the ideas and methods of pure mathematics has rendered it possible, by forming a mathematical theory of dynamics, to bring to light many truths which could not have been discovered without mathematical training, so, if we are to form dynamical theories of other sciences, we must have our minds imbued with these dynamical truths as well as with mathematical methods.

In forming the ideas and words relating to any science, which, like electricity, deals with forces and their effects, we must keep constantly in mind the ideas appropriate to the fundamental science of dynamics, so that we may, during the first development of the science, avoid inconsistency with what is already established, and also that when our views become clearer, the language we have adopted may be a help to us and not a hindrance.

CHAPTER VI.

DYNAMICAL THEORY OF ELECTROMAGNETISM.

568.] We have shewn, in Art. 552, that, when an electric current exists in a conducting circuit, it has a capacity for doing a certain amount of mechanical work, and this independently of any external electromotive force maintaining the current. Now capacity for performing work is nothing else than energy, in whatever way it arises, and all energy is the same in kind, however it may differ in form. The energy of an electric current is either of that form which consists in the actual motion of matter, or of that which consists in the capacity for being set in motion, arising from forces acting between bodies placed in certain positions relative to each other.

The first kind of energy, that of motion, is called Kinetic energy, and when once understood it appears so fundamental a fact of nature that we can hardly conceive the possibility of resolving it into anything else. The second kind of energy, that depending on position, is called Potential energy, and is due to the action of what we call forces, that is to say, tendencies towards change of relative position. With respect to these forces, though we may accept their existence as a demonstrated fact, yet we always feel that every explanation of the mechanism by which bodies are set in motion forms a real addition to our knowledge.

569.] The electric current cannot be conceived except as a kinetic phenomenon. Even Faraday, who constantly endeavoured to emancipate his mind from the influence of those suggestions which the words 'electric current' and 'electric fluid' are too apt to carry with them, speaks of the electric current as 'something progressive, and not a mere arrangement'*.

* *Exp. Res.*, 283.

The effects of the current, such as electrolysis, and the transfer of electrification from one body to another, are all progressive actions which require time for their accomplishment, and are therefore of the nature of motions.

As to the velocity of the current, we have shewn that we know nothing about it, it may be the tenth of an inch in an hour, or a hundred thousand miles in a second *. So far are we from knowing its absolute value in any case, that we do not even know whether what we call the positive direction is the actual direction of the motion or the reverse.

But all that we assume here is that the electric current involves motion of some kind. That which is the cause of electric currents has been called Electromotive Force. This name has long been used with great advantage, and has never led to any inconsistency in the language of science. Electromotive force is always to be understood to act on electricity only, not on the bodies in which the electricity resides. It is never to be confounded with ordinary mechanical force, which acts on bodies only, not on the electricity in them. If we ever come to know the formal relation between electricity and ordinary matter, we shall probably also know the relation between electromotive force and ordinary force.

570.] When ordinary force acts on a body, and when the body yields to the force, the work done by the force is measured by the product of the force into the amount by which the body yields. Thus, in the case of water forced through a pipe, the work done at any section is measured by the fluid pressure at the section multiplied into the quantity of water which crosses the section.

In the same way the work done by an electromotive force is measured by the product of the electromotive force into the quantity of electricity which crosses a section of the conductor under the action of the electromotive force.

The work done by an electromotive force is of exactly the same kind as the work done by an ordinary force, and both are measured by the same standards or units.

Part of the work done by an electromotive force acting on a conducting circuit is spent in overcoming the resistance of the circuit, and this part of the work is thereby converted into heat. Another part of the work is spent in producing the electromagnetic phenomena observed by Ampère, in which conductors are made to move by electromagnetic forces. The rest of the work

* *Exp. Res.*, 1648.

is spent in increasing the kinetic energy of the current, and the effects of this part of the action are shewn in the phenomena of the induction of currents observed by Faraday.

We therefore know enough about electric currents to recognise, in a system of material conductors carrying currents, a dynamical system which is the seat of energy, part of which may be kinetic and part potential.

The nature of the connexions of the parts of this system is unknown to us, but as we have dynamical methods of investigation which do not require a knowledge of the mechanism of the system, we shall apply them to this case.

We shall first examine the consequences of assuming the most general form for the function which expresses the kinetic energy of the system.

571.] Let the system consist of a number of conducting circuits, the form and position of which are determined by the values of a system of variables x_1, x_2, &c., the number of which is equal to the number of degrees of freedom of the system.

If the whole kinetic energy of the system were that due to the motion of these conductors, it would be expressed in the form

$$T = \frac{1}{2}(x_1 x_1)\dot{x}_1^2 + \&c. + (x_1 x_2)\dot{x}_1 \dot{x}_2 + \&c.,$$

where the symbols $(x_1, x_1$, &c.) denote the quantities which we have called moments of inertia, and $(x_1, x_2$, &c.) denote the products of inertia.

If X' is the impressed force, tending to increase the coordinate x, which is required to produce the actual motion, then, by Lagrange's equation,

$$\frac{d}{dt}\frac{dT}{d\dot{x}} - \frac{dT}{dx} = X.$$

When T denotes the energy due to the visible motion only, we shall indicate it by the suffix $_m$, thus, T_m.

But in a system of conductors carrying electric currents, part of the kinetic energy is due to the existence of these currents. Let the motion of the electricity, and of anything whose motion is governed by that of the electricity, be determined by another set of coordinates y_1, y_2, &c., then T will be a homogeneous function of squares and products of all the velocities of the two sets of coordinates. We may therefore divide T into three portions, in the first of which, T_m, the velocities of the coordinates x only occur, while in the second, T_e, the velocities of the coordinates y only occur, and in the third, T_{me}, each term contains the product of the velocities of two coordinates of which one is x and the other y.

We have therefore $T = T_m + T_e + T_{me}$,

where
$$T_m = \tfrac{1}{2}(x_1 x_1)\dot{x}_1^2 + \&c. + (x_1 x_2)\dot{x}_1 \dot{x}_2 + \&c.,$$
$$T_e = \tfrac{1}{2}(y_1 y_1)\dot{y}_1^2 + \&c. + (y_1 y_2)\dot{y}_1 \dot{y}_2 + \&c.,$$
$$T_{me} = (x_1 y_1)\dot{x}_1 \dot{y}_1 + \&c.$$

572.] In the general dynamical theory, the coefficients of every term may be functions of all the coordinates, both x and y. In the case of electric currents, however, it is easy to see that the coordinates of the class y do not enter into the coefficients.

For, if all the electric currents are maintained constant, and the conductors at rest, the whole state of the field will remain constant. But in this case the coordinates y are variable, though the velocities $\dot{y}$ are constant. Hence the coordinates y cannot enter into the expression for T, or into any other expression of what actually takes place.

Besides this, in virtue of the equation of continuity, if the conductors are of the nature of linear circuits, only one variable is required to express the strength of the current in each conductor. Let the velocities $\dot{y}_1$, $\dot{y}_2$, &c. represent the strengths of the currents in the several conductors.

All this would be true, if, instead of electric currents, we had currents of an incompressible fluid running in flexible tubes. In this case the velocities of these currents would enter into the expression for T, but the coefficients would depend only on the variables x, which determine the form and position of the tubes.

In the case of the fluid, the motion of the fluid in one tube does not directly affect that of any other tube, or of the fluid in it. Hence, in the value of T_e, only the squares of the velocities $\dot{y}$, and not their products, occur, and in T_{me} any velocity $\dot{y}$ is associated only with those velocities of the form $\dot{x}$ which belong to its own tube.

In the case of electrical currents we know that this restriction does not hold, for the currents in different circuits act on each other. Hence we must admit the existence of terms involving products of the form $\dot{y}_1 \dot{y}_2$, and this involves the existence of something in motion, whose motion depends on the strength of both electric currents $\dot{y}_1$ and $\dot{y}_2$. This moving matter, whatever it is, is not confined to the interior of the conductors carrying the two currents, but probably extends throughout the whole space surrounding them.

573.] Let us next consider the form which Lagrange's equations of motion assume in this case. Let X' be the impressed force corresponding to the coordinate x, one of those which determine the form and position of the conducting circuits. This is a force in the ordinary sense, a tendency towards change of position. It is given by the equation

$$X' = \frac{d}{dt}\frac{dT}{d\dot{x}} - \frac{dT}{dx}.$$

We may consider this force as the sum of three parts, corresponding to the three parts into which we divided the kinetic energy of the system, and we may distinguish them by the same suffixes. Thus $$X' = X'_m + X'_e + X'_{me}.$$

The part X'_m is that which depends on ordinary dynamical considerations, and we need not attend to it.

Since T_e does not contain $\dot{x}$, the first term of the expression for X'_e is zero, and its value is reduced to

$$X'_e = -\frac{dT_e}{dx}.$$

This is the expression for the mechanical force which must be applied to a conductor to balance the electromagnetic force, and it asserts that it is measured by the rate of *diminution* of the purely electrokinetic energy due to the variation of the coordinate x. The electromagnetic force, X_e, which brings this external mechanical force into play, is equal and opposite to it, and is therefore measured by the rate of *increase* of the electrokinetic energy corresponding to an increase of the coordinate x. The value of X_e, since it depends on squares and products of the currents, remains the same if we reverse the directions of all the currents.

The third part of X' is

$$X'_{me} = \frac{d}{dt}\frac{dT_{me}}{d\dot{x}} - \frac{dT_{me}}{dx}.$$

The quantity T_{me} contains only products of the form $\dot{x}\dot{y}$, so that $\frac{dT_{me}}{d\dot{x}}$ is a linear function of the strengths of the currents $\dot{y}$. The first term, therefore, depends on the rate of variation of the strengths of the currents, and indicates a mechanical force on the conductor, which is zero when the currents are constant, and which is positive or negative according as the currents are increasing or decreasing in strength.

The second term depends, not on the variation of the currents, but on their actual strength. As it is a linear function with respect to these currents, it changes sign when the currents change sign. Since every term involves a velocity $\dot{x}$, it is zero when the conductors are at rest.

We may therefore investigate these terms separately. If the conductors are at rest, we have only the first term to deal with. If the currents are constant, we have only the second.

574.] As it is of great importance to determine whether any part of the kinetic energy is of the form T_{me}, consisting of products of ordinary velocities and strengths of electric currents, it is desirable that experiments should be made on this subject with great care.

The determination of the forces acting on bodies in rapid motion is difficult. Let us therefore attend to the first term, which depends on the variation of the strength of the current.

If any part of the kinetic energy depends on the product of an ordinary velocity and the strength of a current, it will probably be most easily observed when the velocity and the current are in the same or in opposite directions. We therefore take a circular coil of a great many windings, and suspend it by a fine vertical wire, so that its windings are horizontal, and the coil is capable of rotating about a vertical axis, either in the same direction as the current in the coil, or in the opposite direction.

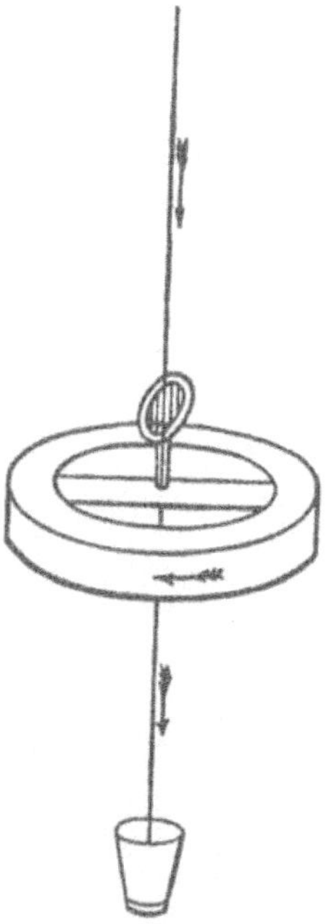

Fig. 33.

We shall suppose the current to be conveyed into the coil by means of the suspending wire, and, after passing round the windings, to complete its circuit by passing downwards through a wire in the same line with the suspending wire and dipping into a cup of mercury.

Since the action of the horizontal component of terrestrial magnetism would tend to turn this coil round a horizontal axis when the current flows through it, we shall suppose that the horizontal component of terrestrial magnetism is exactly neutralized by means of fixed magnets, or that the experiment is made at the magnetic pole. A vertical mirror is attached to the coil to detect any motion in azimuth.

Now let a current be made to pass through the coil in the direction N.E.S.W. If electricity were a fluid like water, flowing along the wire, then, at the moment of starting the current, and as long as its velocity is increasing, a force would require to be supplied to produce the angular momentum of the fluid in passing round the coil, and as this must be supplied by the elasticity of the suspending wire, the coil would at first rotate in the opposite direction or W.S.E.N., and this would be detected by means of the mirror. On stopping the current there would be another movement of the mirror, this time in the same direction as that of the current.

No phenomenon of this kind has yet been observed. Such an action, if it existed, might be easily distinguished from the already known actions of the current by the following peculiarities.

(1) It would occur only when the strength of the current varies, as when contact is made or broken, and not when the current is constant.

All the known *mechanical* actions of the current depend on the strength of the currents, and not on the rate of variation. The electromotive action in the case of induced currents cannot be confounded with this electromagnetic action.

(2) The direction of this action would be reversed when that of all the currents in the field is reversed.

All the known mechanical actions of the current remain the same when all the currents are reversed, since they depend on squares and products of these currents.

If any action of this kind were discovered, we should be able to regard one of the so-called kinds of electricity, either the positive or the negative kind, as a real substance, and we should be able to describe the electric current as a true motion of this substance in a particular direction. In fact, if electrical motions were in any way comparable with the motions of ordinary matter, terms of the form T_{me} would exist, and their existence would be manifested by the mechanical force X_{me}.

According to Fechner's hypothesis, that an electric current consists of two equal currents of positive and negative electricity, flowing in opposite directions through the same conductor, the terms of the second class T_{me} would vanish, each term belonging to the positive current being accompanied by an equal term of opposite sign belonging to the negative current, and the phenomena depending on these terms would have no existence.

It appears to me, however, that while we derive great advantage from the recognition of the many analogies between the electric current and a current of a material fluid, we must carefully avoid making any assumption not warranted by experimental evidence, and that there is, as yet, no experimental evidence to shew whether the electric current is really a current of a material substance, or a double current, or whether its velocity is great or small as measured in feet per second.

A knowledge of these things would amount to at least the beginnings of a complete dynamical theory of electricity, in which we should regard electrical action, not, as in this treatise, as a phenomenon due to an unknown cause, subject only to the general laws of dynamics, but as the result of known motions of known portions of matter, in which not only the total effects and final results, but the whole intermediate mechanism and details of the motion, are taken as the objects of study.

575.] The experimental investigation of the second term of X_{me}, namely $\frac{dT_{me}}{dx}$, is more difficult, as it involves the observation of the effect of forces on a body in rapid motion.

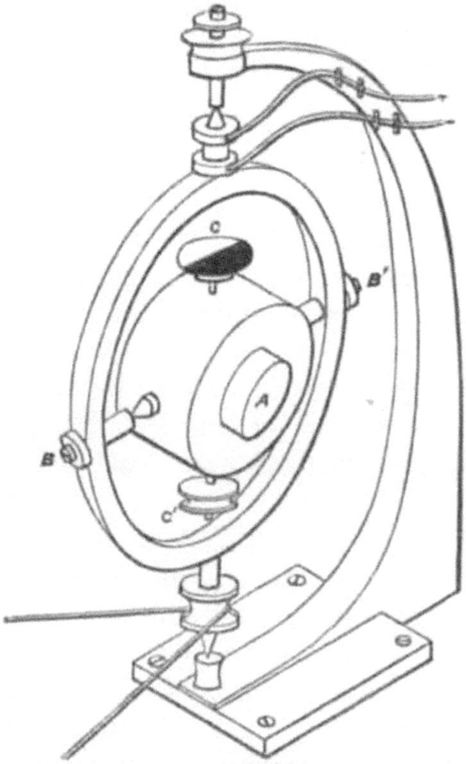

Fig. 34.

The apparatus shewn in Fig. 34, which I had constructed in 1861, is intended to test the existence of a force of this kind.

The electromagnet A is capable of rotating about the horizontal axis BB', within a ring which itself revolves about a vertical axis.

Let A, B, C be the moments of inertia of the electromagnet about the axis of the coil, the horizontal axis BB', and a third axis CC' respectively.

Let θ be the angle which CC' makes with the vertical, ϕ the azimuth of the axis BB', and ψ a variable on which the motion of electricity in the coil depends.

Then the kinetic energy of the electromagnet may be written

$$2T = A\dot{\phi}^2 \sin^2\theta + B\dot{\theta}^2 + C\dot{\phi}^2\cos^2\theta + E(\dot{\phi}\sin\theta + \dot{\psi})^2,$$

where E is a quantity which may be called the moment of inertia of the electricity in the coil.

If Θ is the moment of the impressed force tending to increase θ, we have, by the equations of dynamics,

$$\Theta = B\frac{d^2\theta}{dt^2} - \{(A-C)\dot{\phi}^2\sin\theta\cos\theta + E\dot{\phi}\cos\theta(\dot{\phi}\sin\theta + \dot{\psi})\}.$$

By making Ψ, the impressed force tending to increase ψ, equal to zero, we obtain

$$\dot{\phi}\sin\theta + \dot{\psi} = \gamma,$$

a constant, which we may consider as representing the strength of the current in the coil.

If C is somewhat greater than A, Θ will be zero, and the equilibrium about the axis BB' will be stable when

$$\sin\theta = \frac{E\gamma}{(C-A)\dot{\phi}}.$$

This value of θ depends on that of γ, the electric current, and is positive or negative according to the direction of the current.

The current is passed through the coil by its bearings at B and B', which are connected with the battery by means of springs rubbing on metal rings placed on the vertical axis.

To determine the value of θ, a disk of paper is placed at C, divided by a diameter parallel to BB' into two parts, one of which is painted red and the other green.

When the instrument is in motion a red circle is seen at C when θ is positive, the radius of which indicates roughly the value of θ. When θ is negative, a green circle is seen at C.

By means of nuts working on screws attached to the electromagnet, the axis CC' is adjusted to be a principal axis having its moment of inertia just exceeding that round the axis A, so as to make the instrument very sensible to the action of the force if it exists.

The chief difficulty in the experiments arose from the disturbing action of the earth's magnetic force, which caused the electromagnet to act like a dip-needle. The results obtained were on this account very rough, but no evidence of any change in θ could be obtained even when an iron core was inserted in the coil, so as to make it a powerful electromagnet.

If, therefore, a magnet contains matter in rapid rotation, the angular momentum of this rotation must be very small compared with any quantities which we can measure, and we have as yet no evidence of the existence of the terms T_{me} derived from their mechanical action.

576.] Let us next consider the forces acting on the currents of electricity, that is, the electromotive forces.

Let Y be the effective electromotive force due to induction, the electromotive force which must act on the circuit from without to balance it is $Y' = -Y$, and, by Lagrange's equation,

$$Y = -Y' = -\frac{d}{dt}\frac{dT}{d\dot{y}} + \frac{dT}{dy}.$$

Since there are no terms in T involving the coordinate y, the second term is zero, and Y is reduced to its first term. Hence, electromotive force cannot exist in a system at rest, and with constant currents.

Again, if we divide Y into three parts, Y_m, Y_e, and Y_{me}, corresponding to the three parts of T, we find that, since T_m does not contain $\dot{y}$, $Y_m = 0$.

We also find $$Y_e = -\frac{d}{dt}\frac{dT_e}{d\dot{y}}.$$

Here $\frac{dT_e}{d\dot{y}}$ is a linear function of the currents, and this part of the electromotive force is equal to the rate of change of this function. This is the electromotive force of induction discovered by Faraday. We shall consider it more at length afterwards.

577.] From the part of T, depending on velocities multiplied by currents, we find $$Y_{me} = -\frac{d}{dt}\frac{dT_{me}}{d\dot{y}}.$$

Now $\frac{dT_{me}}{d\dot{y}}$ is a linear function of the velocities of the conductors. If, therefore, any terms of T_{me} have an actual existence, it would be possible to produce an electromotive force independently of all existing currents by simply altering the velocities of the conductors.

For instance, in the case of the suspended coil at Art. 559, if, when the coil is at rest, we suddenly set it in rotation about the vertical axis, an electromotive force would be called into action proportional to the acceleration of this motion. It would vanish when the motion became uniform, and be reversed when the motion was retarded.

Now few scientific observations can be made with greater precision than that which determines the existence or non-existence of a current by means of a galvanometer. The delicacy of this method far exceeds that of most of the arrangements for measuring the mechanical force acting on a body. If, therefore, any currents could be produced in this way they would be detected, even if they were very feeble. They would be distinguished from ordinary currents of induction by the following characteristics.

(1) They would depend entirely on the motions of the conductors, and in no degree on the strength of currents or magnetic forces already in the field.

(2) They would depend not on the absolute velocities of the conductors, but on their accelerations, and on squares and products of velocities, and they would change sign when the acceleration becomes a retardation, though the absolute velocity is the same.

Now in all the cases actually observed, the induced currents depend altogether on the strength and the variation of currents in the field, and cannot be excited in a field devoid of magnetic force and of currents. In so far as they depend on the motion of conductors, they depend on the absolute velocity, and not on the change of velocity of these motions.

We have thus three methods of detecting the existence of the terms of the form T_{me}, none of which have hitherto led to any positive result. I have pointed them out with the greater care because it appears to me important that we should attain the greatest amount of certitude within our reach on a point bearing so strongly on the true theory of electricity.

Since, however, no evidence has yet been obtained of such terms, I shall now proceed on the assumption that they do not exist, or at least that they produce no sensible effect, an assumption which will considerably simplify our dynamical theory. We shall have occasion, however, in discussing the relation of magnetism to light, to shew that the motion which constitutes light may enter as a factor into terms involving the motion which constitutes magnetism.

CHAPTER VII.

THEORY OF ELECTRIC CIRCUITS.

578.] We may now confine our attention to that part of the kinetic energy of the system which depends on squares and products of the strengths of the electric currents. We may call this the Electrokinetic Energy of the system. The part depending on the motion of the conductors belongs to ordinary dynamics, and we have shewn that the part depending on products of velocities and currents does not exist.

Let A_1, A_2, &c. denote the different conducting circuits. Let their form and relative position be expressed in terms of the variables x_1, x_2, &c., the number of which is equal to the number of degrees of freedom of the mechanical system. We shall call these the Geometrical Variables.

Let y_1 denote the quantity of electricity which has crossed a given section of the conductor A_1 since the beginning of the time t. The strength of the current will be denoted by $\dot{y}_1$, the fluxion of this quantity.

We shall call $\dot{y}_1$ the actual current, and y_1 the integral current. There is one variable of this kind for each circuit in the system.

Let T denote the electrokinetic energy of the system. It is a homogeneous function of the second degree with respect to the strengths of the currents, and is of the form

$$T = \tfrac{1}{2} L_1 \dot{y}_1^2 + \tfrac{1}{2} L_2 \dot{y}_2^2 + \&\text{c.} + M_{12} \dot{y}_1 \dot{y}_2 + \&\text{c.}, \qquad (1)$$

where the coefficients L, M, &c. are functions of the geometrical variables x_1, x_2, &c. The electrical variables y_1, y_2 do not enter into the expression.

We may call L_1, L_2, &c. the electric moments of inertia of the circuits A_1, A_2, &c., and M_{12} the electric product of inertia of the two circuits A_1 and A_2. When we wish to avoid the language of the dynamical theory, we shall call L_1 the coefficient of self-induction of the circuit A_1, and M_{12} the coefficient of mutual induction of the circuits A_1 and A_2. M_{12} is also called the potential of the circuit A_1 with respect to A_2. These quantities depend only on the form and relative position of the circuits. We shall find that in the electromagnetic system of measurement they are quantities of the dimension of a line. See Art. 627.

By differentiating T with respect to $\dot{y}_1$ we obtain the quantity p_1, which, in the dynamical theory, may be called the momentum corresponding to y_1. In the electric theory we shall call p_1 the electrokinetic momentum of the circuit A_1. Its value is

$$p_1 = L_1 \dot{y}_1 + M_{12} \dot{y}_2 + \&c.$$

The electrokinetic momentum of the circuit A_1 is therefore made up of the product of its own current into its coefficient of self-induction, together with the sum of the products of the currents in the other circuits, each into the coefficient of mutual induction of A_1 and that other circuit.

Electromotive Force.

579.] Let E be the impressed electromotive force in the circuit A, arising from some cause, such as a voltaic or thermoelectric battery, which would produce a current independently of magneto-electric induction.

Let R be the resistance of the circuit, then, by Ohm's law, an electromotive force $R\dot{y}$ is required to overcome the resistance, leaving an electromotive force $E - R\dot{y}$ available for changing the momentum of the circuit. Calling this force Y', we have, by the general equations,

$$Y' = \frac{dp}{dt} - \frac{dT}{dy},$$

but since T does not involve y, the last term disappears.

Hence, the equation of electromotive force is

$$E - R\dot{y} = Y' = \frac{dp}{dt},$$

or

$$E = R\dot{y} + \frac{dp}{dt}:$$

The impressed electromotive force E is therefore the sum of two parts. The first, $R\dot{y}$, is required to maintain the current $\dot{y}$ against the resistance R. The second part is required to increase the electromagnetic momentum p. This is the electromotive force which must be supplied from sources independent of magneto-electric induction. The electromotive force arising from magneto-electric induction alone is evidently $-\frac{dp}{dt}$, or, *the rate of decrease of the electrokinetic momentum of the circuit.*

Electromagnetic Force.

580.] Let X' be the impressed mechanical force arising from external causes, and tending to increase the variable x. By the general equations

$$X' = \frac{d}{dt}\frac{dT}{d\dot{x}} - \frac{dT}{dx}.$$

Since the expression for the electrokinetic energy does not contain the velocity ($\dot{x}$), the first term of the second member disappears, and we find

$$X' = -\frac{dT}{dx}.$$

Here X' is the external force required to balance the forces arising from electrical causes. It is usual to consider this force as the reaction against the electromagnetic force, which we shall call X, and which is equal and opposite to X'.

Hence

$$X = \frac{dT}{dx},$$

or, *the electromagnetic force tending to increase any variable is equal to the rate of increase of the electrokinetic energy per unit increase of that variable, the currents being maintained constant.*

If the currents are maintained constant by a battery during a displacement in which a quantity, W, of work is done by electromotive force, the electrokinetic energy of the system will be at the same time increased by W. Hence the battery will be drawn upon for a double quantity of energy, or $2W$, in addition to that which is spent in generating heat in the circuit. This was first pointed out by Sir W. Thomson*. Compare this result with the electrostatic property in Art. 93.

Case of Two Circuits.

581.] Let A_1 be called the Primary Circuit, and A_2 the Secondary Circuit. The electrokinetic energy of the system may be written

$$T = \tfrac{1}{2}L\dot{y}_1^2 + M\dot{y}_1\dot{y}_2 + N\dot{y}_2^2,$$

where L and N are the coefficients of self-induction of the primary

* Nichol's *Cyclopaedia of Physical Science*, ed. 1860, Article, 'Magnetism, Dynamical Relations of.'

and secondary circuits respectively, and M is the coefficient of their mutual induction.

Let us suppose that no electromotive force acts on the secondary circuit except that due to the induction of the primary current. We have then

$$E_2 = R_2 \dot{y}_2 + \frac{d}{dt}(M\dot{y}_1 + N\dot{y}_2) = 0.$$

Integrating this equation with respect to t, we have

$$R y_2 + M y_1 + N \dot{y}_2 = C, \text{ a constant,}$$

where y_2 is the integral current in the secondary circuit.

The method of measuring an integral current of short duration will be described in Art. 748, and it is easy in most cases to ensure that the duration of the secondary current shall be very short.

Let the values of the variable quantities in the equation at the end of the time t be accented, then, if y_2 is the integral current, or the whole quantity of electricity which flows through a section of the secondary circuit during the time t,

$$R_2 y_2 = M\dot{y}_1 + N\dot{y}_2 - (M'\dot{y}_1' + N'\dot{y}_2').$$

If the secondary current arises entirely from induction, its initial value $\dot{y}_2$ must be zero if the primary current is constant, and the conductors at rest before the beginning of the time t.

If the time t is sufficient to allow the secondary current to die away, $\dot{y}_2'$, its final value, is also zero, so that the equation becomes

$$R_2 y_2 = M\dot{y}_1 - M'\dot{y}_1'.$$

The integral current of the secondary circuit depends in this case on the initial and final values of $M\dot{y}_1$.

Induced Currents.

582.] Let us begin by supposing the primary circuit broken, or $\dot{y}_1 = 0$, and let a current $\dot{y}_1'$ be established in it when contact is made.

The equation which determines the secondary integral current is

$$R_2 y_2 = -M\dot{y}_1'.$$

When the circuits are placed side by side, and in the same direction, M is a positive quantity. Hence, when contact is made in the primary circuit, a negative current is induced in the secondary circuit.

When the contact is broken in the primary circuit, the primary current ceases, and the induced current is y_2 where

$$R_2 y_2 = M\dot{y}_1.$$

The secondary current is in this case positive.

If the primary current is maintained constant, and the form or relative position of the circuits altered so that M becomes M', the integral secondary current is y_2, where

$$R_2 y_2 = (M - M')\dot{y}_1.$$

In the case of two circuits placed side by side and in the same direction M diminishes as the distance between the circuits increases. Hence, the induced current is positive when this distance is increased and negative when it is diminished.

These are the elementary cases of induced currents described in Art. 530.

Mechanical Action between the Two Circuits.

583.] Let x be any one of the geometrical variables on which the form and relative position of the circuits depend, the electromagnetic force tending to increase x is

$$X = \tfrac{1}{2}\dot{y}_1^2 \frac{dL}{dx} + \dot{y}_1 \dot{y}_2 \frac{dM}{dx} + \tfrac{1}{2}\dot{y}_2^2 \frac{dN}{dx}.$$

If the motion of the system corresponding to the variation of x is such that each circuit moves as a rigid body, L and N will be independent of x, and the equation will be reduced to the form

$$X = \dot{y}_1 \dot{y}_2 \frac{dM}{dx}.$$

Hence, if the primary and secondary currents are of the same sign, the force X, which acts between the circuits, will tend to move them so as to increase M.

If the circuits are placed side by side, and the currents flow in the same direction, M will be increased by their being brought nearer together. Hence the force X is in this case an attraction.

584.] The whole of the phenomena of the mutual action of two circuits, whether the induction of currents or the mechanical force between them, depend on the quantity M, which we have called the coefficient of mutual induction. The method of calculating this quantity from the geometrical relations of the circuits is given in Art. 524, but in the investigations of the next chapter we shall not assume a knowledge of the mathematical form of this quantity. We shall consider it as deduced from experiments on induction, as, for instance, by observing the integral current when the secondary circuit is suddenly moved from a given position to an infinite distance, or to any position in which we know that $M = 0$.

CHAPTER VIII.

EXPLORATION OF THE FIELD BY MEANS OF THE SECONDARY CIRCUIT.

585.] We have proved in Arts. 582, 583, 584 that the electromagnetic action between the primary and the secondary circuit depends on the quantity denoted by M, which is a function of the form and relative position of the two circuits.

Although this quantity M is in fact the same as the potential of the two circuits, the mathematical form and properties of which we deduced in Arts. 423, 492, 521, 539 from magnetic and electromagnetic phenomena, we shall here make no reference to these results, but begin again from a new foundation, without any assumptions except those of the dynamical theory as stated in Chapter VII.

The electrokinetic momentum of the secondary circuit consists of two parts (Art. 578), one, Mi_1, depending on the primary current i_1, while the other, Ni_2, depends on the secondary current i_2. We are now to investigate the first of these parts, which we shall denote by p, where

$$p = Mi_1. \tag{1}$$

We shall also suppose the primary circuit fixed, and the primary current constant. The quantity p, the electrokinetic momentum of the secondary circuit, will in this case depend only on the form and position of the secondary circuit, so that if any closed curve be taken for the secondary circuit, and if the direction along this curve, which is to be reckoned positive, be chosen, the value of p for this closed curve is determinate. If the opposite direction along the curve had been chosen as the positive direction, the sign of the quantity p would have been reversed.

586.] Since the quantity p depends on the form and position of the circuit, we may suppose that each portion of the circuit contributes something to the value of p, and that the part contributed by each portion of the circuit depends on the form and position of that portion only, and not on the position of other parts of the circuit.

This assumption is legitimate, because we are not now considering a *current*, the parts of which may, and indeed do, act on one another, but a mere *circuit*, that is, a closed curve along which a current *may* flow, and this is a purely geometrical figure, the parts of which cannot be conceived to have any physical action on each other.

We may therefore assume that the part contributed by the element ds of the circuit is $J\,ds$, where J is a quantity depending on the position and direction of the element ds. Hence, the value of p may be expressed as a line-integral

$$p = \int J\,ds, \tag{2}$$

where the integration is to be extended once round the circuit.

587.] We have next to determine the form of the quantity J. In the first place, if ds is reversed in direction, J is reversed in sign. Hence, if two circuits $ABCE$ and $AECD$ have the arc AEC common, but reckoned in opposite directions in the two circuits, the sum of the values of p for the two circuits $ABCE$ and $AECD$ will be equal to the value of p for the circuit $ABCD$, which is made up of the two circuits.

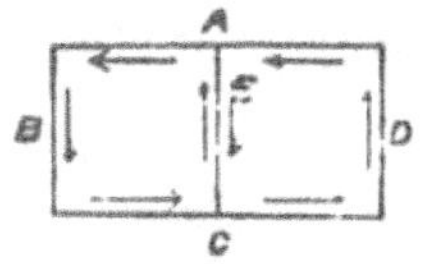

Fig. 35.

For the parts of the line-integral depending on the arc AEC are equal but of opposite sign in the two partial circuits, so that they destroy each other when the sum is taken, leaving only those parts of the line-integral which depend on the external boundary of $ABCD$.

In the same way we may shew that if a surface bounded by a closed curve be divided into any number of parts, and if the boundary of each of these parts be considered as a circuit, the positive direction round every circuit being the same as that round the external closed curve, then the value of p for the closed curve is equal to the sum of the values of p for all the circuits. See Art. 483.

588.] Let us now consider a portion of a surface, the dimensions of which are so small with respect to the principal radii of curvature of the surface that the variation of the direction of the normal within this portion may be neglected. We shall also suppose that if any very small circuit be carried parallel to itself from one part of this surface to another, the value of p for the small circuit is not sensibly altered. This will evidently be the case if the dimensions of the portion of surface are small enough compared with its distance from the primary circuit.

If any closed curve be drawn on this portion of the surface, the value of p will be proportional to its area.

For the areas of any two circuits may be divided into small elements all of the same dimensions, and having the same value of p. The areas of the two circuits are as the numbers of these elements which they contain, and the values of p for the two circuits are also in the same proportion.

Hence, the value of p for the circuit which bounds any element dS of a surface is of the form $I\,dS$,
where I is a quantity depending on the position of dS and on the direction of its normal. We have therefore a new expression for p,

$$p = \iint I\,dS, \tag{3}$$

where the double integral is extended over any surface bounded by the circuit.

589.] Let $ABCD$ be a circuit, of which AC is an elementary portion, so small that it may be considered straight. Let APB and CQB be small equal areas in the same plane, then the value of p will be the same for the small circuits APB and CQB, or

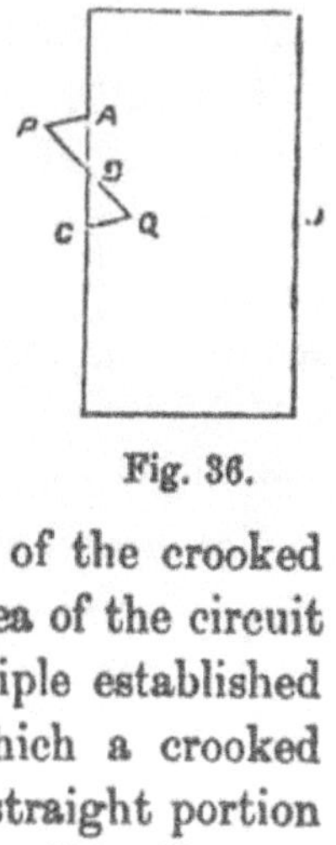

Fig. 36.

$$p(APB) = p(CQB).$$

Hence
$$\begin{aligned} p(APBQCD) &= p(ABQCD) + p(APB), \\ &= p(ABQCD) + p(CQB), \\ &= p(ABCD), \end{aligned}$$

or the value of p is not altered by the substitution of the crooked line $APQC$ for the straight line AC, provided the area of the circuit is not sensibly altered. This, in fact, is the principle established by Ampère's second experiment (Art. 506), in which a crooked portion of a circuit is shewn to be equivalent to a straight portion provided no part of the crooked portion is at a sensible distance from the straight portion.

If therefore we substitute for the element ds three small elements, dx, dy, and dz, drawn in succession, so as to form a continuous path from the beginning to the end of the element ds, and if $F\,dx$, $G\,dy$, and $H\,dz$ denote the elements of the line-integral corresponding to dx, dy, and dz respectively, then

$$J\,ds = F\,dx + G\,dy + H\,dz. \tag{4}$$

590.] We are now able to determine the mode in which the quantity J depends on the direction of the element ds. For, by (4),

$$J = F\frac{dx}{ds} + G\frac{dy}{ds} + H\frac{dz}{ds}. \tag{5}$$

This is the expression for the resolved part, in the direction of ds, of a vector, the components of which, resolved in the directions of the axes of x, y, and z, are F, G, and H respectively.

If this vector be denoted by $\mathfrak{A}$, and the vector from the origin to a point of the circuit by ρ, the element of the circuit will be $d\rho$, and the quaternion expression for J will be

$$-S\,\mathfrak{A}\,d\rho.$$

We may now write equation (2) in the form

$$p = \int\left(F\frac{dx}{ds} + G\frac{dy}{ds} + H\frac{dz}{ds}\right)ds, \tag{6}$$

$$\text{or } p = -\int S\,\mathfrak{A}\,d\rho. \tag{7}$$

The vector $\mathfrak{A}$ and its constituents F, G, H depend on the position of ds in the field, and not on the direction in which it is drawn. They are therefore functions of x, y, z, the coordinates of ds, and not of l, m, n, its direction-cosines.

The vector $\mathfrak{A}$ represents in direction and magnitude the time-integral of the electromotive force which a particle placed at the point (x, y, z) would experience if the primary current were suddenly stopped. We shall therefore call it the Electrokinetic Momentum *at the point* (x, y, z). It is identical with the quantity which we investigated in Art. 405 under the name of the vector-potential of magnetic induction.

The electrokinetic momentum of any finite line or circuit is the line-integral, extended along the line or circuit, of the resolved part of the electrokinetic momentum at each point of the same.

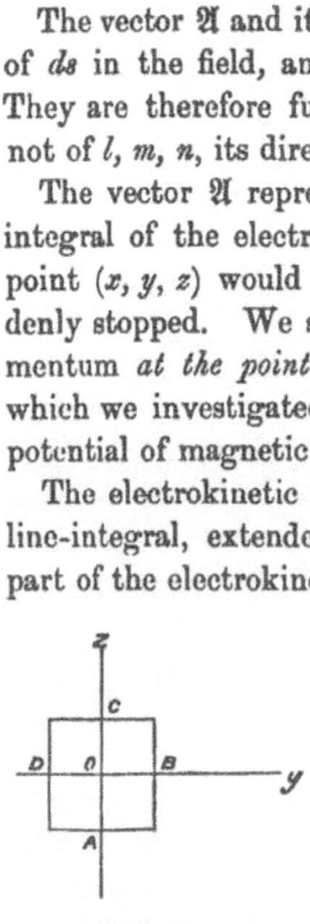

Fig. 37.

591.] Let us next determine the value of p for the elementary rectangle $ABCD$, of which the sides are dy and dz, the positive direction being from the direction of the axis of y to that of z.

Let the coordinates of O, the centre of gravity of the element, be x_0, y_0, z_0, and let G_0, H_0 be the values of G and of H at this point.

The coordinates of A, the middle point of the first side of the rectangle, are y_0 and $z_0 - \frac{1}{2}dz$. The corresponding value of G is

$$G = G_0 - \frac{1}{2}\frac{dG}{dz}dz + \&c., \tag{8}$$

and the part of the value of p which arises from the side A is approximately

$$G_0\, dy - \frac{1}{2}\frac{dG}{dz}dy\, dz. \tag{9}$$

Similarly, for B, $\quad H_0\, dz + \frac{1}{2}\frac{dH}{dy}dy\, dz.$

For C, $\quad -G_0\, dy - \frac{1}{2}\frac{dG}{dz}dy\, dz.$

For D, $\quad -H_0\, dz + \frac{1}{2}\frac{dH}{dy}dy\, dz.$

Adding these four quantities, we find the value of p for the rectangle

$$p = \left(\frac{dH}{dy} - \frac{dG}{dz}\right)dy\, dz. \tag{10}$$

If we now assume three new quantities, a, b, c, such that

$$\left.\begin{aligned} a &= \frac{dH}{dy} - \frac{dG}{dz}, \\ b &= \frac{dF}{dz} - \frac{dH}{dx}, \\ c &= \frac{dG}{dx} - \frac{dF}{dy}, \end{aligned}\right\} \tag{A}$$

and consider these as the constituents of a new vector $\mathfrak{B}$, then, by Theorem IV, Art. 24, we may express the line-integral of $\mathfrak{A}$ round any circuit in the form of the surface-integral of $\mathfrak{B}$ over a surface bounded by the circuit, thus

$$p = \int\left(F\frac{dx}{ds} + G\frac{dy}{ds} + H\frac{dz}{ds}\right)ds = \iint(la + mb + nc)\,dS, \tag{11}$$

or

$$p = \int T\mathfrak{A}\cos\epsilon\, ds = \iint T\mathfrak{B}\cos\eta\, dS, \tag{12}$$

where ϵ is the angle between $\mathfrak{A}$ and ds, and η that between $\mathfrak{B}$ and the normal to dS, whose direction-cosines are l, m, n, and $T\mathfrak{A}$, $T\mathfrak{B}$ denote the numerical values of $\mathfrak{A}$ and $\mathfrak{B}$.

Comparing this result with equation (3), it is evident that the quantity I in that equation is equal to $\mathfrak{B}\cos\eta$, or the resolved part of $\mathfrak{B}$ normal to dS.

592.] We have already seen (Arts. 490, 541) that, according to Faraday's theory, the phenomena of electromagnetic force and induction in a circuit depend on the variation of the number of lines of magnetic induction which pass through the circuit. Now the number of these lines is expressed mathematically by the surface-integral of the magnetic induction through any surface bounded by the circuit. Hence, we must regard the vector $\mathfrak{B}$ and its components a, b, c as representing what we are already acquainted with as the magnetic induction and its components.

In the present investigation we propose to deduce the properties of this vector from the dynamical principles stated in the last chapter, with as few appeals to experiment as possible.

In identifying this vector, which has appeared as the result of a mathematical investigation, with the magnetic induction, the properties of which we learned from experiments on magnets, we do not depart from this method, for we introduce no new fact into the theory, we only give a name to a mathematical quantity, and the propriety of so doing is to be judged by the agreement of the relations of the mathematical quantity with those of the physical quantity indicated by the name.

The vector $\mathfrak{B}$, since it occurs in a surface-integral, belongs evidently to the category of fluxes described in Art. 13. The vector $\mathfrak{A}$, on the other hand, belongs to the category of forces, since it appears in a line-integral.

593.] We must here recall to mind the conventions about positive and negative quantities and directions, some of which were stated in Art. 23. We adopt the right-handed system of axes, so that if a right-handed screw is placed in the direction of the axis of x, and a nut on this screw is turned in the positive direction of rotation, that is, from the direction of y to that of z, it will move along the screw in the positive direction of x.

We also consider vitreous electricity and austral magnetism as positive. The positive direction of an electric current, or of a line of electric induction, is the direction in which positive electricity moves or tends to move, and the positive direction of a line of magnetic induction is the direction in which a compass needle points with the end which turns to the north. See Fig. 24, Art. 498, and Fig. 25, Art. 501.

The student is recommended to select whatever method appears to him most effectual in order to fix these conventions securely in his memory, for it is far more difficult to remember a rule which determines in which of two previously indifferent ways a statement is to be made, than a rule which selects one way out of many.

594.] We have next to deduce from dynamical principles the expressions for the electromagnetic force acting on a conductor carrying an electric current through the magnetic field, and for the electromotive force acting on the electricity within a body moving in the magnetic field. The mathematical method which we shall adopt may be compared with the experimental method used by Faraday * in exploring the field by means of a wire, and with what we have already done at Art. 490, by a method founded on experiments. What we have now to do is to determine the effect on the value of p, the electrokinetic momentum of the secondary circuit, due to given alterations of the form of that circuit.

Let AA', BB' be two parallel straight conductors connected by the conducting arc C, which may be of any form, and by a straight

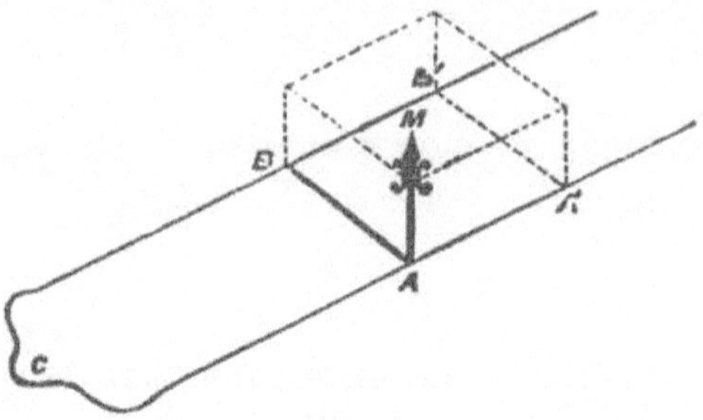

Fig. 38.

conductor AB, which is capable of sliding parallel to itself along the conducting rails AA' and BB'.

Let the circuit thus formed be considered as the secondary circuit, and let the direction ABC be assumed as the positive direction round it.

Let the sliding piece move parallel to itself from the position AB to the position $A'B'$. We have to determine the variation of p, the electrokinetic momentum of the circuit, due to this displacement of the sliding piece.

The secondary circuit is changed from ABC to $A'B'C$, hence, by Art. 587,

$$p(A'B'C) - p(ABC) = p(AA'B'B). \tag{13}$$

We have therefore to determine the value of p for the parallelogram $AA'B'B$. If this parallelogram is so small that we may neglect the variations of the direction and magnitude of the magnetic induction at different points of its plane, the value of p is, by Art. 591, $\mathfrak{B} \cos \eta \,.\, AA'B'B$, where $\mathfrak{B}$ is the magnetic induction,

* *Exp. Res.*, 3082, 3087, 3113.

and η the angle which it makes with the positive direction of the normal to the parallelogram $AA'B'B$.

We may represent the result geometrically by the volume of the parallelepiped, whose base is the parallelogram $AA'B'B$, and one of whose edges is the line AM, which represents in direction and magnitude the magnetic induction $\mathfrak{B}$. If the parallelogram is in the plane of the paper, and if AM is drawn upwards from the paper, the volume of the parallelepiped is to be taken positively, or more generally, if the directions of the circuit AB, of the magnetic induction AM, and of the displacement AA', form a right-handed system when taken in this cyclical order.

The volume of this parallelepiped represents the increment of the value of p for the secondary circuit due to the displacement of the sliding piece from AB to $A'B'$.

Electromotive Force acting on the Sliding Piece.

595.] The electromotive force produced in the secondary circuit by the motion of the sliding piece is, by Art. 579,

$$E = -\frac{dp}{dt}. \tag{14}$$

If we suppose AA' to be the displacement in unit of time, then AA' will represent the velocity, and the parallelepiped will represent $\frac{dp}{dt}$, and therefore, by equation (14), the electromotive force in the negative direction BA.

Hence, the electromotive force acting on the sliding piece AB, in consequence of its motion through the magnetic field, is represented by the volume of the parallelepiped, whose edges represent in direction and magnitude—the velocity, the magnetic induction, and the sliding piece itself, and is positive when these three directions are in right-handed cyclical order.

Electromagnetic Force acting on the Sliding Piece.

596.] Let i_2 denote the current in the secondary circuit in the positive direction ABC, then the work done by the electromagnetic force on AB while it slides from the position AB to the position $A'B'$ is $(M'-M)\,i_1 i_2$, where M and M' are the values of M_{12} in the initial and final positions of AB. But $(M'-M)\,i_1$ is equal to $p'-p$, and this is represented by the volume of the parallelepiped on AB, AM, and AA'. Hence, if we draw a line parallel to AB

to represent the quantity $AB \,.\, i_2$, the parallelepiped contained by this line, by AM, the magnetic induction, and by AA', the displacement, will represent the work done during this displacement.

For a given distance of displacement this will be greatest when the displacement is perpendicular to the parallelogram whose sides are AB and AM. The electromagnetic force is therefore represented by the area of the parallelogram on AB and AM multiplied by i_2, and is in the direction of the normal to this parallelogram, drawn so that AB, AM, and the normal are in right-handed cyclical order.

Four Definitions of a Line of Magnetic Induction.

597.] If the direction AA', in which the motion of the sliding piece takes place, coincides with AM, the direction of the magnetic induction, the motion of the sliding piece will not call electromotive force into action, whatever be the direction of AB, and if AB carries an electric current there will be no tendency to slide along AA'.

Again, if AB, the sliding piece, coincides in direction with AM, the direction of magnetic induction, there will be no electromotive force called into action by any motion of AB, and a current through AB will not cause AB to be acted on by mechanical force.

We may therefore define a line of magnetic induction in four different ways. It is a line such that—

(1) If a conductor be moved along it parallel to itself it will experience no electromotive force.

(2) If a conductor carrying a current be free to move along a line of magnetic induction it will experience no tendency to do so.

(3) If a linear conductor coincide in direction with a line of magnetic induction, and be moved parallel to itself in any direction, it will experience no electromotive force in the direction of its length.

(4) If a linear conductor carrying an electric current coincide in direction with a line of magnetic induction it will not experience any mechanical force.

General Equations of Electromotive Force.

598.] We have seen that E, the electromotive force due to induction acting on the secondary circuit, is equal to $-\frac{dp}{dt}$, where

$$p = \int \left(F\frac{dx}{ds} + G\frac{dy}{ds} + H\frac{dz}{ds}\right) ds. \tag{1}$$

To determine the value of E, let us differentiate the quantity under the integral sign with respect to t, remembering that if the secondary circuit is in motion, x, y, and z are functions of the time. We obtain

$$E = -\int \left(\frac{dF}{dt}\frac{dx}{ds} + \frac{dG}{dt}\frac{dy}{ds} + \frac{dH}{dt}\frac{dz}{ds} \right) ds$$

$$-\int \left(\frac{dF}{dx}\frac{dx}{ds} + \frac{dG}{dx}\frac{dy}{ds} + \frac{dH}{dx}\frac{dz}{ds} \right) \frac{dx}{dt}\, ds$$

$$-\int \left(\frac{dF}{dy}\frac{dx}{ds} + \frac{dG}{dy}\frac{dy}{ds} + \frac{dH}{dy}\frac{dz}{ds} \right) \frac{dy}{dt}\, ds$$

$$-\int \left(\frac{dF}{dz}\frac{dx}{ds} + \frac{dG}{dz}\frac{dy}{ds} + \frac{dH}{dz}\frac{dz}{ds} \right) \frac{dz}{dt}\, ds$$

$$-\int \left(F\frac{d^2x}{ds\,dt} + G\frac{d^2y}{ds\,dt} + H\frac{d^2z}{ds\,dt} \right) ds. \tag{2}$$

Now consider the second term of the integral, and substitute from equations (A), Art. 591, the values of $\frac{dG}{dx}$ and $\frac{dH}{dx}$. This term then becomes,

$$-\int \left(c\frac{dy}{ds} - b\frac{dz}{ds} + \frac{dF}{dx}\frac{dx}{ds} + \frac{dF}{dy}\frac{dy}{ds} + \frac{dF}{dz}\frac{dz}{ds} \right) \frac{dx}{dt}\, ds,$$

which we may write

$$-\int \left(c\frac{dy}{ds} - b\frac{dz}{ds} + \frac{dF}{ds} \right) \frac{dx}{dt}\, ds.$$

Treating the third and fourth terms in the same way, and collecting the terms in $\frac{dx}{ds}$, $\frac{dy}{ds}$, and $\frac{dz}{ds}$, remembering that

$$\int \left(\frac{dF}{ds}\frac{dx}{dt} + F\frac{d^2x}{ds\,dt} \right) ds = F\frac{dx}{dt}, \tag{3}$$

and therefore that the integral, when taken round the closed curve, vanishes,

$$E = \int \left(c\frac{dy}{dt} - b\frac{dz}{dt} - \frac{dF}{dt} \right) \frac{dx}{ds}\, ds$$

$$+\int \left(a\frac{dz}{dt} - c\frac{dx}{dt} - \frac{dG}{dt} \right) \frac{dy}{ds}\, ds$$

$$+\int \left(b\frac{dx}{dt} - a\frac{dy}{dt} - \frac{dH}{dt} \right) \frac{dz}{ds}\, ds. \tag{4}$$

We may write this expression in the form

$$E = \int \left(P\frac{dx}{ds} + Q\frac{dy}{ds} + R\frac{dz}{ds}\right) ds, \tag{5}$$

where

$$\left.\begin{aligned} P &= c\frac{dy}{dt} - b\frac{dz}{dt} - \frac{dF}{dt} - \frac{d\Psi}{dx}, \\ Q &= a\frac{dz}{dt} - c\frac{dx}{dt} - \frac{dG}{dt} - \frac{d\Psi}{dy}, \\ R &= b\frac{dx}{dt} - a\frac{dy}{dt} - \frac{dH}{dt} - \frac{d\Psi}{dz}. \end{aligned}\right\} \text{Equations of Electromotive Force.} \tag{B}$$

The terms involving the new quantity Ψ are introduced for the sake of giving generality to the expressions for P, Q, R. They disappear from the integral when extended round the closed circuit. The quantity Ψ is therefore indeterminate as far as regards the problem now before us, in which the total electromotive force round the circuit is to be determined. We shall find, however, that when we know all the circumstances of the problem, we can assign a definite value to Ψ, and that it represents, according to a certain definition, the *electric potential* at the point x, y, z.

The quantity under the integral sign in equation (5) represents the electromotive force acting on the element ds of the circuit.

If we denote by $T\mathfrak{E}$, the numerical value of the resultant of P, Q, and R, and by ϵ, the angle between the direction of this resultant and that of the element ds, we may write equation (5),

$$E = \int T\mathfrak{E} \cos\epsilon \, ds. \tag{6}$$

The vector $\mathfrak{E}$ is the electromotive force *at* the moving element ds. Its direction and magnitude depend on the position and motion of ds, and on the variation of the magnetic field, but not on the direction of ds. Hence we may now disregard the circumstance that ds forms part of a circuit, and consider it simply as a portion of a moving body, acted on by the electromotive force $\mathfrak{E}$. The electromotive force at a point has already been defined in Art. 68. It is also called the resultant electrical force, being the force which would be experienced by a unit of positive electricity placed at that point. We have now obtained the most general value of this quantity in the case of a body moving in a magnetic field due to a variable electric system.

If the body is a conductor, the electromotive force will produce a current; if it is a dielectric, the electromotive force will produce only electric displacement.

The electromotive force at a point, or on a particle, must be carefully distinguished from the electromotive force along an arc of a curve, the latter quantity being the line-integral of the former. See Art. 69.

599.] The electromotive force, the components of which are defined by equations (B), depends on three circumstances. The first of these is the motion of the particle through the magnetic field. The part of the force depending on this motion is expressed by the first two terms on the right of each equation. It depends on the velocity of the particle transverse to the lines of magnetic induction. If $\mathfrak{G}$ is a vector representing the velocity, and $\mathfrak{B}$ another representing the magnetic induction, then if $\mathfrak{E}_1$ is the part of the electromotive force depending on the motion,

$$\mathfrak{E}_1 = V.\,\mathfrak{G}\,\mathfrak{B}, \tag{7}$$

or, the electromotive force is the vector part of the product of the magnetic induction multiplied by the velocity, that is to say, the magnitude of the electromotive force is represented by the area of the parallelogram, whose sides represent the velocity and the magnetic induction, and its direction is the normal to this parallelogram, drawn so that the velocity, the magnetic induction, and the electromotive force are in right-handed cyclical order.

The third term in each of the equations (B) depends on the time-variation of the magnetic field. This may be due either to the time-variation of the electric current in the primary circuit, or to motion of the primary circuit. Let $\mathfrak{E}_2$ be the part of the electromotive force which depends on these terms. Its components are

$$-\frac{dF}{dt}, \quad -\frac{dG}{dt}, \quad \text{and} \quad -\frac{dH}{dt},$$

and these are the components of the vector, $-\frac{d\mathfrak{A}}{dt}$ or $\dot{\mathfrak{A}}$. Hence,

$$\mathfrak{E}_2 = -\dot{\mathfrak{A}}. \tag{8}$$

The last term of each equation (B) is due to the variation of the function Ψ in different parts of the field. We may write the third part of the electromotive force, which is due to this cause,

$$\mathfrak{E}_3 = -\nabla\Psi. \tag{9}$$

The electromotive force, as defined by equations (B), may therefore be written in the quaternion form,

$$\mathfrak{E} = V.\,\mathfrak{G}\,\mathfrak{B} - \dot{\mathfrak{A}} - \nabla\Psi. \tag{10}$$

On the Modification of the Equations of Electromotive Force when the Axes to which they are referred are moving in Space.

600.] Let x', y', z' be the coordinates of a point referred to a system of rectangular axes moving in space, and let x, y, z be the coordinates of the same point referred to fixed axes.

Let the components of the velocity of the origin of the moving system be u, v, w, and those of its angular velocity ω_1, ω_2, ω_3 referred to the fixed system of axes, and let us choose the fixed axes so as to coincide at the given instant with the moving ones, then the only quantities which will be different for the two systems of axes will be those differentiated with respect to the time. If $\frac{\delta x}{\delta t}$ denotes a component velocity of a point moving in rigid connexion with the moving axes, and $\frac{dx}{dt}$ and $\frac{dx'}{dt}$ that of any moving point, having the same instantaneous position, referred to the fixed and the moving axes respectively, then

$$\frac{dx}{dt} = \frac{\delta x}{\delta t} + \frac{dx'}{dt}, \tag{1}$$

with similar equations for the other components.

By the theory of the motion of a body of invariable form,

$$\left.\begin{aligned} \frac{\delta x}{\delta t} &= u + \omega_2 z - \omega_3 y, \\ \frac{\delta y}{\delta t} &= v + \omega_3 x - \omega_1 z, \\ \frac{\delta z}{\delta t} &= w + \omega_1 y - \omega_2 x. \end{aligned}\right\} \tag{2}$$

Since F is a component of a directed quantity parallel to x, if $\frac{dF'}{dt}$ be the value of $\frac{dF}{dt}$ referred to the moving axes,

$$\frac{dF'}{dt} = \frac{dF}{dx}\frac{\delta x}{\delta t} + \frac{dF}{dy}\frac{\delta y}{\delta t} + \frac{dF}{dz}\frac{\delta z}{\delta t} + G\omega_3 - H\omega_2 + \frac{dF}{dt}. \tag{3}$$

Substituting for $\frac{dF}{dy}$ and $\frac{dF}{dz}$ their values as deduced from the equations (A) of magnetic induction, and remembering that, by (2),

$$\frac{d}{dx}\frac{\delta x}{\delta t} = 0, \qquad \frac{d}{dx}\frac{\delta y}{\delta t} = \omega_3, \qquad \frac{d}{dx}\frac{\delta z}{\delta t} = -\omega_2, \tag{4}$$

$$\frac{dF'}{dt} = \frac{dF}{dx}\frac{\delta x}{\delta t} + F\frac{d}{dx}\frac{\delta x}{\delta t} + \frac{dG}{dx}\frac{\delta y}{\delta t} + G\frac{d}{dy}\frac{\delta y}{\delta t} + \frac{dH}{dx}\frac{\delta z}{\delta t} + H\frac{d}{dx}\frac{\delta z}{\delta t}$$

$$- c\frac{\delta y}{\delta t} + b\frac{\delta z}{\delta t} + \frac{dF}{dt}. \tag{5}$$

If we now put
$$-\Psi' = F\frac{\delta x}{\delta t} + G\frac{\delta y}{\delta t} + H\frac{\delta z}{\delta t}, \tag{6}$$
$$\frac{dF'}{dt} = -\frac{d\Psi'}{dx} - c\frac{\delta y}{\delta t} + b\frac{\delta z}{\delta t} + \frac{dF}{dt}. \tag{7}$$

The equation for P, the component of the electromotive force parallel to x, is, by (B),
$$P = c\frac{dy}{dt} - b\frac{dz}{dt} - \frac{dF}{dt} - \frac{d\Psi}{dx}, \tag{8}$$
referred to the fixed axes. Substituting the values of the quantities as referred to the moving axes, we have
$$P' = c\frac{dy'}{dt} - b\frac{dz'}{dt} - \frac{dF'}{dt} - \frac{d(\Psi+\Psi')}{dx}, \tag{9}$$
for the value of P referred to the moving axes.

601.] It appears from this that the electromotive force is expressed by a formula of the same type, whether the motions of the conductors be referred to fixed axes or to axes moving in space, the only difference between the formulae being that in the case of moving axes the electric potential Ψ must be changed into $\Psi+\Psi'$.

In all cases in which a current is produced in a conducting circuit, the electromotive force is the line-integral
$$E = \int\left(P\frac{dx}{ds} + Q\frac{dy}{ds} + R\frac{dz}{ds}\right)ds, \tag{10}$$
taken round the curve. The value of Ψ disappears from this integral, so that the introduction of Ψ' has no influence on its value. In all phenomena, therefore, relating to closed circuits and the currents in them, it is indifferent whether the axes to which we refer the system be at rest or in motion. See Art. 668.

On the Electromagnetic Force acting on a Conductor which carries an Electric Current through a Magnetic Field.

602.] We have seen in the general investigation, Art. 583, that if x_1 is one of the variables which determine the position and form of the secondary circuit, and if X_1 is the force acting on the secondary circuit tending to increase this variable, then
$$X_1 = \frac{dM}{dx_1} i_1 i_2. \tag{1}$$

Since i_1 is independent of x_1, we may write
$$M i_1 = p = \int\left(F\frac{dx}{ds} + G\frac{dy}{ds} + H\frac{dz}{ds}\right)ds, \tag{2}$$

and we have for the value of X_1,

$$X_1 = i_2 \frac{d}{dx_1}\int\left(F\frac{dx}{ds} + G\frac{dy}{ds} + H\frac{dz}{ds}\right)ds. \qquad (3)$$

Now let us suppose that the displacement consists in moving every point of the circuit through a distance δx in the direction of x, δx being any continuous function of s, so that the different parts of the circuit move independently of each other, while the circuit remains continuous and closed.

Also let X be the total force in the direction of x acting on the part of the circuit from $s = 0$ to $s = s$, then the part corresponding to the element ds will be $\frac{dX}{ds}ds$. We shall then have the following expression for the work done by the force during the displacement,

$$\int\frac{dX}{ds}\delta x\, ds = i_2\int\frac{d}{d\,\delta x}\left(F\frac{dx}{ds} + G\frac{dy}{ds} + H\frac{dz}{ds}\right)\delta x\, ds, \qquad (4)$$

where the integration is to be extended round the closed curve, remembering that δx is an arbitrary function of s. We may therefore perform the differentiation with respect to δx in the same way that we differentiated with respect to t in Art. 598, remembering that

$$\frac{dx}{d\,\delta x} = 1, \quad \frac{dy}{d\,\delta x} = 0, \text{ and } \frac{dz}{d\,\delta x} = 0. \qquad (5)$$

We thus find

$$\int\frac{dX}{ds}\delta x\, ds = i_2\int\left(c\frac{dy}{ds} - b\frac{dz}{ds}\right)\delta x\, ds + \int\frac{d}{ds}(F\,\delta x)\, ds. \qquad (6)$$

The last term vanishes when the integration is extended round the closed curve, and since the equation must hold for all forms of the function δx, we must have

$$\frac{dX}{ds} = i_2\left(c\frac{dy}{ds} - b\frac{dz}{ds}\right), \qquad (7)$$

an equation which gives the force parallel to x on any element of the circuit. The forces parallel to y and z are

$$\frac{dY}{ds} = i_2\left(a\frac{dz}{ds} - c\frac{dx}{ds}\right), \qquad (8)$$

$$\frac{dZ}{ds} = i_2\left(b\frac{dx}{ds} - a\frac{dy}{dx}\right). \qquad (9)$$

The resultant force on the element is given in direction and magnitude by the quaternion expression $i_2 V d\rho\,\mathfrak{B}$, where i_2 is the numerical measure of the current, and $d\rho$ and $\mathfrak{B}$ are vectors representing the element of the circuit and the magnetic induction, and the multiplication is to be understood in the Hamiltonian sense.

603.] If the conductor is to be treated not as a line but as a body, we must express the force on the element of length, and the current through the complete section, in terms of symbols denoting the force per unit of volume, and the current per unit of area.

Let X, Y, Z now represent the components of the force referred to unit of volume, and u, v, w those of the current referred to unit of area. Then, if S represents the section of the conductor, which we shall suppose small, the volume of the element ds will be $S\,ds$, and $u = \frac{i_2}{S}\frac{dx}{ds}$. Hence, equation (7) will become

$$\frac{X S\,ds}{ds} = S(vc - wb), \tag{10}$$

or $X = vc - wb$.
Similarly $Y = wa - uc$,
and $Z = ub - va$.

$$\left.\begin{aligned} X &= vc - wb. \\ Y &= wa - uc, \\ Z &= ub - va. \end{aligned}\right\} \quad \text{(Equations of Electromagnetic Force.)} \qquad \text{(C)}$$

Here X, Y, Z are the components of the electromagnetic force on an element of a conductor divided by the volume of that element; u, v, w are the components of the electric current through the element referred to unit of area, and a, b, c are the components of the magnetic induction at the element, which are also referred to unit of area.

If the vector $\mathfrak{F}$ represents in magnitude and direction the force acting on unit of volume of the conductor, and if $\mathfrak{C}$ represents the electric current flowing through it,

$$\mathfrak{F} = V.\mathfrak{C}\mathfrak{B}. \tag{11}$$

CHAPTER IX.

GENERAL EQUATIONS OF THE ELECTROMAGNETIC FIELD.

604.] In our theoretical discussion of electrodynamics we began by assuming that a system of circuits carrying electric currents is a dynamical system, in which the currents may be regarded as velocities, and in which the coordinates corresponding to these velocities do not themselves appear in the equations. It follows from this that the kinetic energy of the system, so far as it depends on the currents, is a homogeneous quadratic function of the currents, in which the coefficients depend only on the form and relative position of the circuits. Assuming these coefficients to be known, by experiment or otherwise, we deduced, by purely dynamical reasoning, the laws of the induction of currents, and of electromagnetic attraction. In this investigation we introduced the conceptions of the electrokinetic energy of a system of currents, of the electromagnetic momentum of a circuit, and of the mutual potential of two circuits.

We then proceeded to explore the field by means of various configurations of the secondary circuit, and were thus led to the conception of a vector $\mathfrak{A}$, having a determinate magnitude and direction at any given point of the field. We called this vector the electromagnetic momentum at that point. This quantity may be considered as the time-integral of the electromotive force which would be produced at that point by the sudden removal of all the currents from the field. It is identical with the quantity already investigated in Art. 405 as the vector-potential of magnetic induction. Its components parallel to x, y, and z are F, G, and H. The electromagnetic momentum of a circuit is the line-integral of $\mathfrak{A}$ round the circuit.

We then, by means of Theorem IV, Art. 24, transformed the

line-integral of $\mathfrak{A}$ into the surface-integral of another vector, $\mathfrak{B}$, whose components are a, b, c, and we found that the phenomena of induction due to motion of a conductor, and those of electromagnetic force can be expressed in terms of $\mathfrak{B}$. We gave to $\mathfrak{B}$ the name of the Magnetic induction, since its properties are identical with those of the lines of magnetic induction as investigated by Faraday.

We also established three sets of equations: the first set, (A), are those of magnetic induction, expressing it in terms of the electromagnetic momentum. The second set, (B), are those of electromotive force, expressing it in terms of the motion of the conductor across the lines of magnetic induction, and of the rate of variation of the electromagnetic momentum. The third set, (C), are the equations of electromagnetic force, expressing it in terms of the current and the magnetic induction.

The current in all these cases is to be understood as the actual current, which includes not only the current of conduction, but the current due to variation of the electric displacement.

The magnetic induction $\mathfrak{B}$ is the quantity which we have already considered in Art. 400. In an unmagnetized body it is identical with the force on a unit magnetic pole, but if the body is magnetized, either permanently or by induction, it is the force which would be exerted on a unit pole, if placed in a narrow crevasse in the body, the walls of which are perpendicular to the direction of magnetization. The components of $\mathfrak{B}$ are a, b, c.

It follows from the equations (A), by which a, b, c are defined, that

$$\frac{da}{dx}+\frac{db}{dy}+\frac{dc}{dz}=0.$$

This was shewn at Art. 403 to be a property of the magnetic induction.

605.] We have defined the magnetic force within a magnet, as distinguished from the magnetic induction, to be the force on a unit pole placed in a narrow crevasse cut parallel to the direction of magnetization. This quantity is denoted by $\mathfrak{H}$, and its components by α, β, γ. See Art. 398.

If $\mathfrak{I}$ is the intensity of magnetization, and A, B, C its components, then, by Art. 400,

$$\left.\begin{aligned} a &= \alpha+4\pi A,\\ b &= \beta+4\pi B,\\ c &= \gamma+4\pi C. \end{aligned}\right\} \quad \text{(Equations of Magnetization.)} \qquad \text{(D)}$$

We may call these the equations of magnetization, and they indicate that in the electromagnetic system the magnetic induction $\mathfrak{B}$, considered as a vector, is the sum, in the Hamiltonian sense, of two vectors, the magnetic force $\mathfrak{H}$, and the magnetization $\mathfrak{J}$ multiplied by 4π, or

$$\mathfrak{B} = \mathfrak{H} + 4\pi\mathfrak{J}.$$

In certain substances, the magnetization depends on the magnetic force, and this is expressed by the system of equations of induced magnetism given at Arts. 426 and 435.

606.] Up to this point of our investigation we have deduced everything from purely dynamical considerations, without any reference to quantitative experiments in electricity or magnetism. The only use we have made of experimental knowledge is to recognise, in the abstract quantities deduced from the theory, the concrete quantities discovered by experiment, and to denote them by names which indicate their physical relations rather than their mathematical generation.

In this way we have pointed out the existence of the electromagnetic momentum $\mathfrak{A}$ as a vector whose direction and magnitude vary from one part of space to another, and from this we have deduced, by a mathematical process, the magnetic induction, $\mathfrak{B}$, as a derived vector. We have not, however, obtained any data for determining either $\mathfrak{A}$ or $\mathfrak{B}$ from the distribution of currents in the field. For this purpose we must find the mathematical connexion between these quantities and the currents.

We begin by admitting the existence of permanent magnets, the mutual action of which satisfies the principle of the conservation of energy. We make no assumption with respect to the laws of magnetic force except that which follows from this principle, namely, that the force acting on a magnetic pole must be capable of being derived from a potential.

We then observe the action between currents and magnets, and we find that a current acts on a magnet in a manner apparently the same as another magnet would act if its strength, form, and position were properly adjusted, and that the magnet acts on the current in the same way as another current. These observations need not be supposed to be accompanied with actual measurements of the forces. They are not therefore to be considered as furnishing numerical data, but are useful only in suggesting questions for our consideration.

The question these observations suggest is, whether the magnetic field produced by electric currents, as it is similar to that produced by permanent magnets in many respects, resembles it also in being related to a potential?

The evidence that an electric circuit produces, in the space surrounding it, magnetic effects precisely the same as those produced by a magnetic shell bounded by the circuit, has been stated in Arts. 482–485.

We know that in the case of the magnetic shell there is a potential, which has a determinate value for all points outside the substance of the shell, but that the values of the potential at two neighbouring points, on opposite sides of the shell, differ by a finite quantity.

If the magnetic field in the neighbourhood of an electric current resembles that in the neighbourhood of a magnetic shell, the magnetic potential, as found by a line-integration of the magnetic force, will be the same for any two lines of integration, provided one of these lines can be transformed into the other by continuous motion without cutting the electric current.

If, however, one line of integration cannot be transformed into the other without cutting the current, the line-integral of the magnetic force along the one line will differ from that along the other by a quantity depending on the strength of the current. The magnetic potential due to an electric current is therefore a function having an infinite series of values with a common difference, the particular value depending on the course of the line of integration. Within the substance of the conductor, there is no such thing as a magnetic potential.

607.] Assuming that the magnetic action of a current has a magnetic potential of this kind, we proceed to express this result mathematically.

In the first place, the line-integral of the magnetic force round any closed curve is zero, provided the closed curve does not surround the electric current.

In the next place, if the current passes once, and only once, through the closed curve in the positive direction, the line-integral has a determinate value, which may be used as a measure of the strength of the current. For if the closed curve alters its form in any continuous manner without cutting the current, the line-integral will remain the same.

In electromagnetic measure, the line-integral of the magnetic force round a closed curve is numerically equal to the current through the closed curve multiplied by 4π.

If we take for the closed curve the parallelogram whose sides are dy and dz, the line-integral of the magnetic force round the parallelogram is

$$\left(\frac{d\gamma}{dy} - \frac{d\beta}{dz}\right) dy\, dz,$$

and if u, v, w are the components of the flow of electricity, the current through the parallelogram is

$$u\, dy\, dz.$$

Multiplying this by 4π, and equating the result to the line-integral, we obtain the equation

$$4\pi u = \frac{d\gamma}{dy} - \frac{d\beta}{dz},$$

with the similar equations

$$\left.\begin{aligned} 4\pi v &= \frac{d\alpha}{dz} - \frac{d\gamma}{dx}, \\ 4\pi w &= \frac{d\beta}{dx} - \frac{d\alpha}{dy}, \end{aligned}\right\} \quad \text{(Equations of Electric Currents.)} \qquad \text{(E)}$$

which determine the magnitude and direction of the electric currents when the magnetic force at every point is given.

When there is no current, these equations are equivalent to the condition that

$$\alpha\, dx + \beta\, dy + \gamma\, dz = -D\Omega,$$

or that the magnetic force is derivable from a magnetic potential in all points of the field where there are no currents.

By differentiating the equations (E) with respect to x, y, and z respectively, and adding the results, we obtain the equation

$$\frac{du}{dx} + \frac{dv}{dy} + \frac{dw}{dz} = 0,$$

which indicates that the current whose components are u, v, w is subject to the condition of motion of an incompressible fluid, and that it must necessarily flow in closed circuits.

This equation is true only if we take u, v, and w as the components of that electric flow which is due to the variation of electric displacement as well as to true conduction.

We have very little experimental evidence relating to the direct electromagnetic action of currents due to the variation of electric displacement in dielectrics, but the extreme difficulty of reconciling the laws of electromagnetism with the existence of electric currents which are not closed is one reason among many why we must admit the existence of transient currents due to the variation of displacement. Their importance will be seen when we come to the electromagnetic theory of light.

608.] We have now determined the relations of the principal quantities concerned in the phenomena discovered by Örsted, Ampère, and Faraday. To connect these with the phenomena described in the former parts of this treatise, some additional relations are necessary.

When electromotive force acts on a material body, it produces in it two electrical effects, called by Faraday Induction and Conduction, the first being most conspicuous in dielectrics, and the second in conductors.

In this treatise, static electric induction is measured by what we have called the electric displacement, a directed quantity or vector which we have denoted by $\mathfrak{D}$, and its components by f, g, h.

In isotropic substances, the displacement is in the same direction as the electromotive force which produces it, and is proportional to it, at least for small values of this force. This may be expressed by the equation

$$\mathfrak{D} = \frac{1}{4\pi} K \mathfrak{E}, \qquad \text{(Equation of Electric Displacement.)} \qquad \text{(F)}$$

where K is the dielectric capacity of the substance. See Art. 69.

In substances which are not isotropic, the components f, g, h of the electric displacement $\mathfrak{D}$ are linear functions of the components P, Q, R of the electromotive force $\mathfrak{E}$.

The form of the equations of electric displacement is similar to that of the equations of conduction as given in Art. 298.

These relations may be expressed by saying that K is, in isotropic bodies, a scalar quantity, but in other bodies it is a linear and vector function, operating on the vector $\mathfrak{E}$.

609.] The other effect of electromotive force is conduction. The laws of conduction as the result of electromotive force were established by Ohm, and are explained in the second part of this treatise, Art. 241. They may be summed up in the equation

$$\mathfrak{K} = C \mathfrak{E}, \qquad \text{(Equation of Conductivity.)} \qquad \text{(G)}$$

where $\mathfrak{E}$ is the intensity of the electromotive force at the point, $\mathfrak{K}$ is the density of the current of conduction, the components of which are p, q, r, and C is the conductivity of the substance, which, in the case of isotropic substances, is a simple scalar quantity, but in other substances becomes a linear and vector function operating on the vector $\mathfrak{E}$. The form of this function is given in Cartesian coordinates in Art. 298.

610.] One of the chief peculiarities of this treatise is the doctrine which it asserts, that the true electric current $\mathfrak{C}$, that on which the electromagnetic phenomena depend, is not the same thing as $\mathfrak{K}$, the current of conduction, but that the time-variation of $\mathfrak{D}$, the electric displacement, must be taken into account in estimating the total movement of electricity, so that we must write,

$$\mathfrak{C} = \mathfrak{K} + \dot{\mathfrak{D}}, \quad \text{(Equation of True Currents.)} \qquad \text{(H)}$$

or, in terms of the components,

$$\left.\begin{aligned} u &= p + \frac{df}{dt}, \\ v &= q + \frac{dg}{dt}, \\ w &= r + \frac{dh}{dt}. \end{aligned}\right\} \qquad \text{(H*)}$$

611.] Since both $\mathfrak{K}$ and $\mathfrak{D}$ depend on the electromotive force $\mathfrak{E}$, we may express the true current $\mathfrak{C}$ in terms of the electromotive force, thus

$$\mathfrak{C} = \left(C + \frac{1}{4\pi} K \frac{d}{dt}\right)\mathfrak{E}, \qquad \text{(I)}$$

or, in the case in which C and K are constants,

$$\left.\begin{aligned} u &= CP + \frac{1}{4\pi} K \frac{dP}{dt}, \\ v &= CQ + \frac{1}{4\pi} K \frac{dQ}{dt}, \\ w &= CR + \frac{1}{4\pi} K \frac{dR}{dt}. \end{aligned}\right\} \qquad \text{(I*)}$$

612.] The volume-density of the free electricity at any point is found from the components of electric displacement by the equation

$$\rho = \frac{df}{dx} + \frac{dg}{dy} + \frac{dh}{dz}. \qquad \text{(J)}$$

613.] The surface-density of electricity is

$$\sigma = lf + mg + nh + l'f' + m'g' + n'h', \qquad \text{(K)}$$

where l, m, n are the direction-cosines of the normal drawn from the surface into the medium in which f, g, h are the components of the displacement, and l', m', n' are those of the normal drawn from the surface into the medium in which they are f', g', h'.

614.] When the magnetization of the medium is entirely induced by the magnetic force acting on it, we may write the equation of induced magnetization,

$$\mathfrak{B} = \mu \mathfrak{H}, \qquad \text{(L)}$$

where μ is the coefficient of magnetic permeability, which may be considered a scalar quantity, or a linear and vector function operating on $\mathfrak{H}$, according as the medium is isotropic or not.

615.] These may be regarded as the principal relations among the quantities we have been considering. They may be combined so as to eliminate some of these quantities, but our object at present is not to obtain compactness in the mathematical formulae, but to express every relation of which we have any knowledge. To eliminate a quantity which expresses a useful idea would be rather a loss than a gain in this stage of our enquiry.

There is one result, however, which we may obtain by combining equations (A) and (E), and which is of very great importance.

If we suppose that no magnets exist in the field except in the form of electric circuits, the distinction which we have hitherto maintained between the magnetic force and the magnetic induction vanishes, because it is only in magnetized matter that these quantities differ from each other.

According to Ampère's hypothesis, which will be explained in Art. 833, the properties of what we call magnetized matter are due to molecular electric circuits, so that it is only when we regard the substance in large masses that our theory of magnetization is applicable, and if our mathematical methods are supposed capable of taking account of what goes on within the individual molecules, they will discover nothing but electric circuits, and we shall find the magnetic force and the magnetic induction everywhere identical. In order, however, to be able to make use of the electrostatic or of the electromagnetic system of measurement at pleasure we shall retain the coefficient μ, remembering that its value is unity in the electromagnetic system.

616.] The components of the magnetic induction are by equations (A), Art. 591,

$$\left.\begin{aligned} a &= \frac{dH}{dy} - \frac{dG}{dz}, \\ b &= \frac{dF}{dz} - \frac{dH}{dx}, \\ c &= \frac{dG}{dx} - \frac{dF}{dy}. \end{aligned}\right\}$$

The components of the electric current are by equations (E), Art. 607,

$$\left.\begin{aligned} 4\pi u &= \frac{d\gamma}{dy} - \frac{d\beta}{dz}, \\ 4\pi v &= \frac{d\alpha}{dz} - \frac{d\gamma}{dx}, \\ 4\pi w &= \frac{d\beta}{dx} - \frac{d\alpha}{dy}. \end{aligned}\right\}$$

According to our hypothesis a, b, c are identical with $\mu\alpha, \mu\beta, \mu\gamma$ respectively. We therefore obtain

$$4\pi\mu u = \frac{d^2G}{dx\,dy} - \frac{d^2F}{dy^2} - \frac{d^2F}{dz^2} + \frac{d^2H}{dz\,dx}. \tag{1}$$

If we write
$$J = \frac{dF}{dx} + \frac{dG}{dy} + \frac{dH}{dz}, \tag{2}$$

and *
$$\nabla^2 = -\left(\frac{d^2}{dx^2} + \frac{d^2}{dy^2} + \frac{d^2}{dz^2}\right), \tag{3}$$

we may write equation (1),

$$\left.\begin{aligned} 4\pi\mu u &= \frac{dJ}{dx} + \nabla^2 F. \\ 4\pi\mu v &= \frac{dJ}{dy} + \nabla^2 G, \\ 4\pi\mu w &= \frac{dJ}{dz} + \nabla^2 H. \end{aligned}\right\} \tag{4}$$

Similarly,

If we write
$$\left.\begin{aligned} F' &= \frac{1}{\mu}\iiint \frac{u}{r}\,dx\,dy\,dz, \\ G' &= \frac{1}{\mu}\iiint \frac{v}{r}\,dx\,dy\,dz, \\ H' &= \frac{1}{\mu}\iiint \frac{w}{r}\,dx\,dy\,dz, \end{aligned}\right\} \tag{5}$$

$$\chi = \frac{4\pi}{\mu}\iiint \frac{J}{r}\,dx\,dy\,dz, \tag{6}$$

where r is the distance of the given point from the element xyz, and the integrations are to be extended over all space, then

$$\left.\begin{aligned} F &= F' + \frac{d\chi}{dx}, \\ G &= G' + \frac{d\chi}{dy}, \\ H &= H' + \frac{d\chi}{dz}. \end{aligned}\right\} \tag{7}$$

The quantity χ disappears from the equations (A), and it is not related to any physical phenomenon. If we suppose it to be zero everywhere, J will also be zero everywhere, and equations (5), omitting the accents, will give the true values of the components of $\mathfrak{A}$.

* The negative sign is employed here in order to make our expressions consistent with those in which Quaternions are employed.

617.] We may therefore adopt, as a definition of $\mathfrak{A}$, that it is the vector-potential of the electric current, standing in the same relation to the electric current that the scalar potential stands to the matter of which it is the potential, and obtained by a similar process of integration, which may be thus described.—

From a given point let a vector be drawn, representing in magnitude and direction a given element of an electric current, divided by the numerical value of the distance of the element from the given point. Let this be done for every element of the electric current. The resultant of all the vectors thus found is the potential of the whole current. Since the current is a vector quantity, its potential is also a vector. See Art. 422.

When the distribution of electric currents is given, there is one, and only one, distribution of the values of $\mathfrak{A}$, such that $\mathfrak{A}$ is everywhere finite and continuous, and satisfies the equations

$$\nabla^2 \mathfrak{A} = 4\pi\mu\mathfrak{C}, \qquad S.\nabla\mathfrak{A} = 0,$$

and vanishes at an infinite distance from the electric system. This value is that given by equations (5), which may be written

$$\mathfrak{A} = \frac{1}{\mu}\iiint \frac{\mathfrak{C}}{r}\, dx\, dy\, dz.$$

Quaternion Expressions for the Electromagnetic Equations.

618.] In this treatise we have endeavoured to avoid any process demanding from the reader a knowledge of the Calculus of Quaternions. At the same time we have not scrupled to introduce the idea of a vector when it was necessary to do so. When we have had occasion to denote a vector by a symbol, we have used a German letter, the number of different vectors being so great that Hamilton's favourite symbols would have been exhausted at once. Whenever therefore, a German letter is used it denotes a Hamiltonian vector, and indicates not only its magnitude but its direction. The constituents of a vector are denoted by Roman or Greek letters.

The principal vectors which we have to consider are :—

	Symbol of Vector.	Constituents.
The radius vector of a point..................	ρ	$x\ y\ z$
The electromagnetic momentum at a point	$\mathfrak{A}$	$F\ G\ H$
The magnetic induction	$\mathfrak{B}$	$a\ b\ c$
The (total) electric current	$\mathfrak{C}$	$u\ v\ w$
The electric displacement....................	$\mathfrak{D}$	$f\ g\ h$

	Symbol of Vector.	Constituents.
The electromotive force	$\mathfrak{E}$	$P\ Q\ R$
The mechanical force	$\mathfrak{F}$	$X\ Y\ Z$
The velocity of a point........................	$\mathfrak{G}$ or $\dot{\rho}$	$\dot{x}\ \dot{y}\ \dot{z}$
The magnetic force	$\mathfrak{H}$	$\alpha\ \beta\ \gamma$
The intensity of magnetization	$\mathfrak{I}$	$A\ B\ C$
The current of conduction	$\mathfrak{K}$	$p\ q\ r$

We have also the following scalar functions :—

The electric potential Ψ.

The magnetic potential (where it exists) Ω.

The electric density e.

The density of magnetic 'matter' m.

Besides these we have the following quantities, indicating physical properties of the medium at each point :—

C, the conductivity for electric currents.

K, the dielectric inductive capacity.

μ, the magnetic inductive capacity.

These quantities are, in isotropic media, mere scalar functions of ρ, but in general they are linear and vector operators on the vector functions to which they are applied. K and μ are certainly always self-conjugate, and C is probably so also.

619.] The equations (A) of magnetic induction, of which the first is,

$$a = \frac{dH}{dy} - \frac{dG}{dz},$$

may now be written $\qquad \mathfrak{B} = V\nabla\mathfrak{A},$

where ∇ is the operator

$$i\frac{d}{dx} + j\frac{d}{dy} + k\frac{d}{dz},$$

and V indicates that the vector part of the result of this operation is to be taken.

Since $\mathfrak{A}$ is subject to the condition $S\nabla\mathfrak{A} = 0$, $\nabla\mathfrak{A}$ is a pure vector, and the symbol V is unnecessary.

The equations (B) of electromotive force, of which the first is

$$P = c\dot{y} - b\dot{z} - \frac{dF}{dt} - \frac{d\Psi}{dx},$$

become $\qquad \mathfrak{E} = V\mathfrak{G}\mathfrak{B} - \dot{\mathfrak{A}} - \nabla\Psi.$

The equations (C) of mechanical force, of which the first is

$$X = cv - bw - e\frac{d\Psi}{dx} - m\frac{d\Omega}{dx},$$

become $\qquad \mathfrak{F} = V\mathfrak{C}\mathfrak{B} - e\nabla\Psi - m\nabla\Omega.$

The equations (D) of magnetization, of which the first is

$$a = \alpha + 4\pi A,$$

become

$$\mathfrak{B} = \mathfrak{H} + 4\pi\mathfrak{J}.$$

The equations (E) of electric currents, of which the first is

$$4\pi u = \frac{d\gamma}{dy} - \frac{d\beta}{dz},$$

become

$$4\pi\mathfrak{C} = V\nabla\mathfrak{H}.$$

The equation of the current of conduction is, by Ohm's Law,

$$\mathfrak{K} = C\mathfrak{E}.$$

That of electric displacement is

$$\mathfrak{D} = \frac{1}{4\pi}K\mathfrak{E}.$$

The equation of the total current, arising from the variation of the electric displacement as well as from conduction, is

$$\mathfrak{C} = \mathfrak{K} + \dot{\mathfrak{D}}.$$

When the magnetization arises from magnetic induction,

$$\mathfrak{B} = \mu\mathfrak{H}.$$

We have also, to determine the electric volume-density,

$$e = S\nabla\mathfrak{D}.$$

To determine the magnetic volume-density,

$$m = S\nabla\mathfrak{J}.$$

When the magnetic force can be derived from a potential

$$\mathfrak{H} = -\nabla\Omega.$$

CHAPTER X.

DIMENSIONS OF ELECTRIC UNITS.

620.] EVERY electromagnetic quantity may be defined with reference to the fundamental units of Length, Mass, and Time. If we begin with the definition of the unit of electricity, as given in Art. 65, we may obtain definitions of the units of every other electromagnetic quantity, in virtue of the equations into which they enter along with quantities of electricity. The system of units thus obtained is called the Electrostatic System.

If, on the other hand, we begin with the definition of the unit magnetic pole, as given in Art. 374, we obtain a different system of units of the same set of quantities. This system of units is not consistent with the former system, and is called the Electromagnetic System.

We shall begin by stating those relations between the different units which are common to both systems, and we shall then form a table of the dimensions of the units according to each system.

621.] We shall arrange the primary quantities which we have to consider in pairs. In the first three pairs, the product of the two quantities in each pair is a quantity of energy or work. In the second three pairs, the product of each pair is a quantity of energy referred to unit of volume.

FIRST THREE PAIRS.

Electrostatic Pair.

	Symbol.
(1) Quantity of electricity	e
(2) Line-integral of electromotive force, or electric potential	E

Magnetic Pair.

		Symbol.
(3)	Quantity of free magnetism, or strength of a pole .	m
(4)	Magnetic potential	Ω

Electrokinetic Pair.

(5)	Electrokinetic momentum of a circuit	p
(6)	Electric current	C

SECOND THREE PAIRS.

Electrostatic Pair.

(7)	Electric displacement (measured by surface-density) .	$\mathfrak{D}$
(8)	Electromotive force at a point	$\mathfrak{E}$

Magnetic Pair.

(9)	Magnetic induction	$\mathfrak{B}$
(10)	Magnetic force	$\mathfrak{H}$

Electrokinetic Pair.

(11)	Intensity of electric current at a point	$\mathfrak{C}$
(12)	Vector potential of electric currents	$\mathfrak{A}$

622.] The following relations exist between these quantities. In the first place, since the dimensions of energy are $\left[\frac{L^2M}{T^2}\right]$, and those of energy referred to unit of volume $\left[\frac{M}{LT^2}\right]$, we have the following equations of dimensions:

$$[eE] = [m\,\Omega] = [p\,C] = \left[\frac{L^2M}{T^2}\right], \tag{1}$$

$$[\mathfrak{D}\,\mathfrak{E}] = [\mathfrak{B}\,\mathfrak{H}] = [\mathfrak{C}\,\mathfrak{A}] = \left[\frac{M}{LT^2}\right]. \tag{2}$$

Secondly, since e, p and $\mathfrak{A}$ are the time-integrals of C, E, and $\mathfrak{E}$ respectively,

$$\left[\frac{e}{C}\right] = \left[\frac{p}{E}\right] = \left[\frac{\mathfrak{A}}{\mathfrak{E}}\right] = [T]. \tag{3}$$

Thirdly, since E, Ω, and p are the line-integrals of $\mathfrak{E}$, $\mathfrak{H}$, and $\mathfrak{A}$ respectively,

$$\left[\frac{E}{\mathfrak{E}}\right] = \left[\frac{\Omega}{\mathfrak{H}}\right] = \left[\frac{p}{\mathfrak{A}}\right] = [L]. \tag{4}$$

Finally, since e, C, and m are the surface-integrals of $\mathfrak{D}$, $\mathfrak{C}$, and $\mathfrak{B}$ respectively,

$$\left[\frac{e}{\mathfrak{D}}\right] = \left[\frac{C}{\mathfrak{C}}\right] = \left[\frac{m}{\mathfrak{B}}\right] = [L^2]. \tag{5}$$

623.] These fifteen equations are not independent, and in order to deduce the dimensions of the twelve units involved, we require one additional equation. If, however, we take either e or m as an independent unit, we can deduce the dimensions of the rest in terms of either of these.

$$
\begin{array}{lll}
(1) & [e] & = [e] \qquad = \left[\dfrac{L^2 M}{m T}\right]. \\[2ex]
(2) & [E] & = \left[\dfrac{L^2 M}{e T^2}\right] = \left[\dfrac{m}{T}\right]. \\[2ex]
(3) \text{ and } (5) & [p] = [m] & = \left[\dfrac{L^2 M}{e T}\right] = [m]. \\[2ex]
(4) \text{ and } (6) & [C] = [\Omega] & = \left[\dfrac{e}{T}\right] \qquad = \left[\dfrac{L^2 M}{m T^2}\right]. \\[2ex]
(7) & [\mathfrak{D}] & = \left[\dfrac{e}{L^2}\right] \qquad = \left[\dfrac{M}{m T}\right]. \\[2ex]
(8) & [\mathfrak{E}] & = \left[\dfrac{L M}{e T^2}\right] = \left[\dfrac{m}{L T}\right]. \\[2ex]
(9) & [\mathfrak{B}] & = \left[\dfrac{M}{e T}\right] = \left[\dfrac{m}{L^2}\right]. \\[2ex]
(10) & [\mathfrak{H}] & = \left[\dfrac{e}{L T}\right] = \left[\dfrac{L M}{m T^2}\right]. \\[2ex]
(11) & [\mathfrak{C}] & = \left[\dfrac{e}{L^2 T}\right] = \left[\dfrac{M}{m T^2}\right]. \\[2ex]
(12) & [\mathfrak{A}] & = \left[\dfrac{L M}{e T}\right] = \left[\dfrac{m}{L}\right].
\end{array}
$$

624.] The relations of the first ten of these quantities may be exhibited by means of the following arrangement:—

e,	$\mathfrak{D}$,	$\mathfrak{H}$,	C and Ω.	E	$\mathfrak{E}$,	$\mathfrak{B}$,	m and p.
m and p,	$\mathfrak{B}$,	$\mathfrak{E}$,	E.	C and Ω,	$\mathfrak{H}$,	$\mathfrak{D}$,	e.

The quantities in the first line are derived from e by the same operations as the corresponding quantities in the second line are derived from m. It will be seen that the order of the quantities in the first line is exactly the reverse of the order in the second line. The first four of each line have the first symbol in the numerator. The second four in each line have it in the denominator.

All the relations given above are true whatever system of units we adopt.

625.] The only systems of any scientific value are the electrostatic and the electromagnetic system. The electrostatic system is founded on the definition of the unit of electricity, Arts. 41, 42, and may be deduced from the equation

$$\mathfrak{E} = \frac{e}{L^2},$$

which expresses that the resultant force $\mathfrak{E}$ at any point, due to the action of a quantity of electricity e at a distance L, is found by dividing e by L^2. Substituting the equations of dimension (1) and (8), we find

$$\left[\frac{LM}{eT^2}\right] = \left[\frac{e}{L^2}\right], \qquad \left[\frac{m}{LT}\right] = \left[\frac{M}{mT}\right],$$

whence $[e] = [L^{\frac{3}{2}} M^{\frac{1}{2}} T^{-1}], \quad m = [L^{\frac{1}{2}} M^{\frac{1}{2}}],$

in the electrostatic system.

The electromagnetic system is founded on a precisely similar definition of the unit of strength of a magnetic pole, Art. 374, leading to the equation

$$\mathfrak{H} = \frac{m}{L^2}.$$

whence $$\left[\frac{e}{LT}\right] = \left[\frac{M}{eT}\right], \qquad \left[\frac{LM}{mT^2}\right] = \left[\frac{m}{L^2}\right],$$

and $[e] = [L^{\frac{1}{2}} M^{\frac{1}{2}}], \quad [m] = [L^{\frac{3}{2}} M^{\frac{1}{2}} T^{-1}],$

in the electromagnetic system. From these results we find the dimensions of the other quantities.

626.] *Table of Dimensions.*

	Symbol	Dimensions in Electrostatic System	Dimensions in Electromagnetic System
Quantity of electricity	e	$[L^{\frac{3}{2}} M^{\frac{1}{2}} T^{-1}]$	$[L^{\frac{1}{2}} M^{\frac{1}{2}}]$.
Line-integral of electromotive force	E	$[L^{\frac{1}{2}} M^{\frac{1}{2}} T^{-1}]$	$[L^{\frac{3}{2}} M^{\frac{1}{2}} T^{-2}]$.
Quantity of magnetism; Electrokinetic momentum of a circuit	m; p	$[L^{\frac{1}{2}} M^{\frac{1}{2}}]$	$[L^{\frac{3}{2}} M^{\frac{1}{2}} T^{-1}]$.
Electric current; Magnetic potential	C; Ω	$[L^{\frac{3}{2}} M^{\frac{1}{2}} T^{-2}]$	$[L^{\frac{1}{2}} M^{\frac{1}{2}} T^{-1}]$.
Electric displacement; Surface-density	$\mathfrak{D}$	$[L^{-\frac{1}{2}} M^{\frac{1}{2}} T^{-1}]$	$[L^{-\frac{3}{2}} M^{\frac{1}{2}}]$.
Electromotive force at a point	$\mathfrak{E}$	$[L^{-\frac{1}{2}} M^{\frac{1}{2}} T^{-1}]$	$[L^{\frac{1}{2}} M^{\frac{1}{2}} T^{-2}]$.
Magnetic induction	$\mathfrak{B}$	$[L^{-\frac{3}{2}} M^{\frac{1}{2}}]$	$[L^{-\frac{1}{2}} M^{\frac{1}{2}} T^{-1}]$.
Magnetic force	$\mathfrak{H}$	$[L^{\frac{1}{2}} M^{\frac{1}{2}} T^{-2}]$	$[L^{-\frac{1}{2}} M^{\frac{1}{2}} T^{-1}]$.
Strength of current at a point	$\mathfrak{C}$	$[L^{-\frac{1}{2}} M^{\frac{1}{2}} T^{-2}]$	$[L^{-\frac{3}{2}} M^{\frac{1}{2}} T^{-1}]$.
Vector potential	$\mathfrak{A}$	$[L^{-\frac{1}{2}} M^{\frac{1}{2}}]$	$[L^{\frac{1}{2}} M^{\frac{1}{2}} T^{-1}]$.

627.] We have already considered the products of the pairs of these quantities in the order in which they stand. Their ratios are in certain cases of scientific importance. Thus

	Symbol.	Electrostatic System.	Electromagnetic System.
$\frac{e}{E}$ = capacity of an accumulator . .	q	$[L]$	$\left[\frac{T^2}{L}\right]$.
$\frac{p}{C}$ = coefficient of self-induction of a circuit, or electro-magnetic capacity	L	$\left[\frac{T^2}{L}\right]$	$[L]$.
$\frac{\mathfrak{D}}{\mathfrak{E}}$ = specific inductive capacity of dielectric.	K	$[0]$	$\left[\frac{T^2}{L^2}\right]$.
$\frac{\mathfrak{B}}{\mathfrak{H}}$ = magnetic inductive capacity . .	μ	$\left[\frac{T^2}{L^2}\right]$	$[0]$.
$\frac{E}{C}$ = resistance of a conductor	R	$\left[\frac{T}{L}\right]$	$\left[\frac{L}{T}\right]$.
$\frac{\mathfrak{E}}{\mathfrak{C}}$ = specific resistance of a substance . . .	r	$[T]$	$\left[\frac{L^2}{T}\right]$.

628.] If the units of length, mass, and time are the same in the two systems, the number of electrostatic units of electricity contained in one electromagnetic unit is numerically equal to a certain velocity, the absolute value of which does not depend on the magnitude of the fundamental units employed. This velocity is an important physical quantity, which we shall denote by the symbol v.

Number of Electrostatic Units in one Electromagnetic Unit.

For e, C, Ω, $\mathfrak{D}$, $\mathfrak{H}$, $\mathfrak{C}$, ... v.

For m, p, E, $\mathfrak{B}$, $\mathfrak{E}$, $\mathfrak{A}$, ... $\frac{1}{v}$.

For electrostatic capacity, dielectric inductive capacity, and conductivity, v^2.

For electromagnetic capacity, magnetic inductive capacity, and resistance, $\frac{1}{v^2}$.

Several methods of determining the velocity v will be given in Arts. 768–780.

In the electrostatic system the specific dielectric inductive capacity of air is assumed equal to unity. This quantity is therefore represented by $\frac{1}{v^2}$ in the electromagnetic system.

In the electromagnetic system the specific magnetic inductive capacity of air is assumed equal to unity. This quantity is therefore represented by $\frac{1}{v^2}$ in the electrostatic system.

Practical System of Electric Units.

629.] Of the two systems of units, the electromagnetic is of the greater use to those practical electricians who are occupied with electromagnetic telegraphs. If, however, the units of length, time, and mass are those commonly used in other scientific work, such as the mètre or the centimètre, the second, and the gramme, the units of resistance and of electromotive force will be so small that to express the quantities occurring in practice enormous numbers must be used, and the units of quantity and capacity will be so large that only exceedingly small fractions of them can ever occur in practice. Practical electricians have therefore adopted a set of electrical units deduced by the electromagnetic system from a large unit of length and a small unit of mass.

The unit of length used for this purpose is ten million of mètres, or approximately the length of a quadrant of a meridian of the earth.

The unit of time is, as before, one second.

The unit of mass is 10^{-11} gramme, or one hundred millionth part of a milligramme.

The electrical units derived from these fundamental units have been named after eminent electrical discoverers. Thus the practical unit of resistance is called the Ohm, and is represented by the resistance-coil issued by the British Association, and described in Art. 340. It is expressed in the electromagnetic system by a velocity of 10,000,000 metres per second.

The practical unit of electromotive force is called the Volt, and is not very different from that of a Daniell's cell. Mr. Latimer Clark has recently invented a very constant cell, whose electromotive force is almost exactly 1.457 Volts.

The practical unit of capacity is called the Farad. The quantity of electricity which flows through one Ohm under the electromotive force of one Volt during one second, is equal to the charge produced in a condenser whose capacity is one Farad by an electromotive force of one Volt.

The use of these names is found to be more convenient in practice than the constant repetition of the words 'electromagnetic units,' with the additional statement of the particular fundamental units on which they are founded.

When very large quantities are to be measured, a large unit is formed by multiplying the original unit by one million, and placing before its name the prefix *mega*.

In like manner by prefixing *micro* a small unit is formed, one millionth of the original unit.

The following table gives the values of these practical units in the different systems which have been at various times adopted.

Fundamental Units.	Practical System.	B. A. Report, 1863.	Thomson.	Weber.
Length, *Time,* *Mass.*	*Earth's Quadrant,* *Second,* 10^{-11} *Gramme.*	*Metre,* *Second,* *Gramme.*	*Centimetre,* *Second,* *Gramme.*	*Millimetre,* *Second,* *Milligramme.*
Resistance	Ohm	10^7	10^9	10^1
Electromotive force	Volt	10^5	10^8	10^{11}
Capacity	Farad	10^{-7}	10^{-9}	10^{-10}
Quantity	Farad (charged to a Volt.)	10^{-2}	10^{-1}	10

CHAPTER XI.

ON ENERGY AND STRESS IN THE ELECTROMAGNETIC FIELD.

Electrostatic Energy.

630.] THE energy of the system may be divided into the Potential Energy and the Kinetic Energy.

The potential energy due to electrification has been already considered in Art. 85. It may be written

$$W = \tfrac{1}{2}\Sigma(e\Psi), \tag{1}$$

where e is the charge of electricity at a place where the electric potential is Ψ, and the summation is to be extended to every place where there is electrification.

If f, g, h are the components of the electric displacement, the quantity of electricity in the element of volume $dx\,dy\,dz$ is

$$e = \left(\frac{df}{dx} + \frac{dg}{dy} + \frac{dh}{dz}\right) dx\,dy\,dz, \tag{2}$$

and

$$W = \tfrac{1}{2}\iiint\left(\frac{df}{dx} + \frac{dg}{dy} + \frac{dh}{dz}\right)\Psi\, dx\,dy\,dz, \tag{3}$$

where the integration is to be extended throughout all space.

631.] Integrating this expression by parts, and remembering that when the distance, r, from a given point of a finite electrified system becomes infinite, the potential Ψ becomes an infinitely small quantity of the order r^{-1}, and that f, g, h become infinitely small quantities of the order r^{-2}, the expression is reduced to

$$W = -\tfrac{1}{2}\iiint\left(f\frac{d\Psi}{dx} + g\frac{d\Psi}{dy} + h\frac{d\Psi}{dz}\right) dx\,dy\,dz, \tag{4}$$

where the integration is to be extended throughout all space.

If we now write P, Q, R for the components of the electromotive force, instead of $-\frac{d\Psi}{dx}$, $-\frac{d\Psi}{dy}$, and $-\frac{d\Psi}{dz}$, we find

$$W = \tfrac{1}{2}\iiint(Pf + Qg + Rh)\, dx\,dy\,dz. \tag{5}$$

Hence, the electrostatic energy of the whole field will be the same if we suppose that it resides in every part of the field where electrical force and electrical displacement occur, instead of being confined to the places where free electricity is found.

The energy in unit of volume is half the product of the electromotive force and the electric displacement, multiplied by the cosine of the angle which these vectors include.

In Quaternion language it is $-\frac{1}{2} S \mathfrak{E} \mathfrak{D}$.

Magnetic Energy.

632.] We may treat the energy due to magnetization in a similar way. If A, B, C are the components of magnetization and a, β, γ the components of magnetic force, the potential energy of the system of magnets is, by Art. 389,

$$-\frac{1}{2}\iiint(A a+B\beta+C\gamma)\,dx\,dy\,dz, \qquad (6)$$

the integration being extended over the space occupied by magnetized matter. This part of the energy, however, will be included in the kinetic energy in the form in which we shall presently obtain it.

633.] We may transform this expression when there are no electric currents by the following method.

We know that
$$\frac{da}{dx}+\frac{db}{dy}+\frac{dc}{dz}=0. \qquad (7)$$

Hence, by Art. 97, if

$$a=-\frac{d\Omega}{dx}, \quad \beta=-\frac{d\Omega}{dy}, \quad \gamma=-\frac{d\Omega}{dz}, \qquad (8)$$

as is always the case in magnetic phenomena where there are no currents,
$$\iiint(a\,a+b\,\beta)+c\,\gamma)\,dx\,dy\,dz=0, \qquad (9)$$

the integral being extended throughout all space, or

$$\iiint\{(a+4\pi A)a+(\beta+4\pi B)\beta+(\gamma+4\pi C)\gamma\}\,dx\,dy\,dz=0. \qquad (10)$$

Hence, the energy due to a magnetic system

$$-\frac{1}{2}\iiint(A a+B\beta+C\gamma)\,dx\,dy\,dz=\frac{1}{8\pi}\iiint(a^2+\beta^2+\gamma^2)\,dx\,dy\,dz,$$

$$=-\frac{1}{8\pi}\iiint\mathfrak{H}^2\,dx\,dy\,dz. \qquad (11)$$

Electrokinetic Energy.

634.] We have already, in Art. 578, expressed the kinetic energy of a system of currents in the form

$$T = \tfrac{1}{2}\Sigma(p\,i), \tag{12}$$

where p is the electromagnetic momentum of a circuit, and i is the strength of the current flowing round it, and the summation extends to all the circuits.

But we have proved, in Art. 590, that p may be expressed as a line-integral of the form

$$p = \int\left(F\frac{dx}{ds} + G\frac{dy}{ds} + H\frac{dz}{ds}\right)ds, \tag{13}$$

where F, G, H are the components of the electromagnetic momentum, $\mathfrak{A}$, at the point $(x\,y\,z)$, and the integration is to be extended round the closed circuit s. We therefore find

$$T = \tfrac{1}{2}\Sigma i\int\left(F\frac{dx}{ds} + G\frac{dy}{ds} + H\frac{dz}{ds}\right)ds. \tag{14}$$

If u, v, w are the components of the density of the current at any point of the conducting circuit, and if S is the transverse section of the circuit, then we may write

$$i\frac{dx}{ds} = u\,S, \quad i\frac{dy}{ds} = v\,S, \quad i\frac{dz}{ds} = w\,S, \tag{15}$$

and we may also write the volume

$$S\,ds = dx\,dy\,dz,$$

and we now find

$$T = \tfrac{1}{2}\iiint(Fu + Gv + Hw)\,dx\,dy\,dz, \tag{16}$$

where the integration is to be extended to every part of space where there are electric currents.

635.] Let us now substitute for u, v, w their values as given by the equations of electric currents (E), Art. 607, in terms of the components α, β, γ of the magnetic force. We then have

$$T = \frac{1}{8\pi}\iiint\left\{F\left(\frac{d\gamma}{dy} - \frac{d\beta}{dz}\right) + G\left(\frac{d\alpha}{dz} - \frac{d\gamma}{dx}\right) + H\left(\frac{d\beta}{dx} - \frac{d\alpha}{dy}\right)\right\}dx\,dy\,dz, \tag{17}$$

where the integration is extended over a portion of space including all the currents.

If we integrate this by parts, and remember that, at a great distance r from the system, α, β, and γ are of the order of magnitude r^{-3}, we find that when the integration is extended throughout all space, the expression is reduced to

$$T = \frac{1}{8\pi}\iiint\left\{\alpha\left(\frac{dH}{dy} - \frac{dG}{dz}\right) + \beta\left(\frac{dF}{dz} - \frac{dH}{dx}\right) + \gamma\left(\frac{dG}{dx} - \frac{dF}{dy}\right)\right\}dx\,dy\,dz. \tag{18}$$

By the equations (A), Art. 591, of magnetic induction, we may substitute for the quantities in small brackets the components of magnetic induction a, b, c, so that the kinetic energy may be written

$$T = \frac{1}{8\pi}\iiint (a\alpha + b\beta + c\gamma)\,dx\,dy\,dz, \tag{19}$$

where the integration is to be extended throughout every part of space in which the magnetic force and magnetic induction have values differing from zero.

The quantity within brackets in this expression is the product of the magnetic induction into the resolved part of the magnetic force in its own direction.

In the language of quaternions this may be written more simply,

$$-S.\mathfrak{B}\mathfrak{H}.$$

where $\mathfrak{B}$ is the magnetic induction, whose components are a, b, c, and $\mathfrak{H}$ is the magnetic force, whose components are α, β, γ.

636.] The electrokinetic energy of the system may therefore be expressed either as an integral to be taken where there are electric currents, or as an integral to be taken over every part of the field in which magnetic force exists. The first integral, however, is the natural expression of the theory which supposes the currents to act upon each other directly at a distance, while the second is appropriate to the theory which endeavours to explain the action between the currents by means of some intermediate action in the space between them. As in this treatise we have adopted the latter method of investigation, we naturally adopt the second expression as giving the most significant form to the kinetic energy.

According to our hypothesis, we assume the kinetic energy to exist wherever there is magnetic force, that is, in general, in every part of the field. The amount of this energy per unit of volume is $-\frac{1}{8\pi}S\mathfrak{B}\mathfrak{H}$, and this energy exists in the form of some kind of motion of the matter in every portion of space.

When we come to consider Faraday's discovery of the effect of magnetism on polarized light, we shall point out reasons for believing that wherever there are lines of magnetic force, there is a rotatory motion of matter round those lines. See Art. 821.

Magnetic and Electrokinetic Energy compared.

637.] We found in Art. 423 that the mutual potential energy

of two magnetic shells, of strengths ϕ and ϕ', and bounded by the closed curves s and s' respectively, is

$$-\phi\phi'\iint\frac{\cos\epsilon}{r}\,ds\,ds',$$

where ϵ is the angle between the directions of ds and ds', and r is the distance between them.

We also found in Art. 521 that the mutual energy of two circuits s and s', in which currents i and i' flow, is

$$i\,i'\iint\frac{\cos\epsilon}{r}\,ds\,ds'.$$

If i, i' are equal to ϕ, ϕ' respectively, the mechanical action between the magnetic shells is equal to that between the corresponding electric circuits, and in the same direction. In the case of the magnetic shells, the force tends to diminish their mutual potential energy, in the case of the circuits it tends to increase their mutual energy, because this energy is kinetic.

It is impossible, by any arrangement of magnetized matter, to produce a system corresponding in all respects to an electric circuit, for the potential of the magnetic system is single valued at every point of space, whereas that of the electric system is many-valued.

But it is always possible, by a proper arrangement of infinitely small electric circuits, to produce a system corresponding in all respects to any magnetic system, provided the line of integration which we follow in calculating the potential is prevented from passing through any of these small circuits. This will be more fully explained in Art. 833.

The action of magnets at a distance is perfectly identical with that of electric currents. We therefore endeavour to trace both to the same cause, and since we cannot explain electric currents by means of magnets, we must adopt the other alternative, and explain magnets by means of molecular electric currents.

638.] In our investigation of magnetic phenomena, in Part III of this treatise, we made no attempt to account for magnetic action at a distance, but treated this action as a fundamental fact of experience. We therefore assumed that the energy of a magnetic system is potential energy, and that this energy is *diminished* when the parts of the system yield to the magnetic forces which act on them.

If, however, we regard magnets as deriving their properties from electric currents circulating within their molecules, their energy is kinetic, and the force between them is such that it tends to move them in a direction such that if the strengths of the currents were maintained constant the kinetic energy would *increase*.

This mode of explaining magnetism requires us also to abandon the method followed in Part III, in which we regarded the magnet as a continuous and homogeneous body, the minutest part of which has magnetic properties of the same kind as the whole.

We must now regard a magnet as containing a finite, though very great, number of electric circuits, so that it has essentially a molecular, as distinguished from a continuous structure.

If we suppose our mathematical machinery to be so coarse that our line of integration cannot thread a molecular circuit, and that an immense number of magnetic molecules are contained in our element of volume, we shall still arrive at results similar to those of Part III, but if we suppose our machinery of a finer order, and capable of investigating all that goes on in the interior of the molecules, we must give up the old theory of magnetism, and adopt that of Ampère, which admits of no magnets except those which consist of electric currents.

We must also regard both magnetic and electromagnetic energy as kinetic energy, and we must attribute to it the proper sign, as given in Art. 635.

In what follows, though we may occasionally, as in Art. 639, &c., attempt to carry out the old theory of magnetism, we shall find that we obtain a perfectly consistent system only when we abandon that theory and adopt Ampère's theory of molecular currents, as in Art. 644.

The energy of the field therefore consists of two parts only, the electrostatic or potential energy

$$W = \tfrac{1}{2}\iiint (Pf + Qg + Rh)\,dx\,dy\,dz,$$

and the electromagnetic or kinetic energy

$$T = \frac{1}{8\pi}\iiint (a\alpha + b\beta + c\gamma)\,dx\,dy\,dz.$$

ON THE FORCES WHICH ACT ON AN ELEMENT OF A BODY PLACED IN THE ELECTROMAGNETIC FIELD.

Forces acting on a Magnetic Element.

639.] The potential energy of the element $dx\,dy\,dz$ of a body magnetized with an intensity whose components are A, B, C, and placed in a field of magnetic force whose components are α, β, γ, is

$$-(A\alpha+B\beta+C\gamma)\,dx\,dy\,dz.$$

Hence, if the force urging the element to move without rotation in the direction of x is $X_1\,dx\,dy\,dz$,

$$X_1 = A\frac{d\alpha}{dx}+B\frac{d\beta}{dx}+C\frac{d\gamma}{dx}, \tag{1}$$

and if the moment of the couple tending to turn the element about the axis of x from y towards z is $L\,dx\,dy\,dz$,

$$L = B\gamma-C\beta. \tag{2}$$

The forces and the moments corresponding to the axes of y and z may be written down by making the proper substitutions.

640.] If the magnetized body carries an electric current, of which the components are u, v, w, then, by equations *C*, Art. 603, there will be an additional electromagnetic force whose components are X_2, Y_2, Z_2, of which X_2 is

$$X_2 = vc-wb. \tag{3}$$

Hence, the total force, X, arising from the magnetism of the molecule, as well as the current passing through it, is

$$X = A\frac{d\alpha}{dx}+B\frac{d\beta}{dx}+C\frac{d\gamma}{dx}+vc-wb. \tag{4}$$

The quantities a, b, c are the components of magnetic induction, and are related to α, β, γ, the components of magnetic force, by the equations given in Art. 400,

$$\left.\begin{aligned} a &= \alpha+4\pi A,\\ b &= \beta+4\pi B,\\ c &= \gamma+4\pi C.\end{aligned}\right\} \tag{5}$$

The components of the current, u, v, w, can be expressed in terms of α, β, γ by the equations of Art. 607,

$$\left.\begin{aligned} 4\pi u &= \frac{d\gamma}{dy}-\frac{d\beta}{dz},\\ 4\pi v &= \frac{d\alpha}{dz}-\frac{d\gamma}{dx},\\ 4\pi w &= \frac{d\beta}{dx}-\frac{d\alpha}{dy}.\end{aligned}\right\} \tag{6}$$

Hence

$$X = \frac{1}{4\pi}\left\{(a-\alpha)\frac{d\alpha}{dx}+(b-\beta)\frac{d\beta}{dx}+(c-\gamma)\frac{d\gamma}{dx}+b\left(\frac{d\alpha}{dy}-\frac{d\beta}{dx}\right)+c\left(\frac{d\alpha}{dz}-\frac{d\gamma}{dx}\right)\right\},$$

$$= \frac{1}{4\pi}\left\{a\frac{d\alpha}{dx}+b\frac{d\alpha}{dy}+c\frac{d\alpha}{dz}-\frac{1}{2}\frac{d}{dx}(\alpha^2+\beta^2+\gamma^2)\right\}. \tag{7}$$

By Art. 403, $$\frac{da}{dx}+\frac{db}{dy}+\frac{dc}{dz}=0. \tag{8}$$

Multiplying this equation, (8), by a, and dividing by 4π, we may add the result to (7), and we find

$$X=\frac{1}{4\pi}\left\{\frac{d}{dx}\left[a\alpha-\tfrac{1}{2}(a^2+\beta^2+\gamma^2)\right]+\frac{d}{dy}[b\alpha]+\frac{d}{dz}[c\alpha]\right\}, \tag{9}$$

also, by (2), $$L=\frac{1}{4\pi}\left((b-\beta)\gamma-(c-\gamma)\beta\right), \tag{10}$$

$$=\frac{1}{4\pi}(b\gamma-c\beta), \tag{11}$$

where X is the force referred to unit of volume in the direction of x, and L is the moment of the forces about this axis.

On the Explanation of these Forces by the Hypothesis of a Medium in a State of Stress.

641.] Let us denote a stress of any kind referred to unit of area by a symbol of the form P_{hk}, where the first suffix, $_h$, indicates that the normal to the surface on which the stress is supposed to act is parallel to the axis of h, and the second suffix, $_k$, indicates that the direction of the stress with which the part of the body on the positive side of the surface acts on the part on the negative side is parallel to the axis of k.

The directions of h and k may be the same, in which case the stress is a normal stress. They may be oblique to each other, in which case the stress is an oblique stress, or they may be perpendicular to each other, in which case the stress is a tangential stress.

The condition that the stresses shall not produce any tendency to rotation in the elementary portions of the body is

$$P_{hk}=P_{kh}.$$

In the case of a magnetized body, however, there is such a tendency to rotation, and therefore this condition, which holds in the ordinary theory of stress, is not fulfilled.

Let us consider the effect of the stresses on the six sides of the elementary portion of the body $dx\,dy\,dz$, taking the origin of coordinates at its centre of gravity.

On the positive face $dy\,dz$, for which the value of x is $\frac{1}{2}dx$, the forces are—

$$\left.\begin{array}{ll}\text{Parallel to } x, & \left(P_{xx}+\frac{1}{2}\frac{dP_{xx}}{dx}dx\right)dy\,dz = X_{+x},\\ \text{Parallel to } y, & \left(P_{xy}+\frac{1}{2}\frac{dP_{xy}}{dx}dx\right)dy\,dz = Y_{+x},\\ \text{Parallel to } z, & \left(P_{xz}+\frac{1}{2}\frac{dP_{xz}}{dx}dx\right)dy\,dz = Z_{+x}.\end{array}\right\} \qquad (12)$$

The forces acting on the opposite side, $-X_{-x}$, $-Y_{-x}$, and $-Z_{-x}$, may be found from these by changing the sign of dx. We may express in the same way the systems of three forces acting on each of the other faces of the element, the direction of the force being indicated by the capital letter, and the face on which it acts by the suffix.

If $X\,dx\,dy\,dz$ is the whole force parallel to x acting on the element,

$$X\,dx\,dy\,dz = X_{+x}+X_{+y}+X_{+z}+X_{-x}+X_{-y}+X_{-z},$$

$$= \left(\frac{dP_{xx}}{dx}+\frac{dP_{yx}}{dx}+\frac{dP_{zx}}{dx}\right)dx\,dy\,dz,$$

whence
$$X = \frac{d}{dx}P_{xx}+\frac{d}{dy}P_{yx}+\frac{d}{dz}P_{zx}. \qquad (13)$$

If $L\,dx\,dy\,dz$ is the moment of the forces about the axis of x tending to turn the element from y to z,

$$L\,dx\,dy\,dz = \tfrac{1}{2}dy\,(Z_{+y}-Z_{-y})-\tfrac{1}{2}dz\,(Y_{+z}-Y_{-z}),$$

$$= (P_{yz}-P_{zy})\,dx\,dy\,dz,$$

whence
$$L = P_{yz}-P_{zy}. \qquad (14)$$

Comparing the values of X and L given by equations (9) and (11) with those given by (13) and (14), we find that, if we make

$$\left.\begin{array}{ll}P_{xx} = \frac{1}{4\pi}\left(a\alpha-\frac{1}{2}(\alpha^2+\beta^2+\gamma^2)\right), & \\ P_{yy} = \frac{1}{4\pi}\left(b\beta-\frac{1}{2}(\alpha^2+\beta^2+\gamma^2)\right), & \\ P_{zz} = \frac{1}{4\pi}\left(c\gamma-\frac{1}{2}(\alpha^2+\beta^2+\gamma^2)\right), & \\ P_{yz} = \frac{1}{4\pi}b\gamma, & P_{zy} = \frac{1}{4\pi}c\beta, \\ P_{zx} = \frac{1}{4\pi}c\alpha, & P_{xz} = \frac{1}{4\pi}a\gamma, \\ P_{xy} = \frac{1}{4\pi}a\beta, & P_{yx} = \frac{1}{4\pi}b\alpha,\end{array}\right\} \qquad (15)$$

the force arising from a system of stress of which these are the components will be statically equivalent, in its effects on each element of the body, with the forces arising from the magnetization and electric currents.

642.] The nature of the stress of which these are the components may be easily found, by making the axis of x bisect the angle between the directions of the magnetic force and the magnetic induction, and taking the axis of y in the plane of these directions, and measured towards the side of the magnetic force.

If we put $\mathfrak{H}$ for the numerical value of the magnetic force, $\mathfrak{B}$ for that of the magnetic induction, and 2ϵ for the angle between their directions,

$$\left.\begin{array}{lll} \alpha = \mathfrak{H}\cos\epsilon, & \beta = \mathfrak{H}\sin\epsilon, & \gamma = 0, \\ a = \mathfrak{B}\cos\epsilon, & b = -\mathfrak{B}\sin\epsilon, & c = 0; \end{array}\right\} \tag{16}$$

$$\left.\begin{aligned} P_{xx} &= \frac{1}{4\pi}(\mathfrak{B}\mathfrak{H}\cos^2\epsilon - \tfrac{1}{2}\mathfrak{H}^2), \\ P_{yy} &= \frac{1}{4\pi}(-\mathfrak{B}\mathfrak{H}\sin^2\epsilon - \tfrac{1}{2}\mathfrak{H}^2), \\ P_{zz} &= \frac{1}{4\pi}(-\tfrac{1}{2}\mathfrak{H}^2), \\ P_{yz} &= P_{zx} = P_{zy} = P_{xz} = 0, \\ P_{xy} &= \frac{1}{4\pi}\mathfrak{B}\mathfrak{H}\cos\epsilon\sin\epsilon, \\ P_{yx} &= -\frac{1}{4\pi}\mathfrak{B}\mathfrak{H}\cos\epsilon\sin\epsilon. \end{aligned}\right\} \tag{17}$$

Hence, the state of stress may be considered as compounded of—

(1) A pressure equal in all directions $= \dfrac{1}{8\pi}\mathfrak{H}^2$.

(2) A tension along the line bisecting the angle between the directions of the magnetic force and the magnetic induction

$$= \frac{1}{4\pi}\mathfrak{B}\mathfrak{H}\cos^2\epsilon.$$

(3) A pressure along the line bisecting the exterior angle between these directions $= \dfrac{1}{4\pi}\mathfrak{B}\mathfrak{H}\sin^2\epsilon$.

(4) A couple tending to turn every element of the substance in the plane of the two directions *from* the direction of magnetic induction *to* the direction of magnetic force $= \dfrac{1}{4\pi}\mathfrak{B}\mathfrak{H}\sin 2\epsilon$.

When the magnetic induction is in the same direction as the magnetic force, as it always is in fluids and non-magnetized solids, then $\epsilon = 0$, and making the axis of x coincide with the direction of the magnetic force,

$$P_{xx} = \frac{1}{4\pi}(\mathfrak{B}\mathfrak{H} - \tfrac{1}{2}\mathfrak{H}^2), \qquad P_{yy} = P_{zz} = -\frac{1}{8\pi}\mathfrak{H}^2, \tag{18}$$

and the tangential stresses disappear.

The stress in this case is therefore a hydrostatic pressure $\frac{1}{8\pi}\mathfrak{H}^2$, combined with a longitudinal tension $\frac{1}{4\pi}\mathfrak{B}\mathfrak{H}$ along the lines of force.

643.] When there is no magnetization, $\mathfrak{B} = \mathfrak{H}$, and the stress is still further simplified, being a tension along the lines of force equal to $\frac{1}{8\pi}\mathfrak{H}^2$, combined with a pressure in all directions at right angles to the lines of force, numerically equal also to $\frac{1}{8\pi}\mathfrak{H}^2$. The components of stress in this important case are

$$\left.\begin{aligned}
P_{xx} &= \frac{1}{8\pi}(\alpha^2 - \beta^2 - \gamma^2),\\
P_{yy} &= \frac{1}{8\pi}(\beta^2 - \gamma^2 - \alpha^2),\\
P_{zz} &= \frac{1}{8\pi}(\gamma^2 - \alpha^2 - \beta^2),\\
P_{yz} = P_{zy} &= \frac{1}{4\pi}\beta\gamma,\\
P_{zx} = P_{xz} &= \frac{1}{4\pi}\gamma\alpha,\\
P_{xy} = P_{yx} &= \frac{1}{4\pi}\alpha\beta.
\end{aligned}\right\} \tag{19}$$

The force arising from these stresses on an element of the medium referred to unit of volume is

$$\begin{aligned}
X &= \frac{d}{dx}p_{xx} + \frac{d}{dy}p_{yx} + \frac{d}{dz}p_{zx},\\
&= \frac{1}{4\pi}\left\{\alpha\frac{d\alpha}{dx} - \beta\frac{d\beta}{dx} - \gamma\frac{d\gamma}{dx}\right\} + \frac{1}{4\pi}\left\{\alpha\frac{d\beta}{dy} + \beta\frac{d\alpha}{dy}\right\} + \frac{1}{4\pi}\left\{\alpha\frac{d\gamma}{dz} - \gamma\frac{d\alpha}{dz}\right\},\\
&= \frac{1}{4\pi}\alpha\left(\frac{d\alpha}{dx} + \frac{d\beta}{dy} + \frac{d\gamma}{dz}\right) + \frac{1}{4\pi}\gamma\left(\frac{d\alpha}{dz} - \frac{d\gamma}{dx}\right) - \frac{1}{4\pi}\beta\left(\frac{d\beta}{dx} - \frac{d\alpha}{dy}\right).
\end{aligned}$$

Now
$$\begin{aligned}
\frac{d\alpha}{dx} + \frac{d\beta}{dy} + \frac{d\gamma}{dz} &= 4\pi m,\\
\frac{d\alpha}{dz} - \frac{d\gamma}{dx} &= 4\pi v,\\
\frac{d\beta}{dx} - \frac{d\alpha}{dy} &= 4\pi w,
\end{aligned}$$

where m is the density of austral magnetic matter referred to unit of volume, and v and w are the components of electric currents referred to unit of area perpendicular to y and z respectively. Hence,

Similarly
$$\left.\begin{aligned} X &= am + v\gamma - w\beta. \\ Y &= \beta m + w a - u\gamma, \\ Z &= \gamma m + u\beta - v a. \end{aligned}\right\} \quad \text{(Equations of Electromagnetic Force.)} \qquad (20)$$

644.] If we adopt the theories of Ampère and Weber as to the nature of magnetic and diamagnetic bodies, and assume that magnetic and diamagnetic polarity are due to molecular electric currents, we get rid of imaginary magnetic matter, and find that everywhere $m = 0$, and
$$\frac{da}{dx} + \frac{d\beta}{dy} + \frac{d\gamma}{dz} = 0, \qquad (21)$$
so that the equations of electromagnetic force become,
$$\left.\begin{aligned} X &= v\gamma - w\beta, \\ Y &= w a - u\gamma, \\ Z &= u\beta - v a. \end{aligned}\right\} \qquad (22)$$

These are the components of the mechanical force referred to unit of volume of the substance. The components of the magnetic force are a, β, γ, and those of the electric current are u, v, w. These equations are identical with those already established. (Equations (C), Art. 603.)

645.] In explaining the electromagnetic force by means of a state of stress in a medium, we are only following out the conception of Faraday*, that the lines of magnetic force tend to shorten themselves, and that they repel each other when placed side by side. All that we have done is to express the value of the tension along the lines, and the pressure at right angles to them, in mathematical language, and to prove that the state of stress thus assumed to exist in the medium will actually produce the observed forces on the conductors which carry electric currents.

We have asserted nothing as yet with respect to the mode in which this state of stress is originated and maintained in the medium. We have merely shewn that it is possible to conceive the mutual action of electric currents to depend on a particular kind of stress in the surrounding medium, instead of being a direct and immediate action at a distance.

Any further explanation of the state of stress, by means of the motion of the medium or otherwise, must be regarded as a separate and independent part of the theory, which may stand or fall without affecting our present position. See Art. 832.

* *Exp. Res.*, 3266, 3267, 3268.

In the first part of this treatise, Art. 108, we shewed that the observed electrostatic forces may be conceived as operating through the intervention of a state of stress in the surrounding medium. We have now done the same for the electromagnetic forces, and it remains to be seen whether the conception of a medium capable of supporting these states of stress is consistent with other known phenomena, or whether we shall have to put it aside as unfruitful.

In a field in which electrostatic as well as electromagnetic action is taking place, we must suppose the electrostatic stress described in Part I to be superposed on the electromagnetic stress which we have been considering.

646.] If we suppose the total terrestrial magnetic force to be 10 British units (grain, foot, second), as it is nearly in Britain, then the tension perpendicular to the lines of force is 0.128 grains weight per square foot. The greatest magnetic tension produced by Joule* by means of electromagnets was about 140 pounds weight on the square inch.

* Sturgeon's *Annals of Electricity*, vol. v. p. 187 (1840); or *Philosophical Magazine*, Dec., 1851.

CHAPTER XII.

CURRENT-SHEETS.

647.] A CURRENT-SHEET is an infinitely thin stratum of conducting matter, bounded on both sides by insulating media, so that electric currents may flow in the sheet, but cannot escape from it except at certain points called Electrodes, where currents are made to enter or to leave the sheet.

In order to conduct a finite electric current, a real sheet must have a finite thickness, and ought therefore to be considered a conductor of three dimensions. In many cases, however, it is practically convenient to deduce the electric properties of a real conducting sheet, or of a thin layer of coiled wire, from those of a current-sheet as defined above.

We may therefore regard a surface of any form as a current-sheet. Having selected one side of this surface as the positive side, we shall always suppose any lines drawn on the surface to be looked at from the positive side of the surface. In the case of a closed surface we shall consider the outside as positive. See Art. 294, where, however, the direction of the current is defined as seen from the *negative* side of the sheet.

The Current-function.

648.] Let a fixed point A on the surface be chosen as origin, and let a line be drawn on the surface from A to another point P. Let the quantity of electricity which in unit of time crosses this line from left to right be ϕ, then ϕ is called the Current-function at the point P.

The current-function depends only on the position of the point P, and is the same for any two forms of the line AP, provided this line can be transformed by continuous motion from one form to the other without passing through an electrode. For the two forms of the line will enclose an area within which there is no electrode, and therefore the same quantity of electricity which enters the area across one of the lines must issue across the other.

If s denote the length of the line AP, the current across ds from left to right will be $\frac{d\phi}{ds}ds$.

If ϕ is constant for any curve, there is no current across it. Such a curve is called a Current-line or a Stream-line.

649.] Let ψ be the electric potential at any point of the sheet, then the electromotive force along any element ds of a curve will be

$$-\frac{d\psi}{ds}ds,$$

provided no electromotive force exists except that which arises from differences of potential.

If ψ is constant for any curve, the curve is called an Equipotential Line.

650.] We may now suppose that the position of a point on the sheet is defined by the values of ϕ and ψ at that point. Let ds_1 be the length of the element of the equipotential line ψ intercepted between the two current lines ϕ and $\phi+d\phi$, and let ds_2 be the length of the element of the current line ϕ intercepted between the two equipotential lines ψ and $\psi+d\psi$. We may consider ds_1 and ds_2 as the sides of the element $d\phi\, d\psi$ of the sheet. The electromotive force $-d\psi$ in the direction of ds_2 produces the current $d\phi$ across ds_1.

Let the resistance of a portion of the sheet whose length is ds_2, and whose breadth is ds_1, be $\sigma\frac{ds_2}{ds_1}$,

where σ is the specific resistance of the sheet referred to unit of area, then

$$d\psi = \sigma\frac{ds_2}{ds_1}d\phi,$$

whence

$$\frac{ds_1}{d\phi} = \sigma\frac{ds_2}{d\psi}.$$

651.] If the sheet is of a substance which conducts equally well in all directions, ds_1 is perpendicular to ds_2. In the case of a sheet of uniform resistance σ is constant, and if we make $\psi' = \sigma\psi$, we shall have

$$\frac{ds_1}{ds_2} = \frac{d\phi}{d\psi'},$$

and the stream-lines and equipotential lines will cut the surface into little squares.

It follows from this that if ϕ_1 and ψ_1' are conjugate functions (Art. 183) of ϕ and ψ', the curves ϕ_1 may be stream-lines in the sheet for which the curves ψ_1' are the corresponding equipotential lines. One case, of course, is that in which $\phi_1 = \psi'$ and $\psi_1' = -\phi$. In this case the equipotential lines become current-lines, and the current-lines equipotential lines *.

If we have obtained the solution of the distribution of electric currents in a uniform sheet of any form for any particular case, we may deduce the distribution in any other case by a proper transformation of the conjugate functions, according to the method given in Art. 190.

652.] We have next to determine the magnetic action of a current-sheet in which the current is entirely confined to the sheet, there being no electrodes to convey the current to or from the sheet.

In this case the current-function ϕ has a determinate value at every point, and the stream-lines are closed curves which do not intersect each other, though any one stream-line may intersect itself.

Consider the annular portion of the sheet between the stream-lines ϕ and $\phi + \delta\phi$. This part of the sheet is a conducting circuit in which a current of strength $\delta\phi$ circulates in the positive direction round that part of the sheet for which ϕ is greater than the given value. The magnetic effect of this circuit is the same as that of a magnetic shell of strength $\delta\phi$ at any point not included in the substance of the shell. Let us suppose that the shell coincides with that part of the current-sheet for which ϕ has a greater value than it has at the given stream-line.

By drawing all the successive stream-lines, beginning with that for which ϕ has the greatest value, and ending with that for which its value is least, we shall divide the current-sheet into a series of circuits. Substituting for each circuit its corresponding magnetic shell, we find that the magnetic effect of the current-sheet at any point not included in the thickness of the sheet is the same as that of a complex magnetic shell, whose strength at any point is $C + \phi$, where C is a constant.

If the current-sheet is bounded, then we must make $C + \phi = 0$ at the bounding curve. If the sheet forms a closed or an infinite surface, there is nothing to determine the value of the constant C.

* See Thomson, *Camb. and Dub. Math. Journ.*, vol. iii. p. 286.

653.] The magnetic potential at any point on either side of the current-sheet is given, as in Art. 415, by the expression

$$\Omega = \iint \frac{1}{r^2}\phi\cos\theta\, dS,$$

where r is the distance of the given point from the element of surface dS, and θ is the angle between the direction of r, and that of the normal drawn from the positive side of dS.

This expression gives the magnetic potential for all points not included in the thickness of the current-sheet, and we know that for points within a conductor carrying a current there is no such thing as a magnetic potential.

The value of Ω is discontinuous at the current-sheet, for if Ω_1 is its value at a point just within the current-sheet, and Ω_2 its value at a point close to the first but just outside the current-sheet,

$$\Omega_2 = \Omega_1 + 4\pi\phi,$$

where ϕ is the current-function at that point of the sheet.

The value of the component of magnetic force normal to the sheet is continuous, being the same on both sides of the sheet. The component of the magnetic force parallel to the current-lines is also continuous, but the tangential component perpendicular to the current-lines is discontinuous at the sheet. If s is the length of a curve drawn on the sheet, the component of magnetic force in the direction of ds is, for the negative side, $\frac{d\Omega_1}{ds}$, and for the positive side, $\frac{d\Omega_2}{ds} = \frac{d\Omega_1}{ds} + 4\pi\frac{d\phi}{ds}$.

The component of the magnetic force on the positive side therefore exceeds that on the negative side by $4\pi\frac{d\phi}{ds}$. At a given point this quantity will be a maximum when ds is perpendicular to the current-lines.

On the Induction of Electric Currents in a Sheet of Infinite Conductivity.

654.] It was shewn in Art. 579 that in any circuit

$$E = \frac{dp}{dt} + Ri,$$

where E is the impressed electromotive force, p the electrokinetic momentum of the circuit, R the resistance of the circuit, and i the current round it. If there is no impressed electromotive force and no resistance, then $\frac{dp}{dt} = 0$, or p is constant.

Now p, the electrokinetic momentum of the circuit, was shewn in Art. 588 to be measured by the surface-integral of magnetic induction through the circuit. Hence, in the case of a current-sheet of no resistance, the surface-integral of magnetic induction through any closed curve drawn on the surface must be constant, and this implies that the normal component of magnetic induction remains constant at every point of the current-sheet.

655.] If, therefore, by the motion of magnets or variations of currents in the neighbourhood, the magnetic field is in any way altered, electric currents will be set up in the current-sheet, such that their magnetic effect, combined with that of the magnets or currents in the field, will maintain the normal component of magnetic induction at every point of the sheet unchanged. If at first there is no magnetic action, and no currents in the sheet, then the normal component of magnetic induction will always be zero at every point of the sheet.

The sheet may therefore be regarded as impervious to magnetic induction, and the lines of magnetic induction will be deflected by the sheet exactly in the same way as the lines of flow of an electric current in an infinite and uniform conducting mass would be deflected by the introduction of a sheet of the same form made of a substance of infinite resistance.

If the sheet forms a closed or an infinite surface, no magnetic actions which may take place on one side of the sheet will produce any magnetic effect on the other side.

Theory of a Plane Current-sheet.

656.] We have seen that the external magnetic action of a current-sheet is equivalent to that of a magnetic shell whose strength at any point is numerically equal to ϕ, the current-function. When the sheet is a plane one, we may express all the quantities required for the determination of electromagnetic effects in terms of a single function, P, which is the potential due to a sheet of imaginary matter spread over the plane with a surface-density ϕ. The value of P is of course

$$P = \iint \frac{\phi}{r}\, dx'\, dy', \tag{1}$$

where r is the distance from the point (x, y, z) for which P is calculated, to the point $x', y', 0$ in the plane of the sheet, at which the element $dx'\, dy'$ is taken.

To find the magnetic potential, we may regard the magnetic

shell as consisting of two surfaces parallel to the plane of xy, the first, whose equation is $z = \frac{1}{2}c$, having the surface-density $\frac{\phi}{c}$, and the second, whose equation is $z = -\frac{1}{2}c$, having the surface-density $-\frac{\phi}{c}$.

The potentials due to these surfaces will be

$$\frac{1}{c}P_{\left(z-\frac{c}{2}\right)} \quad \text{and} \quad -\frac{1}{c}P_{\left(z+\frac{c}{2}\right)}.$$

respectively, where the suffixes indicate that $z-\frac{c}{2}$ is put for z in the first expression, and $z+\frac{c}{2}$ for z in the second. Expanding these expressions by Taylor's Theorem, adding them, and then making c infinitely small, we obtain for the magnetic potential due to the sheet at any point external to it,

$$\Omega = -\frac{dP}{dz}. \tag{2}$$

657.] The quantity P is symmetrical with respect to the plane of the sheet, and is therefore the same when $-z$ is substituted for z.

Ω, the magnetic potential, changes sign when $-z$ is put for z.

At the positive surface of the sheet

$$\Omega = -\frac{dP}{dz} = 2\pi\phi. \tag{3}$$

At the negative surface of the sheet

$$\Omega = -\frac{dP}{dz} = -2\pi\phi. \tag{4}$$

Within the sheet, if its magnetic effects arise from the magnetization of its substance, the magnetic potential varies continuously from $2\pi\phi$ at the positive surface to $-2\pi\phi$ at the negative surface.

If the sheet contains electric currents, the magnetic force within it does not satisfy the condition of having a potential. The magnetic force within the sheet is, however, perfectly determinate.

The normal component,

$$\gamma = -\frac{d\Omega}{dz} = \frac{d^2P}{dz^2}, \tag{5}$$

is the same on both sides of the sheet and throughout its substance.

If α and β be the components of the magnetic force parallel to

x and to y at the positive surface, and α', β' those on the negative surface

$$\alpha = -2\pi\frac{d\phi}{dx} = -\alpha', \tag{6}$$

$$\beta = -2\pi\frac{d\phi}{dy} = -\beta'. \tag{7}$$

Within the sheet the components vary continuously from α and β to α' and β'.

The equations

$$\left.\begin{aligned} \frac{dH}{dy} - \frac{dG}{dz} &= -\frac{d\Omega}{dx}, \\ \frac{dF}{dz} - \frac{dH}{dx} &= -\frac{d\Omega}{dy}, \\ \frac{dG}{dx} - \frac{dF}{dy} &= -\frac{d\Omega}{dz}, \end{aligned}\right\} \tag{8}$$

which connect the components F, G, H of the vector-potential due to the current-sheet with the scalar potential Ω, are satisfied if we make

$$F = \frac{dP}{dy}, \qquad G = -\frac{dP}{dx}, \qquad H = 0. \tag{9}$$

We may also obtain these values by direct integration, thus for F,

$$F = \iint\frac{u}{r}\,dx'\,dy' = \iint\frac{1}{r}\frac{d\phi}{dy}\,dx'\,dy',$$

$$= \int\frac{\phi}{r}\,dx' - \iint\phi\frac{d}{dy'}\frac{1}{r}\,dx'\,dy'.$$

Since the integration is to be estimated over the infinite plane sheet, and since the first term vanishes at infinity, the expression is reduced to the second term; and by substituting

$$\frac{d}{dy}\frac{1}{r} \text{ for } -\frac{d}{dy'}\frac{1}{r},$$

and remembering that ϕ depends on x' and y', and not on x, y, z, we obtain

$$F = \frac{d}{dy}\iint\frac{\phi}{r}\,dx'\,dy',$$

$$= \frac{dP}{dy}, \text{ by (1).}$$

If Ω' is the magnetic potential due to any magnetic or electric system external to the sheet, we may write

$$P' = -\int\Omega'\,dz, \tag{10}$$

and we shall then have

$$F' = \frac{dP'}{dy}, \qquad G' = -\frac{dP'}{dx}, \qquad H' = 0, \tag{11}$$

for the components of the vector-potential due to this system.

658.] Let us now determine the electromotive force at any point of the sheet, supposing the sheet fixed.

Let X and Y be the components of the electromotive force parallel to x and to y respectively, then, by Art. 598, we have

$$X = -\frac{d}{dt}(F+F') - \frac{d\psi}{dx}, \tag{12}$$

$$Y = -\frac{d}{dt}(G+G') - \frac{d\psi}{dy}. \tag{13}$$

If the electric resistance of the sheet is uniform and equal to σ,

$$X = \sigma u, \qquad Y = \sigma v, \tag{14}$$

where u and v are the components of the current, and if ϕ is the current-function,

$$u = \frac{d\phi}{dy}, \qquad v = -\frac{d\phi}{dx}. \tag{15}$$

But, by equation (3),

$$2\pi\phi = -\frac{dP}{dz}$$

at the positive surface of the current-sheet. Hence, equations (12) and (13) may be written

$$-\frac{\sigma}{2\pi}\frac{d^2P}{dy\,dz} = -\frac{d^2}{dy\,dt}(P+P') - \frac{d\psi}{dx}, \tag{16}$$

$$\frac{\sigma}{2\pi}\frac{d^2P}{dx\,dz} = \frac{d^2}{dx\,dt}(P+P') - \frac{d\psi}{dy}, \tag{17}$$

where the values of the expressions are those corresponding to the positive surface of the sheet.

If we differentiate the first of these equations with respect to x, and the second with respect to y, and add the results, we obtain

$$\frac{d^2\psi}{dx^2} + \frac{d^2\psi}{dy^2} = 0. \tag{18}$$

The only value of ψ which satisfies this equation, and is finite and continuous at every point of the plane, and vanishes at an infinite distance, is

$$\psi = 0. \tag{19}$$

Hence the induction of electric currents in an infinite plane sheet of uniform conductivity is not accompanied with differences of electric potential in different parts of the sheet.

Substituting this value of ψ, and integrating equations (16), (17), we obtain

$$\frac{\sigma}{2\pi}\frac{dP}{dz} - \frac{dP}{dt} - \frac{dP'}{dt} = f(z, t). \tag{20}$$

Since the values of the currents in the sheet are found by

differentiating with respect to x or y, the arbitrary function of z and t will disappear. We shall therefore leave it out of account.

If we also write for $\frac{\sigma}{2\pi}$, the single symbol R, which represents a certain velocity, the equation between P and P' becomes

$$R\frac{dP}{dz} = \frac{dP}{dt} + \frac{dP'}{dt}. \tag{21}$$

659.] Let us first suppose that there is no external magnetic system acting on the current sheet. We may therefore suppose $P'=0$. The case then becomes that of a system of electric currents in the sheet left to themselves, but acting on one another by their mutual induction, and at the same time losing their energy on account of the resistance of the sheet. The result is expressed by the equation

$$R\frac{dP}{dz} = \frac{dP}{dt}, \tag{22}$$

the solution of which is

$$P = f\,(x, y, (z+Rt)). \tag{23}$$

Hence, the value of P on any point on the positive side of the sheet whose coordinates are x, y, z, and at a time t, is equal to the value of P at the point x, y, $(z+Rt)$ at the instant when $t=0$.

If therefore a system of currents is excited in a uniform plane sheet of infinite extent and then left to itself, its magnetic effect at any point on the positive side of the sheet will be the same as if the system of currents had been maintained constant in the sheet, and the sheet moved in the direction of a normal from its negative side with the constant velocity R. The diminution of the electromagnetic forces, which arises from a decay of the currents in the real case, is accurately represented by the diminution of the force on account of the increasing distance in the imaginary case.

660.] Integrating equation (21) with respect to t, we obtain

$$P+P' = \int R\frac{dP}{dz}\,dt. \tag{24}$$

If we suppose that at first P and P' are both zero, and that a magnet or electromagnet is suddenly magnetized or brought from an infinite distance, so as to change the value of P' suddenly from zero to P', then, since the time-integral in the second member of (24) vanishes with the time, we must have at the first instant

$$P = -P'$$

at the surface of the sheet.

Hence, the system of currents excited in the sheet by the sudden introduction of the system to which P' is due is such that at the surface of the sheet it exactly neutralizes the magnetic effect of this system.

At the surface of the sheet, therefore, and consequently at all points on the negative side of it, the initial system of currents produces an effect exactly equal and opposite to that of the magnetic system on the positive side. We may express this by saying that the effect of the currents is equivalent to that of an *image* of the magnetic system, coinciding in position with that system, but opposite as regards the direction of its magnetization and of its electric currents. Such an image is called a *negative* image.

The effect of the currents in the sheet on a point on the positive side of it is equivalent to that of a positive image of the magnetic system on the negative side of the sheet, the lines joining corresponding points being bisected at right angles by the sheet.

The action at a point on either side of the sheet, due to the currents in the sheet, may therefore be regarded as due to an image of the magnetic system on the side of the sheet opposite to the point, this image being a positive or a negative image according as the point is on the positive or the negative side of the sheet.

661.] If the sheet is of infinite conductivity, $R = 0$, and the second term of (24) is zero, so that the image will represent the effect of the currents in the sheet at any time.

In the case of a real sheet, the resistance R has some finite value. The image just described will therefore represent the effect of the currents only during the first instant after the sudden introduction of the magnetic system. The currents will immediately begin to decay, and the effect of this decay will be accurately represented if we suppose the two images to move from their original positions, in the direction of normals drawn from the sheet, with the constant velocity R.

662.] We are now prepared to investigate the system of currents induced in the sheet by any system, M, of magnets or electromagnets on the positive side of the sheet, the position and strength of which vary in any manner.

Let P', as before, be the function from which the direct action of this system is to be deduced by the equations (3), (9), &c., then $\frac{dP'}{dt}\delta t$ will be the function corresponding to the system represented by $\frac{dM}{dt}\delta t$. This quantity, which is the increment of M in the time δt, may be regarded as itself representing a magnetic system.

If we suppose that at the time t a positive image of the system $\frac{dM}{dt}\delta t$ is formed on the negative side of the sheet, the magnetic action at any point on the positive side of the sheet due to this image will be equivalent to that due to the currents in the sheet excited by the change in M during the first instant after the change, and the image will continue to be equivalent to the currents in the sheet, if, as soon as it is formed, it begins to move in the negative direction of z with the constant velocity R.

If we suppose that in every successive element of the time an image of this kind is formed, and that as soon as it is formed it begins to move away from the sheet with velocity R, we shall obtain the conception of a trail of images, the last of which is in process of formation, while all the rest are moving like a rigid body away from the sheet with velocity R.

663.] If P' denotes any function whatever arising from the action of the magnetic system, we may find P, the corresponding function arising from the currents in the sheet, by the following process, which is merely the symbolical expression for the theory of the trail of images.

Let P_τ denote the value of P (the function arising from the currents in the sheet) at the point $(x, y, z+R\tau)$, and at the time $t-\tau$, and let P'_τ denote the value of P' (the function arising from the magnetic system) at the point $(x, y, -(z+R\tau))$, and at the time $t-\tau$. Then

$$\frac{dP_\tau}{d\tau} = R\frac{dP_\tau}{dz} - \frac{dP_\tau}{dt}, \tag{25}$$

and equation (21) becomes

$$\frac{dP_\tau}{d\tau} = \frac{dP'_\tau}{dt}, \tag{26}$$

and we obtain by integrating with respect to τ from $\tau=0$ to $\tau=\infty$,

$$P = \int_0^\infty \frac{dP'_\tau}{dt}\, d\tau \tag{27}$$

as the value of the function P, whence we obtain all the properties of the current sheet by differentiation, as in equations (3), (9), &c.

664.] As an example of the process here indicated, let us take the case of a single magnetic pole of strength unity, moving with uniform velocity in a straight line.

Let the coordinates of the pole at the time t be

$$\xi = \mathfrak{u}t, \qquad \eta = 0, \qquad \zeta = c + \mathfrak{w}t.$$

The coordinates of the image of the pole formed at the time $t-\tau$ are

$$\xi = \mathfrak{u}(t-\tau), \qquad \eta = 0, \qquad \zeta = -(c + \mathfrak{w}(t-\tau) + R\tau),$$

and if r is the distance of this image from the point (x, y, z),

$$r^2 = (x - \mathfrak{u}(t-\tau))^2 + (z + c + \mathfrak{w}(t-\tau) + R\tau)^2.$$

To obtain the potential due to the trail of images we have to calculate

$$\frac{d}{dt}\int_0^\infty \frac{d\tau}{r}.$$

If we write $\qquad Q^2 = \mathfrak{u}^2 + (R-\mathfrak{w})^2,$

$$\int_0^\infty \frac{d\tau}{r} = \frac{1}{Q}\log\{Qr + \mathfrak{u}(x - \mathfrak{u}t) + (R-\mathfrak{w})(z + c + \mathfrak{w}t)\},$$

the value of r in this expression being found by making $\tau = 0$.

Differentiating this expression with respect to t, and putting $t = 0$, we obtain the magnetic potential due to the trail of images,

$$\Omega = \frac{1}{Q}\,\frac{Q\dfrac{\mathfrak{w}(z+c) - \mathfrak{u}x}{r} - \mathfrak{u}^2 - \mathfrak{w}^2 + R\mathfrak{w}}{Qr + \mathfrak{u}x + (R-\mathfrak{w})(z+c)}.$$

By differentiating this expression with respect to x or z, we obtain the components parallel to x or z respectively of the magnetic force at any point, and by putting $x = 0$, $z = c$, and $r = 2c$ in these expressions, we obtain the following values of the components of the force acting on the moving pole itself,

$$X = -\frac{1}{4c^2}\,\frac{\mathfrak{u}}{Q + R - \mathfrak{w}}\left\{1 + \frac{\mathfrak{w}}{Q} - \frac{\mathfrak{u}^2}{Q(Q + R - \mathfrak{w})}\right\},$$

$$Z = -\frac{1}{4c^2}\left\{\frac{\mathfrak{w}}{Q} - \frac{\mathfrak{u}^2}{Q(Q + R - \mathfrak{w})}\right\}.$$

665.] In these expressions we must remember that the motion is supposed to have been going on for an infinite time before the time considered. Hence we must not take $\mathfrak{w}$ a positive quantity, for in that case the pole must have passed through the sheet within a finite time.

If we make $\mathfrak{u} = 0$, and $\mathfrak{w}$ negative, $X = 0$, and

$$Z = \frac{1}{4c^2}\,\frac{\mathfrak{w}}{R + \mathfrak{w}},$$

or the pole as it approaches the sheet is repelled from it.

If we make $\mathfrak{w} = 0$, we find $Q^2 = \mathfrak{u}^2 + R^2$,

$$X = -\frac{1}{4c^2}\,\frac{\mathfrak{u}R}{Q(Q+R)} \quad \text{and} \quad Z = \frac{1}{4c^2}\,\frac{\mathfrak{u}^2}{Q(Q+R)}.$$

The component X represents a retarding force acting on the pole in the direction opposite to that of its own motion. For a given value of R, X is a maximum when $\mathfrak{u} = 1.27R$.

When the sheet is a non-conductor, $R = \infty$ and $X = 0$.

When the sheet is a perfect conductor, $R = 0$ and $X = 0$.

The component Z represents a repulsion of the pole from the sheet. It increases as the velocity increases, and ultimately becomes $\frac{1}{4c^2}$ when the velocity is infinite. It has the same value when R is zero.

666.] When the magnetic pole moves in a curve parallel to the sheet, the calculation becomes more complicated, but it is easy to see that the effect of the nearest portion of the trail of images is to produce a force acting on the pole in the direction opposite to that of its motion. The effect of the portion of the trail immediately behind this is of the same kind as that of a magnet with its axis parallel to the direction of motion of the pole at some time before. Since the nearest pole of this magnet is of the same name with the moving pole, the force will consist partly of a repulsion, and partly of a force parallel to the former direction of motion, but backwards. This may be resolved into a retarding force, and a force towards the concave side of the path of the moving pole.

667.] Our investigation does not enable us to solve the case in which the system of currents cannot be completely formed, on account of a discontinuity or boundary of the conducting sheet.

It is easy to see, however, that if the pole is moving parallel to the edge of the sheet, the currents on the side next the edge will be enfeebled. Hence the forces due to these currents will be less, and there will not only be a smaller retarding force, but, since the repulsive force is least on the side next the edge, the pole will be attracted towards the edge.

Theory of Arago's Rotating Disk.

668.] Arago discovered* that a magnet placed near a rotating metallic disk experiences a force tending to make it follow the motion of the disk, although when the disk is at rest there is no action between it and the magnet.

This action of a rotating disk was attributed to a new kind

* *Annales de Chimie et de Physique*, 1826.

of induced magnetization, till Faraday* explained it by means of the electric currents induced in the disk on account of its motion through the field of magnetic force.

To determine the distribution of these induced currents, and their effect on the magnet, we might make use of the results already found for a conducting sheet at rest acted on by a moving magnet, availing ourselves of the method given in Art. 600 for treating the electromagnetic equations when referred to moving systems of axes. As this case, however, has a special importance, we shall treat it in a direct manner, beginning by assuming that the poles of the magnet are so far from the edge of the disk that the effect of the limitation of the conducting sheet may be neglected.

Making use of the same notation as in the preceding articles (656–667), we find for the components of the electromotive force parallel to x and y respectively,

$$\left.\begin{aligned}\sigma u &= \gamma\frac{dy}{dt} - \frac{d\psi}{dx},\\ \sigma v &= -\gamma\frac{dx}{dt} - \frac{d\psi}{dy},\end{aligned}\right\} \qquad (1)$$

where γ is the resolved part of the magnetic force normal to the disk.

If we now express u and v in terms of ϕ, the current-function,

$$u = \frac{d\phi}{dy}, \qquad v = -\frac{d\phi}{dx}, \qquad (2)$$

and if the disk is rotating about the axis of z with the angular velocity ω,

$$\frac{dy}{dt} = \omega x, \qquad \frac{dx}{dt} = -\omega y. \qquad (3)$$

Substituting these values in equations (1), we find

$$\sigma\frac{d\phi}{dy} = \gamma\omega x - \frac{d\psi}{dx}, \qquad (4)$$

$$-\sigma\frac{d\phi}{dx} = \gamma\omega y - \frac{d\psi}{dy}. \qquad (5)$$

Multiplying (4) by x and (5) by y, and adding, we obtain

$$\sigma\left(x\frac{d\phi}{dy} - y\frac{d\phi}{dx}\right) = \gamma\omega(x^2+y^2) - \left(x\frac{d\psi}{dx} + y\frac{d\psi}{dy}\right). \qquad (6)$$

Multiplying (4) by y and (5) by $-x$, and adding, we obtain

$$\sigma\left(x\frac{d\phi}{dx} + y\frac{d\phi}{dy}\right) = x\frac{d\psi}{dy} - y\frac{d\psi}{dx}. \qquad (7)$$

* *Exp. Res.*, 81.

If we now express these equations in terms of r and θ, where

$$x = r\cos\theta, \qquad y = r\sin\theta, \tag{8}$$

they become

$$\sigma\frac{d\phi}{d\theta} = \gamma\omega r^2 - r\frac{d\psi}{dr}, \tag{9}$$

$$\sigma r\frac{d\phi}{dr} = \frac{d\psi}{d\theta}. \tag{10}$$

Equation (10) is satisfied if we assume any arbitrary function χ of r and θ, and make

$$\phi = \frac{d\chi}{d\theta}, \tag{11}$$

$$\psi = \sigma r\frac{d\chi}{dr}. \tag{12}$$

Substituting these values in equation (9), it becomes

$$\sigma\left(\frac{d^2\chi}{d\theta^2} + \frac{d}{dr}\left(r\frac{d\chi}{dr}\right)\right) = \gamma\omega r^2. \tag{13}$$

Dividing by σr^2, and restoring the coordinates x and y, this becomes

$$\frac{d^2\chi}{dx^2} + \frac{d^2\chi}{dy^2} = \frac{\omega}{\sigma}\gamma. \tag{14}$$

This is the fundamental equation of the theory, and expresses the relation between the function, χ, and the component, γ, of the magnetic force resolved normal to the disk.

Let Q be the potential, at any point on the positive side of the disk, due to imaginary matter distributed over the disk with the surface-density χ.

At the positive surface of the disk

$$\frac{dQ}{dz} = -2\pi\chi. \tag{15}$$

Hence the first member of equation (14) becomes

$$\frac{d^2\chi}{dx^2} + \frac{d^2\chi}{dy^2} = -\frac{1}{2\pi}\frac{d}{dz}\left(\frac{d^2Q}{dx^2} + \frac{d^2Q}{dy^2}\right). \tag{16}$$

But since Q satisfies Laplace's equation at all points external to the disk,

$$\frac{d^2Q}{dx^2} + \frac{d^2Q}{dy^2} = -\frac{d^2Q}{dz^2}, \tag{17}$$

and equation (14) becomes

$$\frac{\sigma}{2\pi}\frac{d^3Q}{dz^3} = \omega\gamma. \tag{18}$$

Again, since Q is the potential due to the distribution χ, the potential due to the distribution ϕ, or $\frac{d\chi}{d\theta}$, will be $\frac{dQ}{d\theta}$. From this we obtain for the magnetic potential due to the currents in the disk,

$$\Omega_1 = -\frac{d^2Q}{d\theta\, dz}, \tag{19}$$

and for the component of the magnetic force normal to the disk due to the currents,

$$\gamma_1 = -\frac{d\Omega}{dz} = \frac{d^3 Q}{d\theta\, dz^2}. \tag{20}$$

If Ω_2 is the magnetic potential due to external magnets, and if we write

$$P = -\int \Omega_2\, dz, \tag{21}$$

the component of the magnetic force normal to the disk due to the magnets will be

$$\gamma_2 = \frac{d^2 P}{dz^2}. \tag{22}$$

We may now write equation (18), remembering that

$$\gamma = \gamma_1 + \gamma_2,$$

$$\frac{\sigma}{2\pi}\frac{d^3 Q}{dz^3} - \omega\frac{d^3 Q}{d\theta\, dz^2} = \omega\frac{d^2 P}{dz^2}. \tag{23}$$

Integrating twice with respect to z, and writing R for $\frac{\sigma}{2\pi}$,

$$\left(R\frac{d}{dz} - \omega\frac{d}{d\theta}\right) Q = \omega P. \tag{24}$$

If the values of P and Q are expressed in terms of r, θ, and ζ, where

$$\zeta = z - \frac{R}{\omega}\theta, \tag{25}$$

equation (24) becomes, by integration with respect to ζ,

$$Q = \int \frac{\omega}{R} P\, d\zeta. \tag{26}$$

669.] The form of this expression shews that the magnetic action of the currents in the disk is equivalent to that of a trail of images of the magnetic system in the form of a helix.

If the magnetic system consists of a single magnetic pole of strength unity, the helix will lie on the cylinder whose axis is that of the disk, and which passes through the magnetic pole. The helix will begin at the position of the optical image of the pole in the disk. The distance, parallel to the axis between consecutive coils of the helix, will be $2\pi\frac{R}{\omega}$. The magnetic effect of the trail will be the same as if this helix had been magnetized everywhere in the direction of a tangent to the cylinder perpendicular to its axis, with an intensity such that the magnetic moment of any small portion is numerically equal to the length of its projection on the disk.

The calculation of the effect on the magnetic pole would be complicated, but it is easy to see that it will consist of—

(1) A dragging force, parallel to the direction of motion of the disk.

(2) A repulsive force acting from the disk.

(3) A force towards the axis of the disk.

When the pole is near the edge of the disk, the third of these forces may be overcome by the force towards the edge of the disk, indicated in Art. 667.

All these forces were observed by Arago, and described by him in the *Annales de Chimie* for 1826. See also Felici, in Tortolini's *Annals*, iv, p. 173 (1853), and v. p. 35; and E. Jochmann, in *Crelle's Journal*, lxiii, pp. 158 and 329; and Pogg. *Ann.* cxxii, p. 214 (1864). In the latter paper the equations necessary for determining the induction of the currents on themselves are given, but this part of the action is omitted in the subsequent calculation of results. The method of images given here was published in the *Proceedings of the Royal Society* for Feb. 15, 1872.

Spherical Current-Sheet.

670.] Let ϕ be the current-function at any point Q of a spherical current-sheet, and let P be the potential at a given point, due to a sheet of imaginary matter distributed over the sphere with surface-density ϕ, it is required to find the magnetic potential and the vector-potential of the current-sheet in terms of P.

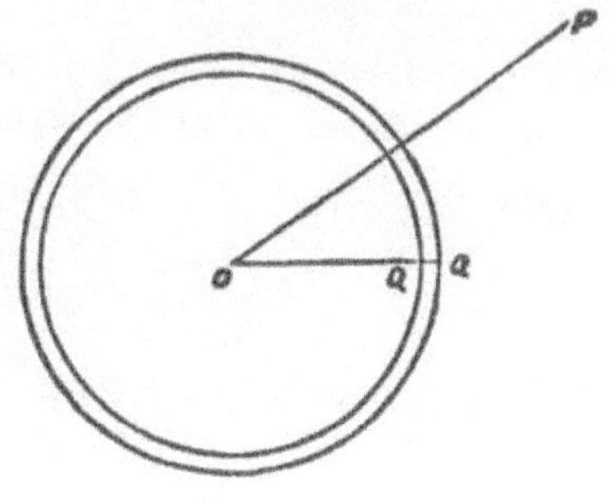

Fig. 39.

Let a denote the radius of the sphere, r the distance of the given point from the centre, and p the reciprocal of the distance of the given point from the point Q on the sphere at which the current-function is ϕ.

The action of the current-sheet at any point not in its substance is identical with that of a magnetic shell whose strength at any point is numerically equal to the current-function.

The mutual potential of the magnetic shell and a unit pole placed at the point P is, by Art. 410,

$$\Omega = \iint \phi \frac{dp}{da} dS.$$

Since p is a homogeneous function of the degree -1 in r and a,

$$a\frac{dp}{da}+r\frac{dp}{dr}=-p,$$

or
$$\frac{dp}{da}=-\frac{1}{a}\frac{d}{dr}(pr),$$

and
$$\Omega=-\iint\frac{\phi}{a}\frac{d}{dr}(pr)\,dS.$$

Since r and a are constant during the surface-integration,

$$\Omega=-\frac{1}{a}\frac{d}{dr}\left(r\iint\phi p\,dS\right).$$

But if P is the potential due to a sheet of imaginary matter of surface-density ϕ,
$$P=\iint\phi p\,dS,$$

and Ω, the magnetic potential of the current-sheet, may be expressed in terms of P in the form

$$\Omega=-\frac{1}{a}\frac{d}{dr}(Pr).$$

671.] We may determine F, the x-component of the vector-potential, from the expression given in Art. 416,

$$F=\iint\phi\left(m\frac{dp}{d\zeta}-n\frac{dp}{d\eta}\right)dS,$$

where ξ, η, ζ are the coordinates of the element dS, and l, m, n are the direction-cosines of the normal.

Since the sheet is a sphere, the direction-cosines of the normal are

$$l=\frac{\xi}{a},\qquad m=\frac{\eta}{a},\qquad n=\frac{\zeta}{a}.$$

But
$$\frac{dp}{d\zeta}=(z-\zeta)p^3=-\frac{dp}{dz},$$

and
$$\frac{dp}{d\eta}=(y-\eta)p^3=-\frac{dp}{dy},$$

so that
$$m\frac{dp}{d\zeta}-n\frac{dp}{d\eta}=(\eta(z-\zeta)-\zeta(y-\eta))\frac{p^3}{a},$$

$$=(z(\eta-y)-y(\zeta-z))\frac{p^3}{a},$$

$$=\frac{z}{a}\frac{dp}{dy}-\frac{y}{a}\frac{dp}{dz};$$

multiplying by $\phi\,dS$, and integrating over the surface of the sphere, we find
$$F=\frac{z}{a}\frac{dP}{dy}-\frac{y}{a}\frac{dP}{dz}.$$

Similarly $$G = \frac{x}{a}\frac{dP}{dz} - \frac{z}{a}\frac{dP}{dx},$$
$$H = \frac{y}{a}\frac{dP}{dx} - \frac{x}{a}\frac{dP}{dy}.$$

The vector $\mathfrak{A}$, whose components are F, G, H, is evidently perpendicular to the radius vector r, and to the vector whose components are $\frac{dP}{dx}$, $\frac{dP}{dy}$, and $\frac{dP}{dz}$. If we determine the lines of intersections of the spherical surface whose radius is r, with the series of equipotential surfaces corresponding to values of P in arithmetical progression, these lines will indicate by their direction the direction of $\mathfrak{A}$, and by their proximity the magnitude of this vector.

In the language of Quaternions,
$$\mathfrak{A} = \frac{1}{a} V\rho\nabla P.$$

672.] If we assume as the value of P within the sphere
$$P = A\left(\frac{r}{a}\right)^i Y_i,$$
where Y_i is a spherical harmonic of degree i, then outside the sphere
$$P' = A\left(\frac{a}{r}\right)^{i+1} Y_i.$$

The current-function ϕ is
$$\phi = \frac{2i+1}{4\pi}\frac{1}{a}AY_i.$$

The magnetic potential within the sphere is
$$\Omega = -(i+1)\frac{1}{a}A\left(\frac{a}{r}\right)^i Y_i,$$
and outside $$\Omega' = i\frac{1}{a}A\left(\frac{a}{r}\right)^{i+1} Y_i.$$

For example, let it be required to produce, by means of a wire coiled into the form of a spherical shell, a uniform magnetic force M within the shell. The magnetic potential within the shell is, in this case, a solid harmonic of the first degree of the form
$$\Omega = Mr\cos\theta,$$
where M is the magnetic force. Hence $A = -\frac{1}{2}a^2 M$, and
$$\phi = \frac{3}{8\pi} Ma\cos\theta.$$

The current-function is therefore proportional to the distance from the equatorial plane of the sphere, and therefore the number of windings of the wire between any two small circles must be proportional to the distance between the planes of these circles.

If N is the whole number of windings, and if γ is the strength of the current in each winding,

$$\phi = \tfrac{1}{2} N\gamma \cos\theta.$$

Hence the magnetic force within the coil is

$$M = \frac{4\pi}{3} \frac{N\gamma}{a}.$$

673.] Let us next find the method of coiling the wire in order to produce within the sphere a magnetic potential of the form of a solid zonal harmonic of the second degree,

$$\Omega = A\frac{r^2}{a^2}(\tfrac{3}{2}\cos^2\theta - \tfrac{1}{2}).$$

Here
$$\phi = \frac{5}{12\pi} A(\tfrac{3}{2}\cos^2\theta - \tfrac{1}{2}).$$

If the whole number of windings is N, the number between the pole and the polar distance θ is $\tfrac{1}{2} N \sin^2\theta$.

The windings are closest at latitude 45°. At the equator the direction of winding changes, and in the other hemisphere the windings are in the contrary direction.

Let γ be the strength of the current in the wire, then within the shell
$$\Omega = \frac{4\pi}{5} N\gamma \frac{r^2}{a^2}(\tfrac{3}{2}\cos^2\theta - \tfrac{1}{2}).$$

Let us now consider a conductor in the form of a plane closed curve placed anywhere within the shell with its plane perpendicular to the axis. To determine its coefficient of induction we have to find the surface-integral of $\frac{d\Omega}{dz}$ over the plane bounded by the curve, putting $\gamma = 1$.

Now
$$\Omega = \frac{4\pi}{5a^2} N(z^2 - \tfrac{1}{2}(x^2+y^2)),$$

and
$$\frac{d\Omega}{dz} = \frac{8\pi}{5a^2} Nz.$$

Hence, if S is the area of the closed curve, its coefficient of induction is
$$M = \frac{8\pi}{5a^2} NSz.$$

If the current in this conductor is γ', there will be, by Art. 583, a force Z, urging it in the direction of z, where

$$Z = \gamma\gamma'\frac{dM}{dz} = \frac{8\pi}{5a^2} NS\gamma\gamma',$$

and, since this is independent of x, y, z, the force is the same in whatever part of the shell the circuit is placed.

674.] The method given by Poisson, and described in Art. 437, may be applied to current-sheets by substituting for the body supposed to be uniformly magnetized in the direction of z with intensity I, a current-sheet having the form of its surface, and for which the current-function is $\phi = Iz.$ (1)

The currents in the sheet will be in planes parallel to that of xy, and the strength of the current round a slice of thickness dz will be $I\,dz$.

The magnetic potential due to this current-sheet at any point outside it will be

$$\Omega = -I\frac{dV}{dz}. \tag{2}$$

At any point inside the sheet it will be

$$\Omega = -4\pi Iz - I\frac{dV}{dz}. \tag{3}$$

The components of the vector-potential are

$$F = -I\frac{dV}{dy}, \qquad G = I\frac{dV}{dx}, \qquad H = 0. \tag{4}$$

These results can be applied to several cases occurring in practice.

675.] (1) A plane electric circuit of any form.

Let V be the potential due to a plane sheet of any form of which the surface-density is unity, then, if for this sheet we substitute either a magnetic shell of strength I or an electric current of strength I round its boundary, the values of Ω and of F, G, H will be those given above.

(2) For a solid sphere of radius a,

$$V = \frac{4\pi}{3}\frac{a^3}{r} \text{ when } r \text{ is greater than } a, \tag{5}$$

and $$V = \frac{2\pi}{3}(3a^2 - r^2) \text{ when } r \text{ is less than } a. \tag{6}$$

Hence, if such a sphere is magnetized parallel to z with intensity I, the magnetic potential will be

$$\Omega = \frac{4\pi}{3} I \frac{a^3}{r^3} z \text{ outside the sphere,} \tag{7}$$

and $$\Omega = \frac{4\pi}{3} Iz \text{ inside the sphere.} \tag{8}$$

If, instead of being magnetized, the sphere is coiled with wire in equidistant circles, the total strength of current between two small circles whose planes are at unit distance being I, then outside the sphere the value of Ω is as before, but within the sphere

$$\Omega = -\frac{8\pi}{3} Iz. \tag{9}$$

This is the case already discussed in Art. 672.

(3) The case of an ellipsoid uniformly magnetized parallel to a given line has been discussed in Art. 437.

If the ellipsoid is coiled with wire in parallel and equidistant planes, the magnetic force within the ellipsoid will be uniform.

(4) *A Cylindric Magnet or Solenoid.*

676.] If the body is a cylinder having any form of section and bounded by planes perpendicular to its generating lines, and if V_1 is the potential at the point (x, y, z) due to a plane area of surface-density unity coinciding with the positive end of the solenoid, and V_2 the potential at the same point due to a plane area of surface-density unity coinciding with the negative end, then, if the cylinder is uniformly and longitudinally magnetized with intensity unity, the potential at the point (x, y, z) will be

$$\Omega = V_1 - V_2. \tag{10}$$

If the cylinder, instead of being a magnetized body, is uniformly lapped with wire, so that there are n windings of wire in unit of length, and if a current, γ, is made to flow through this wire, the magnetic potential outside the solenoid is as before,

$$\Omega = n\gamma(V_1 - V_2), \tag{11}$$

but within the space bounded by the solenoid and its plane ends

$$\Omega = n\gamma(4\pi z + V_1 - V_2). \tag{12}$$

The magnetic potential is discontinuous at the plane ends of the solenoid, but the magnetic force is continuous.

If r_1, r_2, the distances of the centres of inertia of the positive and negative plane end respectively from the point (x, y, z), are very great compared with the transverse dimensions of the solenoid, we may write

$$V_1 = \frac{A}{r_1}, \qquad V_2 = \frac{A}{r_2}, \tag{13}$$

where A is the area of either section.

The magnetic force outside the solenoid is therefore very small, and the force inside the solenoid approximates to a force parallel to the axis in the positive direction and equal to $4\pi n\gamma$.

If the section of the solenoid is a circle of radius a, the values of V_1 and V_2 may be expressed in the series of spherical harmonics given in Thomson and Tait's *Natural Philosophy*, Art. 546, Ex. II.,

$$V = 2\pi\left\{-rQ_1 + a + \tfrac{1}{2}\frac{r^2}{a}Q_2 - \frac{1.1}{2.4}\frac{r^4}{a^3}Q_4 + \frac{1.1.3}{2.4.6}\frac{r^6}{a^5}Q_6 + \&c.\right\} \text{ when } r < a, \tag{14}$$

$$V = 2\pi\left\{\tfrac{1}{2}\frac{a^2}{r} - \frac{1.1}{2.4}\frac{a^4}{r^3}Q_2 + \frac{1.1.3}{2.4.6}\frac{a^6}{r^5}Q_4 - \&c.\right\} \text{ when } r > a. \tag{15}$$

In these expressions r is the distance of the point (x, y, z) from the centre of one of the circular ends of the solenoid, and the zonal harmonics, Q_1, Q_2, &c., are those corresponding to the angle θ which r makes with the axis of the cylinder.

The first of these expressions is discontinuous when $\theta = \frac{\pi}{2}$, but we must remember that within the solenoid we must add to the magnetic force deduced from this expression a longitudinal force $4\pi n \gamma$.

677.] Let us now consider a solenoid so long that in the part of space which we consider, the terms depending on the distance from the ends may be neglected.

The magnetic induction through any closed curve drawn within the solenoid is $4\pi n \gamma A'$, where A' is the area of the projection of the curve on a plane normal to the axis of the solenoid.

If the closed curve is outside the solenoid, then, if it encloses the solenoid, the magnetic induction through it is $4\pi n \gamma A$, where A is the area of the section of the solenoid. If the closed curve does not surround the solenoid, the magnetic induction through it is zero.

If a wire be wound n' times round the solenoid, the coefficient of induction between it and the solenoid is

$$M = 4\pi n n' A. \tag{16}$$

By supposing these windings to coincide with n windings of the solenoid, we find that the coefficient of self-induction of unit of length of the solenoid, taken at a sufficient distance from its extremities, is

$$L = 4\pi n^2 A. \tag{17}$$

Near the ends of a solenoid we must take into account the terms depending on the imaginary distribution of magnetism on the plane ends of the solenoid. The effect of these terms is to make the coefficient of induction between the solenoid and a circuit which surrounds it less than the value $4\pi n A$, which it has when the circuit surrounds a very long solenoid at a great distance from either end.

Let us take the case of two circular and coaxal solenoids of the same length l. Let the radius of the outer solenoid be c_1, and let it be wound with wire so as to have n_1 windings in unit of length. Let the radius of the inner solenoid be c_2, and let the number of windings in unit of length be n_2, then the coefficient of induction between the solenoids, neglecting the effect of the ends, is

$$M = Gg, \tag{18}$$

where

$$G = 4\pi n, \tag{19}$$

and

$$g = \pi c_2^2 l n_2. \tag{20}$$

678.] To determine the effect of the positive end of the solenoids we must calculate the coefficient of induction on the outer solenoid due to the circular disk which forms the end of the inner solenoid. For this purpose we take the second expression for V, as given in equation (15), and differentiate it with respect to r. This gives the magnetic force in the direction of the radius. We then multiply this expression by $2\pi r^2 d\mu$, and integrate it with respect to μ from $\mu = 0$ to $\mu = \dfrac{z}{\sqrt{z_2+c_1^2}}$. This gives the coefficient of induction with respect to a single winding of the outer solenoid at a distance z from the positive end. We then multiply this by dz, and integrate with respect to z from $z = l$ to $z = 0$. Finally, we multiply the result by $n_1 n_2$, and so find the effect of one of the ends in diminishing the coefficient of induction.

We thus find for the value of the coefficient of mutual induction between the two cylinders,

$$M = 4\pi^2 n_1 n_2 c_2^2 (l - 2 c_1 a), \tag{21}$$

where

$$a = \frac{1}{2}\frac{c_2+l-r}{c_2} + \frac{1.3}{2.4}\cdot\frac{1}{2.3}\frac{c_2^2}{c_1^2}\left(1-\frac{c_1^3}{r^3}\right)$$
$$+ \frac{1.3.5}{2.4.6}\cdot\frac{1}{4.5}\frac{c_2^4}{c_1^4}\left(\frac{5}{12} - \frac{2}{3}\frac{c_1^3}{r^3} + 4\frac{c_1^5}{r^5} - \frac{15}{4}\frac{c_1^7}{r^7}\right) + \&c., \tag{22}$$

where r is put, for brevity, for $\sqrt{l^2+c_1^2}$.

It appears from this, that in calculating the mutual induction of two coaxal solenoids, we must use in the expression (20) instead of the true length l the corrected length $l - 2c_1 a$, in which a portion equal to $a c_1$ is supposed to be cut off at each end. When the solenoid is very long compared with its external radius,

$$a = \frac{1}{2} + \frac{1}{16}\frac{c_2^2}{c_1^2} + \frac{5}{768}\frac{c_2^4}{c_1^4} + \&c. \tag{23}$$

679.] When a solenoid consists of a number of layers of wire of such a diameter that there are n layers in unit of length, the number of layers in the thickness dr is $n\,dr$, and we have

$$G = 4\pi\int n^2 dr, \quad \text{and} \quad g = \pi l \int n^2 r^2 dr. \tag{24}$$

If the thickness of the wire is constant, and if the induction take place between an external coil whose outer and inner radii are x and y respectively, and an inner coil whose outer and inner radii are y and z, then, neglecting the effect of the ends,

$$Gg = \frac{4}{3}\pi^2 l n_1^2 n_2^2 (x-y)(y^3-z^3). \tag{25}$$

That this may be a maximum, x and z being given, and y variable,

$$x = \tfrac{4}{3} y - \tfrac{1}{3} \frac{z^3}{y^2}. \tag{26}$$

This equation gives the best relation between the depths of the primary and secondary coil for an induction-machine without an iron core.

If there is an iron core of radius z, then G remains as before, but

$$g = \pi l \int n^2 (r^2 + 4 \pi \kappa z^2)\, dr, \tag{27}$$

$$= \pi l n^2 \left(\frac{y^3 - z^3}{3} + 4 \pi \kappa z^2 (y - z)\right). \tag{28}$$

If y is given, the value of z which gives the maximum value of g is

$$z = \tfrac{2}{3} y \frac{18 \pi \kappa}{18 \pi \kappa + 1}. \tag{29}$$

When, as in the case of iron, κ is a large number, $z = \tfrac{2}{3} y$, nearly.

If we now make x constant, and y and z variable, we obtain the maximum value of Gg when

$$x : y : z :: 4 : 3 : 2. \tag{30}$$

The coefficient of self-induction of a long solenoid whose outer and inner radii are x and y, and having a long iron core whose radius is z, is

$$L = \tfrac{2}{3} \pi^2 l n^4 (x - y)^2 (x^2 + 2xy + 3y^2 + 24 \pi \kappa z^2). \tag{31}$$

680.] We have hitherto supposed the wire to be of uniform thickness. We shall now determine the law according to which the thickness must vary in the different layers in order that, for a given value of the resistance of the primary or the secondary coil, the value of the coefficient of mutual induction may be a maximum.

Let the resistance of unit of length of a wire, such that n windings occupy unit of length of the solenoid, be ρn^2.

The resistance of the whole solenoid is

$$R = 2 \pi l \int n^4 r\, dr. \tag{32}$$

The condition that, with a given value of R, G may be a maximum is $\dfrac{dG}{dr} = C \dfrac{dR}{dr}$, where C is some constant.

This gives n^2 proportional to $\dfrac{1}{r}$, or the diameter of the wire of the exterior coil must be proportional to the square root of the radius.

In order that, for a given value of R, g may be a maximum

$$n^2 = C\left(r + \frac{4 \pi \kappa z^2}{r}\right). \tag{33}$$

Hence, if there is no iron core, the diameter of the wire of the interior coil should be inversely as the square root of the radius, but if there is a core of iron having a high capacity for magnetization, the diameter of the wire should be more nearly directly proportional to the square root of the radius of the layer.

An Endless Solenoid.

681.] If a solid be generated by the revolution of a plane area A about an axis in its own plane, not cutting it, it will have the form of a ring. If this ring be coiled with wire, so that the windings of the coil are in planes passing through the axis of the ring, then, if n is the whole number of windings, the current-function of the layer of wire is $\phi = \frac{1}{2\pi} n\gamma\theta$, where θ is the angle of azimuth about the axis of the ring.

If Ω is the magnetic potential inside the ring and Ω' that outside, then
$$\Omega - \Omega' = 4\pi\phi + C = 2n\gamma\theta + C.$$
Outside the ring Ω' must satisfy Laplace's equation, and must vanish at an infinite distance. From the nature of the problem it must be a function of θ only. The only value of Ω' which fulfils these conditions is zero. Hence
$$\Omega' = 0, \qquad \Omega = 2n\gamma\theta + C.$$

The magnetic force at any point within the ring is perpendicular to the plane passing through the axis, and is equal to $2n\gamma\frac{1}{r}$ where r is the distance from the axis. Outside the ring there is no magnetic force.

If the form of a closed curve be given by the coordinates z, r, and θ of its tracing point as functions of s, its length from a fixed point, the magnetic induction through the closed curve is
$$2n\gamma\int_0^s \frac{z}{r}\frac{dr}{ds}ds$$
taken round the curve, provided the curve is wholly inside the ring. If the curve lies wholly without the ring, but embraces it, the magnetic induction through it is
$$2n\gamma\int_0^{s'} \frac{z'}{r'}\frac{dr'}{ds'}ds' = 2n\gamma a,$$
where the accented coordinates refer not to the closed curve, but to a single winding of the solenoid.

The magnetic induction through any closed curve embracing the ring is therefore the same, and equal to $2n\gamma a$, where a is the linear quantity $\int_0^{s'} \frac{z'}{r'} \frac{dr'}{ds'} ds'$. If the closed curve does not embrace the ring, the magnetic induction through it is zero.

Let a second wire be coiled in any manner round the ring, not necessarily in contact with it, so as to embrace it n' times. The induction through this wire is $2nn'\gamma a$, and therefore M, the coefficient of induction of the one coil on the other, is $M = 2nn'a$.

Since this is quite independent of the particular form or position of the second wire, the wires, if traversed by electric currents, will experience no mechanical force acting between them. By making the second wire coincide with the first, we obtain for the coefficient of self-induction of the ring-coil

$$L = 2n^2 a.$$

CHAPTER XIII.

PARALLEL CURRENTS.

Cylindrical Conductors.

682.] In a very important class of electrical arrangements the current is conducted through round wires of nearly uniform section, and either straight, or such that the radius of curvature of the axis of the wire is very great compared with the radius of the transverse section of the wire. In order to be prepared to deal mathematically with such arrangements, we shall begin with the case in which the circuit consists of two very long parallel conductors, with two pieces joining their ends, and we shall confine our attention to a part of the circuit which is so far from the ends of the conductors that the fact of their not being infinitely long does not introduce any sensible change in the distribution of force.

We shall take the axis of z parallel to the direction of the conductors, then, from the symmetry of the arrangements in the part of the field considered, everything will depend on H, the component of the vector-potential parallel to z.

The components of magnetic induction become, by equations (A),

$$a = \frac{dH}{dy}, \tag{1}$$

$$b = -\frac{dH}{dx}, \tag{2}$$

$$c = 0.$$

For the sake of generality we shall suppose the coefficient of magnetic induction to be μ, so that $a = \mu\,\alpha$, $b = \mu\,\beta$, where α and β are the components of the magnetic force.

The equations (E) of electric currents, Art. 607, give

$$u = 0, \qquad v = 0, \qquad 4\,\pi\,w = \frac{d\beta}{dx} - \frac{d\alpha}{dy}. \tag{3}$$

683.] If the current is a function of r, the distance from the axis of z, and if we write

$$x = r\cos\theta, \quad \text{and} \quad y = r\sin\theta, \tag{4}$$

and β for the magnetic force, in the direction in which θ is measured perpendicular to the plane through the axis of z, we have

$$4\pi w = \frac{d\beta}{dr} + \frac{1}{r}\beta = \frac{1}{r}\frac{d}{dr}(\beta r). \tag{5}$$

If C is the whole current flowing through a section bounded by a circle in the plane xy, whose centre is the origin and whose radius is r,

$$C = \int_0^r 2\pi r w\, dr = \tfrac{1}{2}\beta r. \tag{6}$$

It appears, therefore, that the magnetic force at a given point due to a current arranged in cylindrical strata, whose common axis is the axis of z, depends only on the total strength of the current flowing through the strata which lie between the given point and the axis, and not on the distribution of the current among the different cylindrical strata.

For instance, let the conductor be a uniform wire of radius a, and let the total current through it be C, then, if the current is uniformly distributed through all parts of the section, w will be constant, and

$$C = \pi w a^2. \tag{7}$$

The current flowing through a circular section of radius r, r being less than a, is $C' = \pi w r^2$. Hence at any point within the wire,

$$\beta = \frac{2C'}{r} = 2C\frac{r}{a^2}. \tag{8}$$

Outside the wire

$$\beta = 2\frac{C}{r}. \tag{9}$$

In the substance of the wire there is no magnetic potential, for within a conductor carrying an electric current the magnetic force does not fulfil the condition of having a potential.

Outside the wire the magnetic potential is

$$\Omega = 2C\theta. \tag{10}$$

Let us suppose that instead of a wire the conductor is a metal tube whose external and internal radii are a_1 and a_2, then, if C is the current through the tubular conductor,

$$C = \pi w (a_1^2 - a_2^2). \tag{11}$$

The magnetic force within the tube is zero. In the metal of the tube, where r is between a_1 and a_2,

$$\beta = 2C\frac{1}{a_1^2 - a_2^2}\left(r - \frac{a_2^2}{r}\right), \tag{12}$$

and outside the tube,
$$\beta = 2\frac{C}{r}, \tag{13}$$
the same as when the current flows through a solid wire.

684.] The magnetic induction at any point is $b = \mu\beta$, and since, by equation (2),
$$b = -\frac{dH}{dr}, \tag{14}$$
$$H = -\int \mu\beta\, dr. \tag{15}$$

The value of H outside the tube is
$$A - 2\mu_0 C \log r, \tag{16}$$
where μ_0 is the value of μ in the space outside the tube, and A is a constant, the value of which depends on the position of the return current.

In the substance of the tube,
$$H = A - 2\mu_0 C \log a_1 + \frac{\mu C}{a_1^2 - a_2^2}\left(a_1^2 - r^2 + 2a_2^2 \log\frac{r}{a_1}\right). \tag{17}$$

In the space within the tube H is constant, and
$$H = A - 2\mu_0 C \log a_1 + \mu C\left(1 + \frac{2a_2^2}{a_1^2 - a_2^2}\log\frac{a_2}{a_1}\right). \tag{18}$$

685.] Let the circuit be completed by a return current, flowing in a tube or wire parallel to the first, the axes of the two currents being at a distance b. To determine the kinetic energy of the system we have to calculate the integral
$$T = \tfrac{1}{2}\iiint Hw\, dx\, dy\, dz. \tag{19}$$

If we confine our attention to that part of the system which lies between two planes perpendicular to the axes of the conductors, and distant l from each other, the expression becomes
$$T = \tfrac{1}{2} l \iint Hw\, dx\, dy. \tag{20}$$

If we distinguish by an accent the quantities belonging to the return current, we may write this
$$\frac{2T}{l} = \iint Hw'\, dx'\, dy' + \iint H'w\, dx\, dy + \iint Hw\, dx\, dy + \iint H'w'\, dx'\, dy'. \tag{21}$$

Since the action of the current on any point outside the tube is the same as if the same current had been concentrated at the axis of the tube, the mean value of H for the section of the return current is $A - 2\mu_0 C \log b$, and the mean value of H' for the section of the positive current is $A' - 2\mu_0 C' \log b$.

Hence, in the expression for T, the first two terms may be written

$$AC' - 2\mu_0 CC' \log b, \text{ and } A'C - 2\mu_0 CC' \log b.$$

Integrating the two latter terms in the ordinary way, and adding the results, remembering that $C + C' = 0$, we obtain the value of the kinetic energy T. Writing this $\frac{1}{2}LC^2$, where L is the coefficient of self-induction of the system of two conductors, we find as the value of L for unit of length of the system

$$\frac{L}{l} = 2\mu_0 \log \frac{b^2}{a_1 a_1'} + \frac{1}{2}\mu \left\{\frac{a_1^2 - 3a_2^2}{a_1^2 - a_2^2} + \frac{4a_2^4}{(a_1^2 - a_2^2)^2} \log \frac{a_1}{a_2}\right\}$$
$$+ \frac{1}{2}\mu' \left\{\frac{a_1'^2 - 3a_2'^2}{a_1'^2 - a_2'^2} + \frac{4a_2'^4}{(a_1'^2 - a_2'^2)^2} \log \frac{a_1'}{a_2'}\right\}. \tag{22}$$

If the conductors are solid wires, a_2 and a_2' are zero, and

$$\frac{L}{l} = 2\mu_0 \log \frac{b^2}{a_1 a_1'} + \frac{1}{2}(\mu + \mu'). \tag{23}$$

It is only in the case of iron wires that we need take account of the magnetic induction in calculating their self-induction. In other cases we may make μ_0, μ, and μ' all equal to unity. The smaller the radii of the wires, and the greater the distance between them, the greater is the self-induction.

To find the Repulsion, X, between the Two Portions of Wire.

686.] By Art. 580 we obtain for the force tending to increase b,

$$X = \frac{1}{2}\frac{dL}{db}C^2,$$
$$= 2\mu_0 \frac{l}{b} C^2, \tag{24}$$

which agrees with Ampère's formula, when $\mu_0 = 1$, as in air.

687.] If the length of the wires is great compared with the distance between them, we may use the coefficient of self-induction to determine the tension of the wires arising from the action of the current.

If Z is this tension,

$$Z = \frac{1}{2}\frac{dL}{dl}C^2,$$
$$= C^2 \left\{\mu_0 \log \frac{b^2}{a_1 a_1'} + \frac{\mu}{2}\right\}. \tag{25}$$

In one of Ampère's experiments the parallel conductors consist of two troughs of mercury connected with each other by a floating bridge of wire. When a current is made to enter at the extremity of one of the troughs, to flow along it till it reaches one extremity of the floating wire, to pass into the other trough through the floating bridge, and so to return along the second trough, the floating bridge moves along the troughs so as to lengthen the part of the mercury traversed by the current.

Professor Tait has simplified the electrical conditions of this experiment by substituting for the wire a floating siphon of glass filled with mercury, so that the current flows in mercury throughout its course.

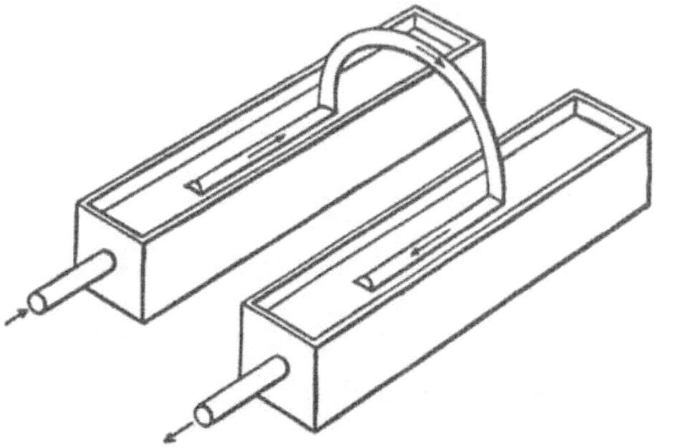

Fig. 40.

This experiment is sometimes adduced to prove that two elements of a current in the same straight line repel one another, and thus to shew that Ampère's formula, which indicates such a repulsion of collinear elements, is more correct than that of Grassmann, which gives no action between two elements in the same straight line; Art. 526.

But it is manifest that since the formulae both of Ampère and of Grassmann give the same results for closed circuits, and since we have in the experiment only a closed circuit, no result of the experiment can favour one more than the other of these theories.

In fact, both formulae lead to the very same value of the repulsion as that already given, in which it appears that b, the distance between the parallel conductors is an important element.

When the length of the conductors is not very great compared with their distance apart, the form of the value of L becomes somewhat more complicated.

688.] As the distance between the conductors is diminished, the value of L diminishes. The limit to this diminution is when the wires are in contact, or when $b = a_1 + a_2$. In this case

$$L = 2\,l \log\left(\frac{(a_1 + a_2)^2}{a_1 a_2} + \tfrac{1}{2}\right). \qquad (26)$$

This is a minimum when $a_1 = a_2$, and then

$$L = 2l(\log 4 + \tfrac{1}{2}),$$
$$= 2l(1.8863),$$
$$= 3.7726\,l. \tag{27}$$

This is the smallest value of the self-induction of a round wire doubled on itself, the whole length of the wire being $2l$.

Since the two parts of the wire must be insulated from each other, the self-induction can never actually reach this limiting value. By using broad flat strips of metal instead of round wires the self-induction may be diminished indefinitely.

On the Electromotive Force required to produce a Current of Varying Intensity along a Cylindrical Conductor.

689.] When the current in a wire is of varying intensity, the electromotive force arising from the induction of the current on itself is different in different parts of the section of the wire, being in general a function of the distance from the axis of the wire as well as of the time. If we suppose the cylindrical conductor to consist of a bundle of wires all forming part of the same circuit, so that the current is compelled to be of uniform strength in every part of the section of the bundle, the method of calculation which we have hitherto used would be strictly applicable. If, however, we consider the cylindrical conductor as a solid mass in which electric currents are free to flow in obedience to electromotive force, the intensity of the current will not be the same at different distances from the axis of the cylinder, and the electromotive forces themselves will depend on the distribution of the current in the different cylindric strata of the wire.

The vector-potential H, the density of the current w, and the electromotive force at any point, must be considered as functions of the time and of the distance from the axis of the wire.

The total current, C, through the section of the wire, and the total electromotive force, E, acting round the circuit, are to be regarded as the variables, the relation between which we have to find.

Let us assume as the value of H,

$$H = S + T_0 + T_1 r^2 + \&c. + T_n r^{2n}, \tag{1}$$

where S, T_0, T_1, &c. are functions of the time.

Then, from the equation

$$\frac{d^2H}{dr^2} + \frac{1}{r}\frac{dH}{dr} = -4\pi w, \tag{2}$$

we find

$$-\pi w = T_1 + \&c + n^2 T_n r^{2n-2}. \tag{3}$$

If ρ denotes the specific resistance of the substance per unit of volume, the electromotive force at any point is ρw, and this may be expressed in terms of the electric potential and the vector potential H by equations (B), Art. 598,

$$\rho w = -\frac{d\Psi}{dz} - \frac{dH}{dt}, \tag{4}$$

or

$$-\rho w = \frac{d\Psi}{dz} + \frac{dS}{dt} + \frac{dT_0}{dt} + \frac{dT_1}{dt} r^2 + \&c. + \frac{dT_n}{dt} r^{2n}. \tag{5}$$

Comparing the coefficients of like powers of r in equations (3) and (5),

$$T_1 = \frac{\pi}{\rho}\left(\frac{d\Psi}{dz} + \frac{dS}{dt} + \frac{dT_0}{dt}\right), \tag{6}$$

$$T_2 = \frac{\pi}{\rho}\frac{dT_1}{dt}, \tag{7}$$

$$T_n = \frac{\pi}{\rho}\frac{1}{n^2}\frac{dT_{n-1}}{dt}. \tag{8}$$

Hence we may write

$$\frac{dS}{dt} = -\frac{d\Psi}{dz}, \tag{9}$$

$$T_0 = T, \qquad T_1 = \frac{\pi}{\rho}\frac{dT}{dt}, \ldots \qquad T_n = \frac{\pi^n}{\rho^n}\frac{1}{(\underline{|n})^2}\frac{d^nT}{dt^n}. \tag{10}$$

690.] To find the total current C, we must integrate w over the section of the wire whose radius is a,

$$C = 2\pi\int_0^a w\, r\, dr. \tag{11}$$

Substituting the value of πw from equation (3), we obtain

$$C = -(T_1 a^2 + \&c. + n\, T_n a^{2n}). \tag{12}$$

The value of H at any point outside the wire depends only on the total current C, and not on the mode in which it is distributed within the wire. Hence we may assume that the value of H at the surface of the wire is AC, where A is a constant to be determined by calculation from the general form of the circuit. Putting $H = AC$ when $r = a$, we obtain

$$AC = S + T_0 + T_1 a^2 + \&c. + T_n a^{2n}. \tag{13}$$

If we now write $\frac{\pi a^2}{\rho} = \alpha$, α is the value of the conductivity of unit of length of the wire, and we have

$$C = -\left(\alpha\frac{dT}{dt} + \frac{2\alpha^2}{1^2 2^2}\frac{d^2T}{dt^2} + \&c. + \frac{n\alpha^n}{(\underline{|n})^2}\frac{d^nT}{dt^n} + \&c.\right), \tag{14}$$

$$AC - S = T + \alpha\frac{dT}{dt} + \frac{\alpha^2}{1^2 2^2}\frac{d^2T}{dt^2} + \&c. + \frac{\alpha^n}{(\underline{|n})^2}\frac{d^nT}{dt^n} + \&c. \tag{15}$$

Eliminating T from these two equations, we find

$$a\left(A\frac{dC}{dt}-\frac{dS}{dt}\right)+C+\tfrac{1}{2}a\frac{dC}{dt}-\tfrac{1}{12}a^2\frac{d^2C}{dt^2}+\tfrac{1}{48}a^3\frac{d^3C}{dt^3}$$
$$-\tfrac{1}{180}a^4\frac{d^4C}{dt^4}+\&c. = 0. \quad (16)$$

If l is the whole length of the circuit, R its resistance, and E the electromotive force due to other causes than the induction of the current on itself,

$$\frac{dS}{dt}=-\frac{E}{l}, \qquad a=\frac{l}{R}, \quad (17)$$

$$E=RC+l(A+\tfrac{1}{2})\frac{dC}{dt}-\tfrac{1}{12}\frac{l^2}{R}\frac{d^2C}{dt^2}+\tfrac{1}{48}\frac{l^3}{R^2}\frac{d^3C}{dt^3}-\tfrac{1}{180}\frac{l^4}{R^3}\frac{d^4C}{dt^4}+\&c. \quad (18)$$

The first term, RC, of the right-hand member of this equation expresses the electromotive force required to overcome the resistance according to Ohm's law.

The second term, $l(A+\frac{1}{2})\frac{dC}{dt}$, expresses the electromotive force which would be employed in increasing the electrokinetic momentum of the circuit, on the hypothesis that the current is of uniform strength at every point of the section of the wire.

The remaining terms express the correction of this value, arising from the fact that the current is not of uniform strength at different distances from the axis of the wire. The actual system of currents has a greater degree of freedom than the hypothetical system, in which the current is constrained to be of uniform strength throughout the section. Hence the electromotive force required to produce a rapid change in the strength of the current is somewhat less than it would be on this hypothesis.

The relation between the time-integral of the electromotive force and the time-integral of the current is

$$\int E\,dt = R\int C\,dt + l(A+\tfrac{1}{2})C-\tfrac{1}{12}\frac{l^2}{R}\frac{dC}{dt}+\&c. \quad (19)$$

If the current before the beginning of the time has a constant value C_0, and if during the time it rises to the value C_1, and remains constant at that value, then the terms involving the differential coefficients of C vanish at both limits, and

$$\int E\,dt = R\int C\,dt + l(A+\tfrac{1}{2})(C_1-C_0), \quad (20)$$

the same value of the electromotive impulse as if the current had been uniform throughout the wire.

*On the Geometrical Mean Distance of Two Figures in a Plane.**

691.] In calculating the electromagnetic action of a current flowing in a straight conductor of any given section on the current in a parallel conductor whose section is also given, we have to find the integral

$$\iiiint \log r \, dx\, dy\, dx'\, dy',$$

where $dx\,dy$ is an element of the area of the first section, $dx'\,dy'$ an element of the second section, and r the distance between these elements, the integration being extended first over every element of the first section, and then over every element of the second.

If we now determine a line R, such that this integral is equal to

$$A_1 A_2 \log R,$$

where A_1 and A_2 are the areas of the two sections, the length of R will be the same whatever unit of length we adopt, and whatever system of logarithms we use. If we suppose the sections divided into elements of equal size, then the logarithm of R, multiplied by the number of pairs of elements, will be equal to the sum of the logarithms of the distances of all the pairs of elements. Here R may be considered as the geometrical mean of all the distances between pairs of elements. It is evident that the value of R must be intermediate between the greatest and the least values of r.

If R_A and R_B are the geometric mean distances of two figures, A and B, from a third, C, and if R_{A+B} is that of the sum of the two figures from C, then

$$(A+B)\log R_{A+B} = A \log R_A + B \log R_B.$$

By means of this relation we can determine R for a compound figure when we know R for the parts of the figure.

692.] EXAMPLES.

(1) Let R be the mean distance from the point O to the line AB. Let OP be perpendicular to AB, then

$$AB(\log R + 1) = AP \log OA + PB \log OB + OP\, \widehat{AOB}.$$

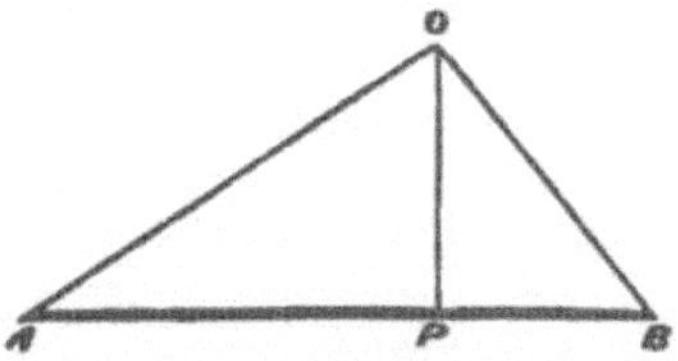

Fig. 41.

* *Trans. R. S. Edin.*, 1871-2.

(2) For two lines (Fig. 42) of lengths a and b drawn perpendicular to the extremities of a line of length c and on the same side of it.

$$\begin{aligned} ab(2\log R+3) = {} & (c^2-(a-b)^2)\log\sqrt{c^2+(a-b)^2}+c^2\log c \\ & +(a^2-c^2)\log\sqrt{a^2+c^2}+(b^2-c^2)\log\sqrt{b^2+c^2} \\ & -c(a-b)\tan^{-1}\frac{a-b}{c}+ac\tan^{-1}\frac{a}{c}+bc\tan^{-1}\frac{b}{c}. \end{aligned}$$

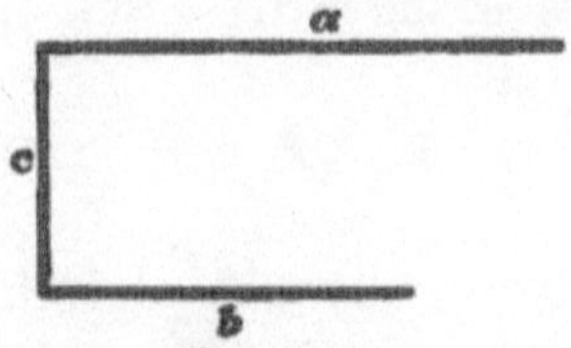

Fig. 42.

(3) For two lines, PQ and RS (Fig. 43), whose directions intersect at O.

$$\begin{aligned} PQ.RS(2\log R+3) = {} & \log PR(2OP.OR\sin^2 O-PR^2\cos O) \\ & +\log QS(2OQ.OS\sin^2 O-QS^2\cos O) \\ & -\log PS(2OP.OS\sin^2 O-PS^2\cos O) \\ & -\log QR(2OQ.OR\sin^2 O-QR^2\cos O) \\ & -\sin O\{OP^2.\widehat{SPR}-OQ^2.\widehat{SQR}+OR^2.\widehat{PRQ}-OS^2.\widehat{PSQ}\}. \end{aligned}$$

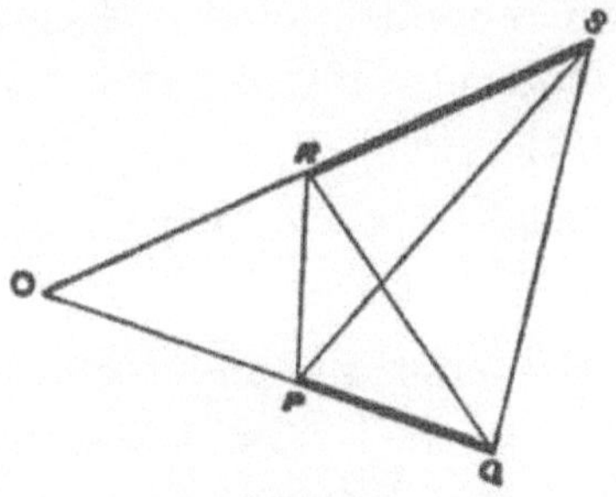

Fig. 43.

(4) For a point O and a rectangle $ABCD$ (Fig. 44). Let OP, OQ, OR, OS, be perpendiculars on the sides, then

$$\begin{aligned} AB.AD(2\log R+3) = {} & 2.OP.OQ\log OA+2.OQ.OR\log OB \\ & +2.OR.OS\log OC+2.OS.OP\log OD \\ & +OP^2.\widehat{DOA}+OQ^2.\widehat{AOB} \\ & +OR^2.\widehat{BOC}+OS^2.\widehat{COD}. \end{aligned}$$

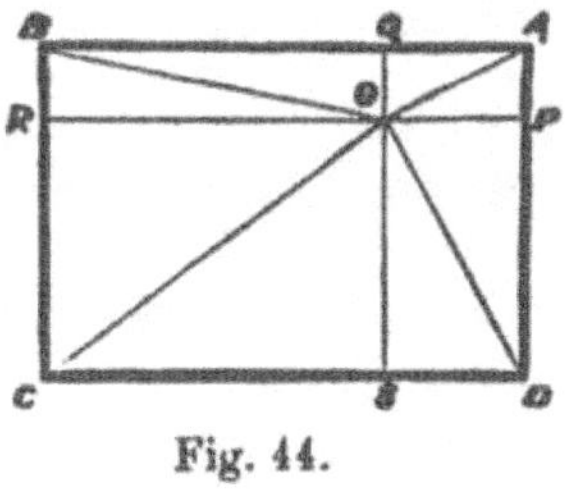

Fig. 44.

(5) It is not necessary that the two figures should be different, for we may find the geometric mean of the distances between every pair of points in the same figure. Thus, for a straight line of length a,

$$\log R = \log a - \tfrac{3}{2},$$

or

$$R = ae^{-\frac{3}{2}},$$

$$R = 0.22313\,a.$$

(6) For a rectangle whose sides are a and b,

$$\log R = \log\sqrt{a^2+b^2} - \tfrac{1}{6}\frac{a^2}{b^2}\log\sqrt{1+\frac{b^2}{a^2}} - \tfrac{1}{6}\frac{b^2}{a^2}\log\sqrt{1+\frac{a^2}{b^2}}$$

$$+ \tfrac{2}{3}\frac{a}{b}\tan^{-1}\frac{b}{a} + \tfrac{2}{3}\frac{b}{a}\tan^{-1}\frac{a}{b} - \tfrac{25}{12}.$$

When the rectangle is a square, whose side is a,

$$\log R = \log a + \tfrac{1}{3}\log 2 + \frac{\pi}{3} - \tfrac{25}{12},$$

$$R = 0.44705\,a.$$

(7) The geometric mean distance of a point from a circular line is equal to the greater of the two quantities, its distance from the centre of the circle, and the radius of the circle.

(8) Hence the geometric mean distance of any figure from a ring bounded by two concentric circles is equal to its geometric mean distance from the centre if it is entirely outside the ring, but if it is entirely within the ring

$$\log R = \frac{a_1^2\log a_1 - a_2^2\log a_2}{a_1^2 - a_2^2} - \tfrac{1}{2},$$

where a_1 and a_2 are the outer and inner radii of the ring. R is in this case independent of the form of the figure within the ring.

(9) The geometric mean distance of all pairs of points in the ring is found from the equation

$$\log R = \log a_1 - \frac{a_2^4}{(a_1^2-a_2^2)^2}\log\frac{a_1}{a_2} + \tfrac{1}{4}\,\frac{3\,a_2^2 - a_1^2}{a_1^2 - a_2^2}.$$

For a circular area of radius a, this becomes

$$\log R = \log a - \tfrac{1}{4},$$

or

$$R = ae^{-\frac{1}{4}},$$

$$R = 0.7788\,a.$$

For a circular line it becomes

$$R = a.$$

693.] In calculating the coefficient of self-induction of a coil of uniform section, the radius of curvature being great compared with the dimensions of the transverse section, we first determine the geometric mean of the distances of every pair of points of the section by the method already described, and then we calculate the coefficient of mutual induction between two linear conductors of the given form, placed at this distance apart.

This will be the coefficient of self-induction when the total current in the coil is unity, and the current is uniform at all points of the section.

But if there are n windings in the coil we must multiply the coefficient already obtained by n^2, and thus we shall obtain the coefficient of self-induction on the supposition that the windings of the conducting wire fill the whole section of the coil.

But the wire is cylindric, and is covered with insulating material, so that the current, instead of being uniformly distributed over the section, is concentrated in certain parts of it, and this increases the coefficient of self-induction. Besides this, the currents in the neighbouring wires have not the same action on the current in a given wire as a uniformly · istributed current.

The corrections arising from these considerations may be determined by the method of the geometric mean distance. They are proportional to the length of the whole wire of the coil, and may be expressed as numerical quantities, by which we must multiply the length of the wire in order to obtain the correction of the coefficient of self-induction.

Let the diameter of the wire be d. It is covered with insulating material, and wound into a coil. We shall suppose that the sections of the wires are in square order, as in Fig. 45, and that the distance between the axis of each wire and that of the next is D, whether in the direction of the breadth or the depth of the coil. D is evidently greater than d.

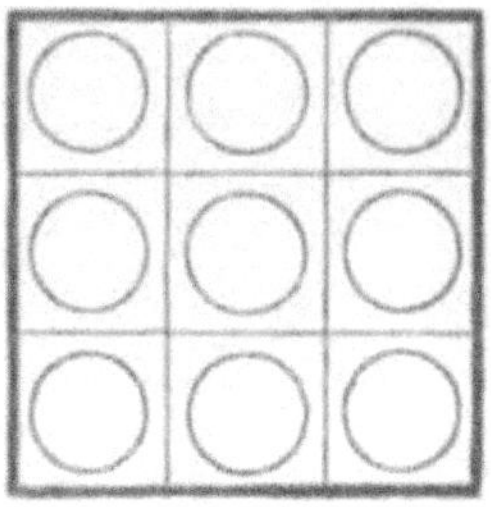

Fig. 45.

We have first to determine th· excess of self-induction of unit of length of a cylindric wire of diameter d over that of unit of length of a square wire of side D, or

$$2 \log \frac{R \text{ for the square}}{R \text{ for the circle}}$$

$$= 2\left(\log \frac{D}{d} + \tfrac{4}{3} \log 2 + \frac{\pi}{3} - \tfrac{11}{6}\right)$$

$$= 2\left(\log \frac{D}{d} + 0.1380606\right).$$

The inductive action of the eight nearest round wires on the wire under consideration is less than that of the corresponding eight square wires on the square wire in the middle by $2 \times (.01971)$.

The corrections for the wires at a greater distance may be neglected, and the total correction may be written

$$2\left(\log_e \frac{D}{d} + 0.11835\right).$$

The final value of the self-induction is therefore

$$L = n^2 M + 2l\left(\log_e \frac{D}{d} + 0.11835\right),$$

where n is the number of windings, and l the length of the wire, M the mutual induction of two circuits of the form of the mean wire of the coil placed at a distance R from each other, where R is the mean geometric distance between pairs of points of the section. D is the distance between consecutive wires, and d the diameter of the wire.

CHAPTER XIV.

CIRCULAR CURRENTS.

Magnetic Potential due to a Circular Current.

694.] THE magnetic potential at a given point, due to a circuit carrying a unit current, is numerically equal to the solid angle subtended by the circuit at that point; see Arts. 409, 485.

When the circuit is circular, the solid angle is that of a cone of the second degree, which, when the given point is on the axis of the circle, becomes a right cone. When the point is not on the axis, the cone is an elliptic cone, and its solid angle is numerically equal to the area of the spherical ellipse which it traces on a sphere whose radius is unity.

This area can be expressed in finite terms by means of elliptic integrals of the third kind. We shall find it more convenient to expand it in the form of an infinite series of spherical harmonics, for the facility with which mathematical operations may be performed on the general term of such a series more than counterbalances the trouble of calculating a number of terms sufficient to ensure practical accuracy.

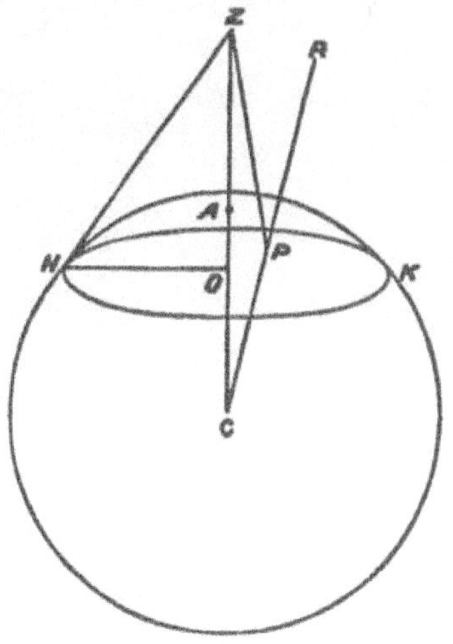

Fig. 46.

For the sake of generality we shall assume the origin at any point on the axis of the circle, that is to say, on the line through the centre perpendicular to the plane of the circle.

Let O (Fig. 46) be the centre of the circle, C the point on the axis which we assume as origin, H a point on the circle.

Describe a sphere with C as centre, and CH as radius. The circle will lie on this sphere, and will form a small circle of the sphere of angular radius a.

Let
$$CH = c,$$
$$OC = b = c\cos a,$$
$$OH = a = c\sin a.$$

Let A be the pole of the sphere, and Z any point on the axis, and let $CZ=z$.

Let R be any point in space, and let $CR = r$, and $ACR = \theta$.

Let P be the point when CR cuts the sphere.

The magnetic potential due to the circular current is equal to that due to a magnetic shell of strength unity bounded by the current. As the form of the surface of the shell is indifferent, provided it is bounded by the circle, we may suppose it to coincide with the surface of the sphere.

We have shewn in Art. 670 that if P is the potential due to a stratum of matter of surface-density unity, spread over the surface of the sphere within the small circle, the potential due to a magnetic shell of strength unity and bounded by the same circle is

$$\omega = \frac{1}{c}\frac{d}{dr}(rP).$$

We have in the first place, therefore, to find P.

Let the given point be on the axis of the circle at Z, then the part of the potential at Z due to an element dS of the spherical surface at P is

$$\frac{dS}{ZP}.$$

This may be expanded in one of the two series of spherical harmonics,

$$\frac{dS}{c}\left\{Q_0 + Q_1\frac{z}{c} + \&c. + Q_i\frac{z^i}{c^i} + \&c.\right\},$$

or

$$\frac{dS}{z}\left\{Q_0 + Q_1\frac{c}{z} + \&c. + Q^i\frac{z^i}{z^i} + \&c.\right\},$$

the first series being convergent when z is less than c, and the second when z is greater than c.

Writing
$$dS = -c^2\,d\mu\,d\phi,$$
and integrating with respect to ϕ between the limits 0 and 2π, and with respect to μ between the limits $\cos a$ and 1, we find

$$P = 2\pi c\left\{\int_\mu^1 Q_0\,d\mu + \&c. + \frac{z^i}{c^i}\int_\mu^1 Q_i\,d\mu\right\}, \tag{1}$$

or
$$P' = 2\pi\frac{c^2}{z}\left\{\int_\mu^1 Q_0\,d\mu + \&c. + \frac{c^i}{z^i}\int_\mu^1 Q_i\,d\mu\right\}. \tag{1'}$$

By the characteristic equation of Q_i,

$$i(i+1)Q_i + \frac{d}{d\mu}\left[(1-\mu^2)\frac{dQ_i}{d\mu}\right] = 0.$$

Hence
$$\int_\mu^1 Q_i\,d\mu = \frac{1-\mu^2}{i(i+1)}\frac{dQ_i}{d\mu}. \tag{2}$$

This expression fails when $i = 0$, but since $Q_0 = 1$,
$$\int_\mu^1 Q_0\,d\mu = 1-\mu. \tag{3}$$

As the function $\frac{dQ_i}{d\mu}$ occurs in every part of this investigation we shall denote it by the abbreviated symbol Q_i'. The values of Q_i' corresponding to several values of i are given in Art. 698.

We are now able to write down the value of P for any point R, whether on the axis or not, by substituting r for z, and multiplying each term by the zonal harmonic of θ of the same order. For P must be capable of expansion in a series of zonal harmonics of θ with proper coefficients. When $\theta = 0$ each of the zonal harmonics becomes equal to unity, and the point R lies on the axis. Hence the coefficients are the terms of the expansion of P for a point on the axis. We thus obtain the two series
$$P = 2\pi c\left\{1-\mu+\&c+\frac{1-\mu^2}{i(i+1)}\frac{r^i}{c^i}Q_i'(a)\,Q_i(\theta)\right\}, \tag{4}$$
or
$$P' = 2\pi\frac{c^2}{r}\left\{1-\mu+\&c.+\frac{1-\mu^2}{i(i+1)}\frac{c^i}{r^i}Q_i'(a)\,Q_i(\theta)\right\}. \tag{4'}$$

695.] We may now find ω, the magnetic potential of the circuit, by the method of Art. 670, from the equation
$$\omega = \frac{1}{c}\frac{d}{dr}(Pr). \tag{5}$$

We thus obtain the two series
$$\omega = -2\pi\left\{1-\cos a+\&c.+\frac{\sin^2 a}{i}\frac{r^i}{c^i}Q_i'(a)Q_i(\theta)+\&c.\right\}, \tag{6}$$
or
$$\omega' = 2\pi\sin^2 a\left\{\tfrac{1}{2}\frac{c^2}{r^2}Q_1'(a)Q_1(\theta)+\&c.+\frac{1}{i+1}\frac{c^{i+1}}{r^{i+1}}Q_i'(a)Q_i(\theta)\right\}. \tag{6'}$$

The series (6) is convergent for all values of r less than c, and the series (6′) is convergent for all values of r greater than c. At the surface of the sphere, where $r = c$, the two series give the same value for ω when θ is greater than a, that is, for points not occupied by the magnetic shell, but when θ is less than a, that is, at points on the magnetic shell,
$$\omega' = \omega + 4\pi. \tag{7}$$

If we assume O, the centre of the circle, as the origin of coordinates, we must put $a = \frac{\pi}{2}$, and the series become

$$\omega = -2\pi\left\{1+\frac{r}{c}Q_1(\theta)+\&c.+(-)^s\frac{1.3.(2s-1)}{2.4.2s}\frac{r^{2s+1}}{c^{2s+1}}Q_{2s+1}(\theta)\right\}, \quad (8)$$

$$\omega' = 2\pi\left\{\tfrac{1}{2}\frac{c^2}{r^2}Q_1(\theta)+\&c.+(-)^s\frac{1.3\ldots(2s+1)}{2.4\ldots(2s+2)}\frac{c^{2s+2}}{r^{2s+2}}Q_{2s+1}(\theta)\right\}, \quad (8')$$

where the orders of all the harmonics are odd *.

On the Potential Energy of two Circular Currents.

696.] Let us begin by supposing the two magnetic shells which are equivalent to the currents to be portions of two concentric spheres, their radii being c_1 and c_2, of which c_1 is the greater (Fig. 47). Let us also suppose that the axes of the two shells coincide, and that a_1 is the angle subtended by the radius of the first shell, and a_2 the angle subtended by the radius of the second shell at the centre C.

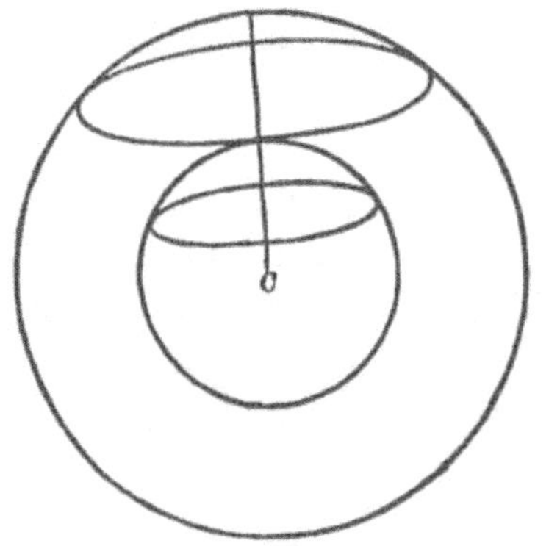

Fig. 47.

Let ω_1 be the potential due to the first shell at any point within it, then the work required to carry the second shell to an infinite distance is the value of the surface-integral

$$M = -\iint\frac{d\omega_1}{dr}\,dS$$

extended over the second shell. Hence

$$M = \int_{\mu_2}^{1}\frac{d\omega_1}{dr}\,2\pi\,c_2^2\,d\mu_2,$$

$$4\pi^2\sin^2 a_1\,c_2^2\left\{\frac{1}{c_1}Q'(a_1)\int_{\mu_2}^{1}Q(a_2)d\mu_2+\&c.+\frac{c_2^{i-1}}{c^i}Q_i'(a_1)\int_{\mu_2}^{1}Q(a_2)d\mu_2\right\},$$

or, substituting the value of the integrals from equation (2), Art. 694,

$$M = 4\pi^2\sin^2 a_1\sin^2 a_2\,c_2^2\left\{\tfrac{1}{2}\frac{c_2}{c_1}Q_1'(a_1)Q_1'(a_2)+\&c.+\frac{1}{i(i+1)}\frac{c_2^i}{c_1^i}Q_i'(a_1)Q_i'(a_2)\right\}.$$

* The value of the solid angle subtended by a circle may be obtained in a more direct way as follows.—

The solid angle subtended by the circle at the point Z in the axis is easily shewn to be

$$\omega = 2\pi\left(1-\frac{z-c\cos a}{HZ}\right).$$

Expanding this expression in spherical harmonics, we find

$$\omega = 2\pi\left\{(\cos a-1)+(Q_1(a)\cos a-Q_0(a))\frac{z}{c}+\&c.+(Q_i(a)\cos a-Q_{i-1}(a))\frac{z^i}{c^i}+\&c.\right\},$$

$$\omega' = 2\pi\left\{(Q_0(a)\cos a-Q_1(a))\frac{c}{z}+\&c.+(Q_i(a)\cos a-Q_{i+1}(a))\frac{c^{i+1}}{z^{i+1}}+\&c.\right\},$$

for the expansions of ω for points on the axis for which z is less than c or greater than c respectively. Remembering the equations (42) and (43) of Art. 132 (vol. i. p. 165), the coefficients in these equations are evidently the same as those we have now obtained in a more convenient form for computation.

697.] Let us next suppose that the axis of one of the shells is turned about C as a centre, so that it now makes an angle θ with the axis of the other shell (Fig. 48). We have only to introduce the zonal harmonics of θ into this expression for M, and we find for the more general value of M,

$$M = 4\pi^2 \sin^2 a_1 \sin^2 a_2 c_2 \left\{ \frac{1}{2} \frac{c_2}{c_1} Q_1'(a_1) Q_1'(a_2) Q_1(\theta) + \&c. \right.$$
$$\left. + \frac{1}{i(i+1)} \frac{c_2^i}{c_1^i} Q_i'(a_1) Q_i'(a_2) Q_i(\theta) \right\}.$$

This is the value of the potential energy due to the mutual action of two circular currents of unit strength, placed so that the normals through the centres of the circles meet in a point C in an angle θ, the distances of the circumferences of the circles from the point C being c_1 and c_2, of which c_1 is the greater.

If any displacement dx alters the value of M, then the force acting in the direction of the displacement is $X = \dfrac{dM}{dx}$.

For instance, if the axis of one of the shells is free to turn about the point C, so as to cause θ to vary, then the moment of the force tending to increase θ is Θ, where

$$\Theta = \frac{dM}{d\theta}.$$

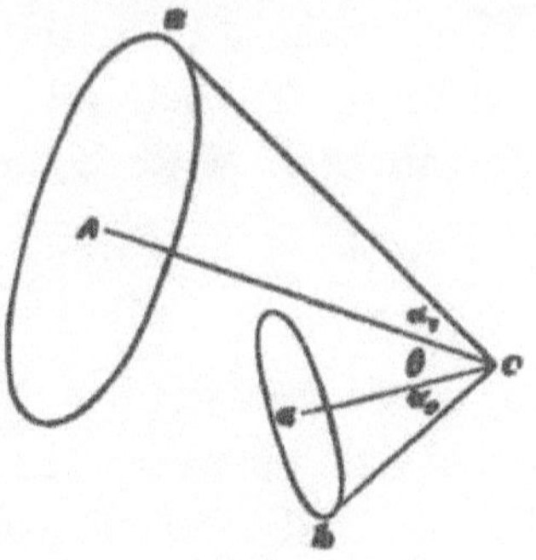

Fig. 48.

Performing the differentiation, and remembering that

$$\frac{dQ_i(\theta)}{d\theta} = -\sin\theta\, Q_i'(\theta),$$

where Q_1' has the same signification as in the former equations,

$$\Theta = -4\pi^2 \sin^2 a_1 \sin^2 a_2 \sin\theta\, c_2 \left\{ \frac{1}{2} \frac{c_2}{c_1} Q_1'(a_1) Q_1'(a_2) Q_1'(\theta) + \&c. \right.$$
$$\left. + \frac{1}{i(i+1)} \frac{c_2^i}{c_1^i} Q_i'(a_1) Q_i'(a_2) Q_i'(\theta) \right\}.$$

698.] As t' e values of Q_i' occur frequently in these calculations the following table of values of the first six degrees may be useful. In this table μ stands for $\cos\theta$, and ν for $\sin\theta$.

$$Q_1' = 1,$$
$$Q_2' = 3\mu,$$
$$Q_3' = \tfrac{3}{2}(5\mu^2 - 1) = 6(\mu^2 - \tfrac{1}{4}\nu^2),$$
$$Q_4' \quad \tfrac{5}{2}\mu(7\mu^2 - 3) = 10\mu(\mu^2 - \tfrac{3}{4}\nu^2),$$
$$Q_5' = \tfrac{15}{8}(21\mu^4 - 14\mu^2 + 1) = 15(\mu^4 - \tfrac{3}{2}\mu^2\nu^2 + \tfrac{1}{8}\nu^4),$$
$$Q_6' = \tfrac{21}{8}\mu(33\mu^4 - 30\mu^2 + 5) = 21\mu(\mu^4 - \tfrac{5}{2}\mu^2\nu^2 + \tfrac{5}{8}\nu^4).$$

699.] It is sometimes convenient to express the series for M in terms of linear quantities as follows:—

Let a be the radius of the smaller circuit, b the distance of its plane from the origin, and $c = \sqrt{a^2+b^2}$.

Let A, B, and C be the corresponding quantities for the larger circuit.

The series for M may then be written,

$$M = 1.2.\pi^2 \frac{A^2}{C^3} a^2 \cos\theta$$

$$+ 2.3.\pi^2 \frac{A^2 B}{C^5} a^2 b (\cos^2\theta - \tfrac{1}{2}\sin^2\theta)$$

$$+ 3.4.\pi^2 \frac{A^2(B^2-\tfrac{1}{4}A^2)}{C^7} a^2 (b^2 - \tfrac{1}{4}a^2)(\cos^3\theta - \tfrac{3}{2}\sin^2\theta\cos\theta)$$

$$+ \&c.$$

If we make $\theta = 0$, the two circles become parallel and on the same axis. To determine the attraction between them we may differentiate M with respect to b. We thus find

$$\frac{dM}{db} = \pi^2 \frac{A^2 a^2}{C^4}\left\{2.3\frac{B}{C} + 2.3.4\frac{B^2-\tfrac{1}{4}A^2}{C^3} b + \&c.\right\}.$$

700.] In calculating the effect of a coil of rectangular section we have to integrate the expressions already found with respect to A, the radius of the coil, and B, the distance of its plane from the origin, and to extend the integration over the breadth and depth of the coil.

In some cases direct integration is the most convenient, but there are others in which the following method of approximation leads to more useful results.

Let P be any function of x and y, and let it be required to find the value of $\overline{P}$ where

$$\overline{P}xy = \int_{-\frac{1}{2}x}^{+\frac{1}{2}x}\int_{-\frac{1}{2}y}^{+\frac{1}{2}y} P\,dx\,dy.$$

In this expression $\overline{P}$ is the mean value of P within the limits of integration.

Let P_0 be the value of P when $x = 0$ and $y = 0$, then, expanding P by Taylor's Theorem,

$$P = P_0 + x\frac{dP_0}{dx} + y\frac{dP_0}{dy} + \tfrac{1}{2}x^2\frac{d^2P}{dx^2} + \&c.$$

Integrating this expression between the limits, and dividing the result by xy, we obtain as the value of $\overline{P}$,

$$\overline{P} = P_0 + \tfrac{1}{24}\left(x^2 \frac{d^2 P_0}{dx^2} + y^2 \frac{d^2 P_0}{dy^2}\right)$$
$$+ \tfrac{1}{960}\left(x^4 \frac{d^4 P_0}{dx^4} + y^4 \frac{d^4 P_0}{dy^4}\right) + \tfrac{1}{576} x^2 y^2 \frac{d^4 P_0}{dx^2\, dy^2} + \&c.$$

In the case of the coil, let the outer and inner radii be $A + \frac{1}{2}\xi$, and $A - \frac{1}{2}\xi$ respectively, and let the distance of the planes of the windings from the origin lie between $B + \frac{1}{2}\eta$ and $B - \frac{1}{2}\eta$, then the breadth of the coil is η, and its depth ξ, these quantities being small compared with A or C.

In order to calculate the magnetic effect of such a coil we may write the successive terms of the series as follows :—

$$G_0 = \pi \frac{B}{C}\left(1 + \tfrac{1}{24}\frac{2A^2 - B^2}{C^4}\xi^2 - \tfrac{1}{8}\frac{A^2}{C^4}\eta^2\right),$$

$$G_1 = 2\pi \frac{A^2}{C^3}\left(1 + \tfrac{1}{24}\left(\frac{2}{A^2} - 15\frac{B^2}{C^4}\right)\xi^2 + \tfrac{1}{8}\frac{4B^2 - A^2}{C^4}\eta^2\right),$$

$$G_2 = 3\pi \frac{A^2 B}{C^5}\left(1 + \tfrac{1}{24}\left(\frac{2}{A^2} - \frac{25}{C^2} + \frac{35A^2}{C^4}\right)\xi^2 + \tfrac{5}{24}\frac{4B^2 - 3A^2}{C^4}\eta^2\right),$$

$$G_3 = 4\pi \frac{A^2(B^2 - \frac{1}{4}A^2)}{C^7} + \frac{\pi}{24}\frac{\xi^2}{C^{11}}\{C^4(8B^2 - 12A^2) + 35A^2B^2(5A^2 - 4B^2)$$
$$+ \frac{\pi}{24}\frac{\eta^2}{C^{11}}\{3A^2C^2(5A^2 - 44B^2) + 63A^2B^2(4B^2 - A^2)\},$$

&c., &c. ;

$$g_1 = \pi a^2 \qquad + \tfrac{1}{12}\pi\xi^2,$$
$$g_2 = 2\pi a^2 b \qquad + \tfrac{1}{8}\pi b \xi^2,$$
$$g_3 = 3\pi a^2(b^2 - \tfrac{1}{4}a^2) + \frac{\pi}{8}\xi^2(2b^2 - 3a^2) + \frac{\pi}{4}\eta^2 a^2,$$

&c., &c.

The quantities G_0, G_1, G_2, &c. belong to the large coil. The value of ω at points for which r is less than C is

$$\omega = -2\pi + 2G_0 - G_1 r Q_1(\theta) - G_2 r^2 Q_2(\theta) - \&c.$$

The quantities g_1, g_2, &c. belong to the small coil. The value of ω' at points for which r is greater than c is

$$\omega' = g_1 \frac{1}{r^2} Q_1(\theta) + g_2 \frac{1}{r^3} Q_2(\theta) + \&c.$$

The potential of the one coil with respect to the other when the total current through the section of each coil is unity is

$$M = G_1 g_1 Q_1(\theta) + G_2 g_2 Q_2(\theta) + \&c.$$

To find M by Elliptic Integrals.

701.] When the distance of the circumferences of the two circles is moderate as compared with the radii of the smaller, the series already given do not converge rapidly. In every case, however, we may find the value of M for two parallel circles by elliptic integrals.

For let b be the length of the line joining the centres of the circles, and let this line be perpendicular to the planes of the two circles, and let A and a be the radii of the circles, then

$$M = \iint \frac{\cos \epsilon}{r}\, ds\, ds',$$

the integration being extended round both curves.

In this case,

$$r^2 = A^2 + a^2 + b^2 - 2\, Aa \cos(\phi - \phi'),$$

$$\epsilon = \phi - \phi', \qquad ds = a\, d\phi, \qquad ds' = A\, d\phi',$$

$$M = \int_0^{2\pi}\int_0^{2\pi} \frac{Aa \cos(\phi - \phi')\, d\phi\, d\phi'}{\sqrt{A^2 + a^2 + b^2 - 2\, Aa \cos(\phi - \phi')}}$$

$$= 2\pi\sqrt{Aa}\left\{\left(c - \frac{2}{c}\right) F + \frac{2}{c} E\right\},$$

where $$c = \frac{\sqrt{Aa}}{\sqrt{(A+a)^2 + b^2}},$$

and F and E are complete elliptic integrals to modulus c.

From this we get, by differentiating with respect to b and remembering that c is a function of b,

$$\frac{dM}{db} = \frac{4\,\pi\, b c^{\frac{1}{2}}}{\sqrt{Aa}\,(1 - c^2)^2}\{E(1 + c^2) - F(1 - c^2)\}.$$

If r_1 and r_2 denote the greatest and least values of r,

$$r_1^2 = (A + a)^2 + b^2, \qquad r_2^2 = (A - a)^2 + b^2,$$

and if an angle γ be taken such that $\cos\gamma = \frac{r_2}{r_1}$,

$$\frac{dM}{db} = \pi\frac{b \sin\gamma}{\sqrt{Aa}}\{2F_\gamma - (1 + \sec^2\gamma)\, E_\gamma\},$$

where F_γ and E_γ denote the complete elliptic integrals of the first and second kind whose modulus is $\sin\gamma$.

If $A = a$, $\cot\gamma = \frac{b}{2a}$, and

$$\frac{dM}{db} = 2\pi\cos\gamma\{2F_\gamma - (1 + \sec^2\gamma)\, E_\gamma\}.$$

The quantity $\frac{dM}{db}$ represents the attraction between two parallel circular currents, the current in each being unity.

Second Expression for M.

An expression for M, which is sometimes more convenient, is got by making $c_1 = \frac{r_1 - r_2}{r_1 + r_2}$, in which case

$$M = 4\pi\sqrt{Aa}\,\frac{1}{\sqrt{c_1}}(F_{c_1} - E_{c_1}).$$

To draw the Lines of Magnetic Force for a Circular Current.

702.] The lines of magnetic force are evidently in planes passing through the axis of the circle, and in each of these lines the value of M is constant.

Calculate the value of $K_\theta = \frac{\sin\theta}{(F_{\sin\theta} - E_{\sin\theta})^2}$ from Legendre's tables for a sufficient number of values of θ.

Draw rectangular axes of x and z on the paper, and, with centre at the point $x = \frac{1}{2}a(\sin\theta + \operatorname{cosec}\theta)$, draw a circle with radius $\frac{1}{2}a(\operatorname{cosec}\theta - \sin\theta)$. For all points of this circle the value of c_1 will be $\sin\theta$. Hence, for all points of this circle,

$$M = 4\pi\sqrt{Aa}\,\frac{1}{\sqrt{K_\theta}}, \quad \text{and} \quad A = \frac{1}{16\pi^2}\,\frac{M^2 K_\theta}{a}.$$

Now A is the value of x for which the value of M was found. Hence, if we draw a line for which $x = A$, it will cut the circle in two points having the given value of M.

Giving M a series of values in arithmetical progression, the values of A will be as a series of squares. Drawing therefore a series of lines parallel to z, for which x has the values found for A, the points where these lines cut the circle will be the points where the corresponding lines of force cut the circle.

If we put $m = 4\pi a$, and $M = nm$, then

$$A = x = n^2 K_\theta\, a.$$

We may call n the index of the line of force.

The forms of these lines are given in Fig. XVIII at the end of this volume. They are copied from a drawing given by Sir W. Thomson in his paper on 'Vortex Motion *.'

703.] If the position of a circle having a given axis is regarded as defined by b, the distance of its centre from a fixed point on the axis, and a, the radius of the circle, then M, the coefficient of induction of the circle with respect to any system whatever

* *Trans. R. S., Edin.*, vol. xxv. p. 217 (1869).

of magnets or currents, is subject to the following equation

$$\frac{d^2M}{da^2}+\frac{d^2M}{db^2}-\frac{1}{a}\frac{dM}{da}=0. \qquad (1)$$

To prove this, let us consider the number of lines of magnetic force cut by the circle when a or b is made to vary.

(1) Let a become $a+\delta a$, b remaining constant. During this variation the circle, in expanding, sweeps over an annular surface in its own plane whose breadth is δa.

If V is the magnetic potential at any point, and if the axis of y be parallel to that of the circle, then the magnetic force perpendicular to the plane of the ring is $\frac{dV}{dy}$.

To find the magnetic induction through the annular surface we have to integrate

$$\int_0^{2\pi} a\,\delta a\frac{dV}{dy}\,d\theta,$$

where θ is the angular position of a point on the ring.

But this quantity represents the variation of M due to the variation of a, or $\frac{dM}{da}\delta a$. Hence

$$\frac{dM}{da}=\int_0^{2\pi} a\frac{dV}{dy}\,d\theta. \qquad (2)$$

(2) Let b become $b+\delta b$, a remaining constant. During this variation the circle sweeps over a cylindric surface of radius a and length δb.

The magnetic force perpendicular to this surface at any point is $\frac{dV}{dr}$ where r is the distance from the axis. Hence

$$\frac{dM}{db}=-\int_0^{2\pi} a\frac{dV}{dr}\,d\theta. \qquad (3)$$

Differentiating equation (2) with respect to a, and (3) with respect to b, we get

$$\frac{d^2M}{da^2}=\int_0^{2\pi}\frac{dV}{dy}\,d\theta+\int_0^{2\pi} a\frac{d^2V}{dr\,dy}\,d\theta, \qquad (4)$$

$$\frac{d^2M}{db^2}=-\int_0^{2\pi} a\frac{d^2V}{dr\,dy}\,d\theta, \qquad (5)$$

Hence

$$\frac{d^2M}{da^2}+\frac{d^2M}{db^2}=\int_0^{2\pi}\frac{dV}{dy}\,d\theta, \qquad (6)$$

$$=\frac{1}{a}\frac{dM}{da}, \text{ by (2).}$$

Transposing the last term we obtain equation (1).

Coefficient of Induction of Two Parallel Circles when the Distance between the Arcs is Small compared with the Radius of either Circle.

704.] We might deduce the value of M in this case from the expansion of the elliptic integral already given when its modulus is nearly unity. The following method, however, is a more direct application of electrical principles.

First Approximation.

Let A and a be the radii of the circles, and b the distance between their planes, then the shortest distance between the arcs is

$$r = \sqrt{(A-a)^2+b^2}.$$

We have to find M_1, the magnetic induction through the circle A, due to a unit current in a on the supposition that r is small compared with A or a.

We shall begin by calculating the magnetic induction through a circle in the plane of a whose radius is $a-c$, c being a quantity small compared with a (Fig. 49).

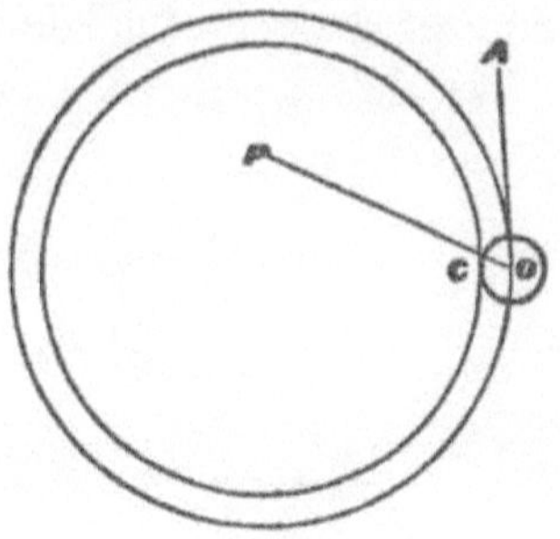

Fig. 49.

Consider a small element ds of the circle a. At a point in the plane of the circle, distant ρ from the middle of ds, measured in a direction making an angle θ with the direction of ds, the magnetic force due to ds is perpendicular to the plane, and equal to

$$\frac{1}{\rho^2}\sin\theta\, ds.$$

If we now calculate the surface-integral of this force over the space which lies within the circle a, but outside of a circle whose centre is ds and whose radius is c, we find it

$$\int_0^{\pi}\int_c^{2a\sin\theta}\frac{1}{\rho^2}\sin\theta\, ds\, d\theta\, d\rho = \{\log 8a - \log c - 2\}\, ds.$$

If c is small, the surface-integral for the part of the annular space outside the small circle c may be neglected.

We then find for the induction through the circle whose radius is $a-c$, by integrating with respect to ds,

$$M_{ac} = 4\pi a\,\{\log 8a - \log c - 2\},$$

provided c is very small compared with a.

Since the magnetic force at any point, the distance of which from a curved wire is small compared with the radius of curvature, is nearly the same as if the wire had been straight, we can calculate the difference between the induction through the circle whose radius is $a-c$, and the circle A by the formula

$$M_{aA}-M_{ac} = 4\pi a\{\log c-\log r\}.$$

Hence we find the value of the induction between A and a to be

$$M_{Aa} = 4\pi a(\log 8a-\log r-2)$$

approximately, provided r is small compared with a.

705.] Since the mutual induction between two windings of the same coil is a very important quantity in the calculation of experimental results, I shall now describe a method by which the approximation to the value of M for this case can be carried to any required degree of accuracy.

We shall assume that the value of M is of the form

$$M = 4\pi\left\{A\log\frac{8a}{r}+B\right\},$$

where $A = \quad a+A_1x+A_2\dfrac{x^2}{a}+A_2'\dfrac{y^2}{a}+A_3\dfrac{x^3}{a^2}+A_3'\dfrac{xy^2}{a^2}+\&c.,$

and $B = -2a+B_1x+B_2\dfrac{x^2}{a}+B_2'\dfrac{y^2}{a}+B_3\dfrac{x^3}{a^2}+B_3'\dfrac{xy^2}{a^2}+\&c.,$

where a and $a+x$ are the radii of the circles, and y the distance between their planes.

We have to determine the values of the coefficients A and B. It is manifest that only even powers of y can occur in these quantities, because, if the sign of y is reversed, the value of M must remain the same.

We get another set of conditions from the reciprocal property of the coefficient of induction, which remains the same whichever circle we take as the primary circuit. The value of M must therefore remain the same when we substitute $a+x$ for a, and $-x$ for x in the above expression.

We thus find the following conditions of reciprocity by equating the coefficients of similar combinations of x and y,

$$A_1 = 1-A_1, \qquad B_1 = 1-2-B_1,$$

$$A_3 = -A_2-A_3, \qquad B_3 = \tfrac{1}{3}-\tfrac{1}{2}A_1+A_2-B_2-B_3,$$

$$A_3' = -A_2'-A_3', \qquad B_3' = A_2'-B_2'-B_3';$$

$$(-)^nA_n = A_2+(n-2)A_3+\frac{(n-2)(n-3)}{1.2}A_4+\&c.+A_n,$$

$$(-)^nB_n = -\frac{1}{n}+\frac{1}{n-1}A_1-\frac{1}{n-2}A_2+\&c.+(-)^nA_{n-1}$$
$$+B_2+(n-2)B_3+\frac{(n-2)(n-3)}{1.2}B_4+\&c.+B_n.$$

From the general equation of M, Art. 703,

$$\frac{d^2M}{dx^2}+\frac{d^2M}{dy^2}-\frac{1}{a+x}\frac{dM}{dx}=0,$$

we obtain another set of conditions,

$$2A_2+2A'_2=A_1,$$
$$2A_2+2A'_2+6A_3+2A'_3=2A_2;$$
$$n(n-1)A_n+(n+1)nA_{n+1}+1.2A'_n+1.2A'_{n+1}=nA_n,$$
$$(n-1)(n-2)A'_n+n(n-1)A'_{n+1}+2.3A''_n+2.3A''_{n+1}=(n-2)A'_n;$$

&c.;

$$4A_2+A_1=2B_2+2B'_2-B_1=4A'_2,$$
$$6A_3+3A_2=2B'_2+6B_3+2B'_3=6A'_3+3A'_2,$$
$$(2n-1)A_n+(2n+2)A_{n+1}$$
$$=n(n-2)B_n+(n+1)nB_{n+1}+1.2B'_n+1.2B'_{n+1}.$$

Solving these equations and substituting the values of the coefficients, the series for M becomes

$$M=4\pi a\log\frac{8a}{r}\left\{1+\frac{1}{2}\frac{x}{a}+\frac{x^2+3y^2}{16a^2}-\frac{x^3+3xy^2}{32a^3}+\&c.\right\}$$
$$+4\pi a\left\{-2-\frac{1}{2}\frac{x}{a}+\frac{3x^2-y^2}{16a^2}-\frac{x^3-6xy^2}{48a^3}+\&c.\right\}.$$

To find the form of a coil for which the coefficient of self-induction is a maximum, the total length and thickness of the wire being given.

706.] Omitting the corrections of Art. 705, we find by Art. 673

$$L=4\pi n^2 a\left(\log\frac{8a}{R}-2\right),$$

where n is the number of windings of the wire, a is the mean radius of the coil, and R is the geometrical mean distance of the transverse section of the coil from itself. See Art. 690. If this section is always similar to itself, R is proportional to its linear dimensions, and n varies as R^2.

Since the total length of the wire is $2\pi an$, a varies inversely as n. Hence

$$\frac{dn}{n}=2\frac{dR}{R},\quad\text{and}\quad\frac{da}{a}=-2\frac{dR}{R},$$

and we find the condition that L may be a maximum

$$\log\frac{8a}{R}=\frac{7}{2}.$$

If the transverse section of the coil is circular, of radius c, then, by Art. 692,

$$\log\frac{R}{c} = -\tfrac{1}{4},$$

$$\text{and } \log\frac{8a}{c} = \tfrac{13}{4},$$

whence $$a = 3.22c;$$

or, the mean radius of the coil should be 3.22 times the radius of the transverse section of the coil in order that such a coil may have the greatest coefficient of self-induction. This result was found by Gauss *.

If the channel in which the coil is wound has a square transverse section, the mean diameter of the coil should be 3.7 times the side of the square section.

* *Werke*, Göttingen edition, 1867, vol. v. p. 622.

CHAPTER XV.

ELECTROMAGNETIC INSTRUMENTS.

Galvanometers.

707.] A GALVANOMETER is an instrument by means of which an electric current is indicated or measured by its magnetic action.

When the instrument is intended to indicate the existence of a feeble current, it is called a Sensitive Galvanometer.

When it is intended to measure a current with the greatest accuracy in terms of standard units, it is called a Standard Galvanometer.

All galvanometers are founded on the principle of Schweigger's Multiplier, in which the current is made to pass through a wire, which is coiled so as to pass many times round an open space, within which a magnet is suspended, so as to produce within this space an electromagnetic force, the intensity of which is indicated by the magnet.

In sensitive galvanometers the coil is so arranged that its windings occupy the positions in which their influence on the magnet is greatest. They are therefore packed closely together in order to be near the magnet.

Standard galvanometers are constructed so that the dimensions and relative positions of all their fixed parts may be accurately known, and that any small uncertainty about the position of the moveable parts may introduce the smallest possible error into the calculations.

In constructing a sensitive galvanometer we aim at making the field of electromagnetic force in which the magnet is suspended as intense as possible. In designing a standard galvanometer we wish to make the field of electromagnetic force near the magnet as uniform as possible, and to know its exact intensity in terms of the strength of the current.

On Standard Galvanometers.

708.] In a standard galvanometer the strength of the current has to be determined from the force which it exerts on the suspended magnet. Now the distribution of the magnetism within the magnet, and the position of its centre when suspended, are not capable of being determined with any great degree of accuracy. Hence it is necessary that the coil should be arranged so as to produce a field of force which is very nearly uniform throughout the whole space occupied by the magnet during its possible motion. The dimensions of the coil must therefore in general be much larger than those of the magnet.

By a proper arrangement of several coils the field of force within them may be made much more uniform than when one coil only is used, and the dimensions of the instrument may be thus reduced and its sensibility increased. The errors of the linear measurements, however, introduce greater uncertainties into the values of the electrical constants for small instruments than for large ones. It is therefore best to determine the electrical constants of small instruments, not by direct measurement of their dimensions, but by an electrical comparison with a large standard instrument, of which the dimensions are more accurately known; see Art. 752.

In all standard galvanometers the coils are circular. The channel in which the coil is to be wound is carefully turned. Its breadth

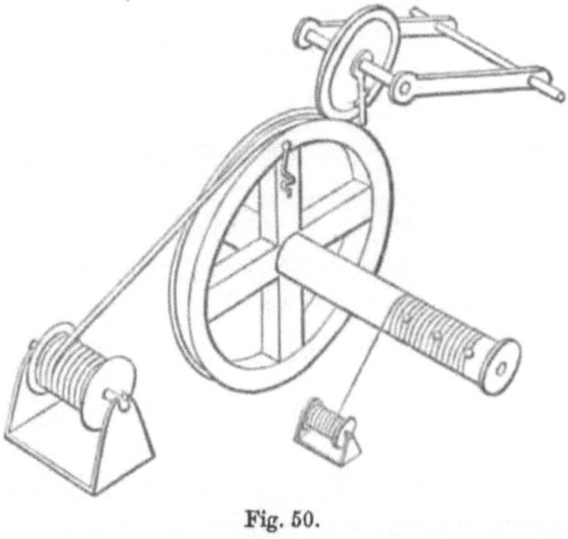

Fig. 50.

is made equal to some multiple, n, of the diameter of the covered wire. A hole is bored in the side of the channel where the wire is to enter, and one end of the covered wire is pushed out through this hole to form the inner connexion of the coil. The channel is placed on a lathe, and a wooden axis is fastened to it; see Fig. 50. The end of a long string is nailed to the wooden axis at the same part of the circumference as the entrance of the wire. The whole is then turned round, and the wire is smoothly and regularly laid on the bottom of the channel till it is completely covered by n windings. During this process the string has been wound n times round the wooden axis, and a nail is driven into the string at the nth turn. The windings of the string should be kept exposed so that they can easily be counted. The external circumference of the first layer of windings is then measured and a new layer is begun, and so on till the proper number of layers has been wound on. The use of the string is to count the number of windings. If for any reason we have to unwind part of the coil, the string is also unwound, so that we do not lose our reckoning of the actual number of windings of the coil. The nails serve to distinguish the number of windings in each layer.

The measure of the circumference of each layer furnishes a test of the regularity of the winding, and enables us to calculate the electrical constants of the coil. For if we take the arithmetic mean of the circumferences of the channel and of the outer layer, and then add to this the circumferences of all the intermediate layers, and divide the sum by the number of layers, we shall obtain the mean circumference, and from this we can deduce the mean radius of the coil. The circumference of each layer may be measured by means of a steel tape, or better by means of a graduated wheel which rolls on the coil as the coil revolves in the process of winding. The value of the divisions of the tape or wheel must be ascertained by comparison with a straight scale.

709.] The moment of the force with which a unit current in the coil acts upon the suspended apparatus may be expressed in the series

$$G_1 g_1 \sin\theta + G_2 g_2 \sin\theta\, Q_2'(\theta) + \&c.,$$

where the coefficients G refer to the coil, and the coefficients g to the suspended apparatus, θ being the angle between the axis of the coil and that of the suspended apparatus; see Art. 700.

When the suspended apparatus is a thin uniformly and longitudinally magnetized bar magnet of length $2l$ and strength unity, suspended by its middle,

$$g_1 = 2l, \quad g_2 = 0, \quad g_3 = 2l^3, \ \&c.$$

The values of the coefficients for a magnet of length $2l$ magnetized in any other way are smaller than when it is magnetized uniformly.

710.] When the apparatus is used as a tangent galvanometer, the coil is fixed with its plane vertical and parallel to the direction of the earth's magnetic force. The equation of equilibrium of the magnet is in this case

$$m g_1 H \cos\theta = m\gamma \sin\theta \{G_1 g_1 + G_2 g_2 Q_1'(\theta) + \&c.\},$$

where $m g_1$ is the magnetic moment of the magnet, H the horizontal component of the terrestrial magnetic force, and γ the strength of the current in the coil. When the length of the magnet is small compared with the radius of the coil the terms after the first in G and g may be neglected, and we find

$$\gamma = \frac{H}{G_1} \cot\theta.$$

The angle usually measured is the deflexion, δ, of the magnet which is the complement of θ, so that $\cot\theta = \tan\delta$.

The current is thus proportional to the tangent of the deviation, and the instrument is therefore called a Tangent Galvanometer.

Another method is to make the whole apparatus moveable about a vertical axis, and to turn it till the magnet is in equilibrium with its axis parallel to the plane of the coil. If the angle between the plane of the coil and the magnetic meridian is δ, the equation of equilibrium is

$$m g_1 H \sin\delta = M\gamma\frac{1}{g}\{G_1 g_1 - \tfrac{3}{2} G_3 g_3 + \&c.\},$$

whence

$$\gamma = \frac{H}{(G_1 - \&c.)} \sin\delta.$$

Since the current is measured by the sine of the deviation, the instrument when used in this way is called a Sine Galvanometer.

The method of sines can be applied only when the current is so steady that we can regard it as constant during the time of adjusting the instrument and bringing the magnet to equilibrium.

711.] We have next to consider the arrangement of the coils of a standard galvanometer.

The simplest form is that in which there is a single coil, and the magnet is suspended at its centre.

Let A be the mean radius of the coil, ξ its depth, η its breadth, and n the number of windings, the values of the coefficients are

$$G_1 = \frac{2\pi n}{A}\left\{1 + \tfrac{1}{12}\frac{\xi^2}{A^2} - \tfrac{1}{8}\frac{\eta^2}{A^2}\right\},$$

$$G_2 = 0,$$

$$G_3 = -\frac{\pi n}{A^3}\left\{1 + \tfrac{1}{2}\frac{\xi^2}{A^2} - \tfrac{5}{8}\frac{\eta^2}{A^2}\right\},$$

$$G_4 = 0, \text{ \&c.}$$

The principal correction is that arising from G_3. The series

$$G_1 g_1 + G_3 g_3\, Q_3'(\theta)$$

becomes

$$G_1 g_1 \left(1 - \tfrac{3}{2}\frac{1}{A^2}\frac{g_3}{g_1}(\cos^2\theta - \tfrac{1}{4}\sin^2\theta)\right).$$

The factor of correction will differ most from unity when the magnet is uniformly magnetized and when $\theta = 0$. In this case it becomes $1 - \frac{1}{2}\frac{l^2}{A^2}$. It vanishes when $\tan\theta = 2$, or when the deflexion is $\tan^{-1}\frac{1}{2}$, or $26°34'$. Some observers, therefore, arrange their experiments so as to make the observed deflexion as near this angle as possible. The best method, however, is to use a magnet so short compared with the radius of the coil that the correction may be altogether neglected.

The suspended magnet is carefully adjusted so that its centre shall coincide as nearly as possible with the centre of the coil. If, however, this adjustment is not perfect, and if the coordinates of the centre of the magnet relative to the centre of the coil are x, y, z, z being measured parallel to the axis of the coil, the factor of correction is

$$\left(1 + \tfrac{3}{2}\frac{x^2 + y^2 - 2z^2}{A^2}\right).$$

When the radius of the coil is large, and the adjustment of the magnet carefully made, we may assume that this correction is insensible.

Gaugain's Arrangement.

712.] In order to get rid of the correction depending on G_3 Gaugain constructed a galvanometer in which this term was rendered zero by suspending the magnet, not at the centre of the coil, but at a point on the axis at a distance from the centre equal to half the radius of the coil. The form of G_3 is

$$G_3 = 4\pi \frac{A^2(B^2 - \frac{1}{4}A^2)}{C^7},$$

and, since in this arrangement $B = \frac{1}{2}A$, $G_3 = 0$.

This arrangement would be an improvement on the first form if we could be sure that the centre of the suspended magnet is exactly at the point thus defined. The position of the centre of the magnet, however, is always uncertain, and this uncertainty introduces a factor of correction of unknown amount depending on G_2 and of the form $\left(1-\frac{9}{8}\frac{z}{A}\right)$, where z is the unknown excess of distance of the centre of the magnet from the plane of the coil. This correction depends on the first power of $\frac{z}{A}$. Hence Gaugain's coil with eccentrically suspended magnet is subject to far greater uncertainty than the old form.

Helmholtz's Arrangement.

713.] Helmholtz converted Gaugain's galvanometer into a trustworthy instrument by placing a second coil, equal to the first, at an equal distance on the other side of the magnet.

By placing the coils symmetrically on both sides of the magnet we get rid at once of all terms of even order.

Let A be the mean radius of either coil, the distance between their mean planes is made equal to A, and the magnet is suspended at the middle point of their common axis. The coefficients are

$$G_1 = \frac{16\pi n}{5\sqrt{5}}\frac{1}{A}\left(1-\frac{1}{60}\frac{\xi^2}{A^2}\right),$$

$$G_2 = 0,$$

$$G_3 = 0.0512\,\frac{\pi n}{3\sqrt{5}\,A^5}\,(31\xi^2-36\eta^2),$$

$$G_4 = 0,$$

$$G_5 = -0.73728\,\frac{\pi n}{\sqrt{5}\,A^5},$$

where n denotes the number of windings in both coils together.

It appears from these results that if the section of the coils be rectangular, the depth being ξ and the breadth η, the value of G_3, as corrected for the finite size of the section, will be small, and will vanish, if ξ is to η as 36 to 31.

It is therefore quite unnecessary to attempt to wind the coils upon a conical surface, as has been done by some instrument makers, for the conditions may be satisfied by coils of rectangular section, which can be constructed with far greater accuracy than coils wound upon an obtuse cone.

The arrangement of the coils in Helmholtz's double galvanometer is represented in Fig. 54, Art. 725.

The field of force due to the double coil is represented in section in Fig. XIX at the end of this volume.

Galvanometer of Four Coils.

714.] By combining four coils we may get rid of the coefficients G_2, G_3, G_4, G_5, and G_6. For by any symmetrical combinations we get rid of the coefficients of even orders Let the four coils be parallel circles belonging to the same sphere, corresponding to angles θ, ϕ, $\pi-\phi$, and $\pi-\theta$.

Let the number of windings on the first and fourth coil be n, and the number on the second and third pn. Then the condition that $G_3 = 0$ for the combination gives

$$n \sin^2\theta\, Q_3'(\theta) + pn \sin^2\phi\, Q_3'(\phi) = 0, \tag{1}$$

and the condition that $G_5 = 0$ gives

$$n \sin^2\theta\, Q_5'(\theta) + pn \sin^2\phi\, Q_5'(\phi) = 0, \tag{2}$$

Putting $\sin^2\theta = x$ and $\sin^2\phi = y$, (3)

and expressing Q_3' and Q_5' (Art. 698) in terms of these quantities, the equations (1) and (2) become

$$4x - 5x^2 + 4py - 5py^2 = 0, \tag{4}$$

$$8x - 28x^2 + 21x^3 + 8py - 28py^2 + 21py^3 = 0. \tag{5}$$

Taking twice (4) from (5), and dividing by 3, we get

$$6x^2 - 7x^3 + 6py^2 - 7py^3 = 0. \tag{6}$$

Hence, from (4) and (6),

$$p = \frac{x}{y}\,\frac{5x-4}{4-5y} = \frac{x^2}{y^2}\,\frac{7x-6}{6-7y},$$

and we obtain

$$y = \tfrac{4}{7}\,\frac{7x-6}{5x-4}, \qquad p = \frac{32}{49x}\,\frac{7x-6}{(5x-4)^3}.$$

Both x and y are the squares of the sines of angles and must therefore lie between 0 and 1. Hence, either x is between 0 and $\frac{4}{7}$, in which case y is between $\frac{6}{7}$ and 1, and p between ∞ and $\frac{49}{32}$, or else x is between $\frac{6}{7}$ and 1, in which case y is between 0 and $\frac{4}{7}$, and p between 0 and $\frac{32}{49}$.

Galvanometer of Three Coils.

715.] The most convenient arrangement is that in which $x = 1$. Two of the coils then coincide and form a great circle of the sphere whose radius is C. The number of windings in this compound coil is 64. The other two coils form small circles of the sphere. The radius of each of them is $\sqrt{\frac{4}{7}}\,C$. The distance of either of them from the plane of the first is $\sqrt{\frac{3}{7}}\,C$. The number of windings on each of these coils is 49.

The value of G_1 is $\dfrac{120}{C}$.

This arrangement of coils is represented in Fig. 51.

Fig. 51.

Since in this three-coiled galvanometer the first term after G_1 which has a finite value is G_7, a large portion of the sphere on whose surface the coils lie forms a field of force sensibly uniform.

If we could wind the wire over the whole of a spherical surface, as described in Art. 627, we should obtain a field of perfectly uniform force. It is practically impossible, however, to distribute the windings on a spherical surface with sufficient accuracy, even if such a coil were not liable to the objection that it forms a closed surface, so that its interior is inaccessible.

By putting the middle coil out of the circuit, and making the current flow in opposite directions through the two side coils, we obtain a field of force which exerts a nearly uniform action in the direction of the axis on a magnet or coil suspended within it, with its axis coinciding with that of the coils; see Art. 673. For in this case all the coefficients of odd orders disappear, and since

$$\mu = \sqrt{\tfrac{3}{7}}, \qquad Q_4' = \tfrac{5}{2}\mu\,(7\mu^2 - 3) = 0.$$

Hence the expression for the magnetic potential near the centre of the coil becomes

$$\omega = \sqrt{\tfrac{3}{7}}\,\pi n\gamma\left\{3\frac{r^2}{C^2}Q_2(\theta) + \tfrac{11}{7}\frac{r^6}{C^6}Q_6(\theta) + \&c.\right\}.$$

On the Proper Thickness of the Wire of a Galvanometer, the External Resistance being given.

716.] Let the form of the channel in which the galvanometer coil is to be wound be given, and let it be required to determine whether it ought to be filled with a long thin wire or with a shorter thick wire.

Let l be the length of the wire, y its radius, $y+b$ the radius of the wire when covered, ρ its specific resistance, g the value of G for unit of length of the wire, and r the part of the resistance which is independent of the galvanometer.

The resistance of the galvanometer wire is

$$R = \frac{\rho}{\pi}\frac{l}{y^2}.$$

The volume of the coil is

$$V = 4\,l(y+b)^2.$$

The electromagnetic force is γG, where γ is the strength of the current and

$$G = gl.$$

If E is the electromotive force acting in the circuit whose resistance is $R+r$,

$$E = \gamma(R+r).$$

The electromagnetic force due to this electromotive force is

$$E\frac{G}{R+r},$$

which we have to make a maximum by the variation of y and l.

Inverting the fraction, we find that

$$\frac{\rho}{\pi g}\frac{1}{y^2} + \frac{r}{gl}$$

is to be made a minimum. Hence

$$2\frac{\rho}{\pi}\frac{dy}{y^3} + \frac{r\,dl}{l^2} = 0.$$

If the volume of the coil remains constant

$$\frac{dl}{l} + 2\frac{dy}{y+b} = 0.$$

Eliminating dl and dy, we obtain

$$\frac{\rho}{\pi}\frac{y+b}{y^3} = \frac{r}{l},$$

or

$$\frac{r}{R} = \frac{y+b}{y}.$$

Hence the thickness of the wire of the galvanometer should be such that the external resistance is to the resistance of the galvanometer coil as the diameter of the covered wire to the diameter of the wire itself.

On Sensitive Galvanometers.

717.] In the construction of a sensitive galvanometer the aim of every part of the arrangement is to produce the greatest possible deflexion of the magnet by means of a given small electromotive force acting between the electrodes of the coil.

The current through the wire produces the greatest effect when it is placed as near as possible to the suspended magnet. The magnet, however, must be left free to oscillate, and therefore there is a certain space which must be left empty within the coil. This defines the internal boundary of the coil.

Outside of this space each winding must be placed so as to have the greatest possible effect on the magnet. As the number of windings increases, the most advantageous positions become filled up, so that at last the increased resistance of a new winding diminishes the effect of the current in the former windings more than the new winding itself adds to it. By making the outer windings of thicker wire than the inner ones we obtain the greatest magnetic effect from a given electromotive force.

718.] We shall suppose that the windings of the galvanometer are circles, the axis of the galvanometer passing through the centres of these circles at right angles to their planes.

Let $r \sin\theta$ be the radius of one of these circles, and $r\cos\theta$ the distance of its centre from the centre of the galvanometer, then, if l is the length of a portion of wire coinciding with this circle, and γ the current which flows in it, the magnetic force at the centre of the galvanometer resolved in the direction of the axis is

$$\gamma l \frac{\sin\theta}{r^2}.$$

If we write $$r^2 = x^2 \sin\theta, \tag{1}$$

this expression becomes $\gamma \frac{l}{x^2}$.

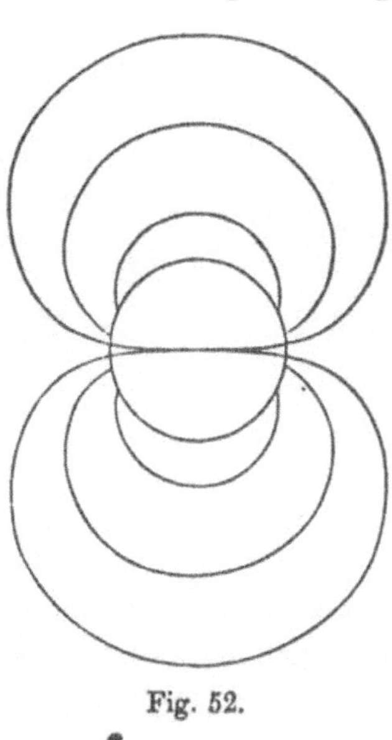

Fig. 52.

Hence, if a surface be constructed similar to those represented in section in Fig. 52, whose polar equation is

$$r^2 = x_1^2 \sin\theta, \tag{2}$$

where x_1 is any constant, a given length of wire bent into the form of a circular arc will produce a greater magnetic effect when it lies within this surface than when it lies outside it.

It follows from this that the outer surface of any layer of wire ought to have a constant value of x, for if x is greater at one place than another a portion of wire might be transferred from the first place to the second, so as to increase the force at the centre of the galvanometer.

The whole force due to the coil is γG, where

$$G = \int \frac{dl}{x^2}, \tag{3}$$

the integration being extended over the whole length of the wire, x being considered as a function of l.

719.] Let y be the radius of the wire, its transverse section will be πy^2. Let ρ be the specific resistance of the material of which the wire is made referred to unit of volume, then the resistance of a length l is $\frac{l\rho}{\pi y^2}$, and the whole resistance of the coil is

$$R = \frac{\rho}{\pi}\int \frac{dl}{y^2}, \tag{4}$$

where y is considered a function of l.

Let Y^2 be the area of the quadrilateral whose angles are the sections of the axes of four neighbouring wires of the coil by a plane through the axis, then $Y^2 l$ is the volume occupied in the coil by a length l of wire together with its insulating covering, and including any vacant space necessarily left between the windings of the coil. Hence the whole volume of the coil is

$$V = \int Y^2\, dl, \tag{5}$$

where Y is considered a function of l.

But since the coil is a figure of revolution

$$V = 2\pi \iint r^2 \sin\theta\, dr\, d\theta, \tag{6}$$

or, expressing r in terms of x, by equation (2),

$$V = 2\pi \iint x^2 (\sin\theta)^{\frac{3}{2}}\, dx\, d\theta. \tag{7}$$

Now $2\pi\int_0^{\pi} (\sin\theta)^{\frac{3}{2}}\, d\theta$ is a numerical quantity, call it N, then

$$V = \tfrac{1}{3} N x^3 - V_0, \tag{8}$$

where V_0 is the volume of the interior space left for the magnet.

Let us now consider a layer of the coil contained between the surfaces x and $x + dx$.

The volume of this layer is

$$dV = Nx^2\,dx = Y^2\,dl, \tag{9}$$

where dl is the length of wire in this layer.

This gives us dl in terms of dx. Substituting this in equations (3) and (4), we find

$$dG = N\frac{dx}{Y^2}, \tag{10}$$

$$dR = N\frac{\rho}{\pi}\frac{x^2\,dx}{Y^2 y^2}, \tag{11}$$

where dG and dR represent the portions of the values of G and of R due to this layer of the coil.

Now if E be the given electromotive force,

$$E = \gamma(R+r),$$

where r is the resistance of the external part of the circuit, independent of the galvanometer, and the force at the centre is

$$\gamma G = E\frac{G}{R+r}.$$

We have therefore to make $\dfrac{G}{R+r}$ a maximum, by properly adjusting the section of the wire in each layer. This also necessarily involves a variation of Y because Y depends on y.

Let G_0 and R_0 be the values of G and of $R+r$ when the given layer is excluded from the calculation. We have then

$$\frac{G}{R+r} = \frac{G_0 + dG}{R_0 + dR}, \tag{12}$$

and to make this a maximum by the variation of the value of y for the given layer we must have

$$\frac{\dfrac{d}{dy}\cdot dG}{\dfrac{d}{dy}\cdot dR} = \frac{G}{R+r}. \tag{13}$$

Since dx is very small and ultimately vanishes, $\dfrac{G_0}{R_0}$ will be sensibly, and ultimately exactly, the same whichever layer is excluded, and we may therefore regard it as constant. We have therefore, by (10) and (11),

$$\frac{x^2}{y^2}\left(1 + \frac{Y}{y}\frac{dy}{dY}\right) = \frac{\rho}{\pi}\frac{R+r}{G} = \text{constant}. \tag{14}$$

If the method of covering the wire and of winding it is such that the proportion between the space occupied by the metal of the wire bears the same proportion to the space between the wires whether the wire is thick or thin, then

$$\frac{Y}{y}\frac{dy}{dY} = 1,$$

and we must make both y and Y proportional to x, that is to say, the diameter of the wire in any layer must be proportional to the linear dimension of that layer.

If the thickness of the insulating covering is constant and equal to b, and if the wires are arranged in square order,

$$Y = 2\,(y+b), \tag{15}$$

and the condition is

$$\frac{x^2\,(2y+b)}{y^3} = \text{constant}. \tag{16}$$

In this case the diameter of the wire increases with the diameter of the layer of which it forms part, but not in so high a ratio.

If we adopt the first of these two hypotheses, which will be nearly true if the wire itself nearly fills up the whole space, then we may put $\qquad y = a\,x, \qquad Y = \beta y,$

where a and β are constant numerical quantities, and

$$G = N\frac{1}{a^2\,\beta^2}\left(\frac{1}{a}-\frac{1}{x}\right),$$

$$R = N\frac{\rho}{\pi}\frac{1}{a^4\,\beta^2}\left(\frac{1}{a}-\frac{1}{x}\right),$$

where a is a constant depending upon the size and form of the free space left inside the coil.

Hence, if we make the thickness of the wire vary in the same ratio as x, we obtain very little advantage by increasing the external size of the coil after the external dimensions have become a large multiple of the internal dimensions.

720.] If increase of resistance is not regarded as a defect, as when the external resistance is far greater than that of the galvanometer, or when our only object is to produce a field of intense force, we may make y and Y constant. We have then

$$G = \frac{N}{Y^2}(x-a),$$

$$R = \tfrac{1}{3}\frac{N}{Yy^2}\frac{\rho}{\pi}(x^3-a_1{}^3),$$

where a is a constant depending on the vacant space inside the coil. In this case the value of G increases uniformly as the dimensions of the coil are increased, so that there is no limit to the value of G except the labour and expense of making the coil.

On Suspended Coils.

721.] In the ordinary galvanometer a suspended magnet is acted on by a fixed coil. But if the coil can be suspended with sufficient delicacy, we may determine the action of the magnet, or of another coil on the suspended coil, by its deflexion from the position of equilibrium.

We cannot, however, introduce the electric current into the coil unless there is metallic connexion between the electrodes of the battery and those of the wire of the coil. This connexion may be made in two different ways, by the Bifilar Suspension, and by wires in opposite directions.

The bifilar suspension has already been described in Art. 459 as applied to magnets. The arrangement of the upper part of the suspension is shewn in Fig. 55. When applied to coils, the two fibres are no longer of silk but of metal, and since the torsion of a metal wire capable of supporting the coil and transmitting the current is much greater than that of a silk fibre, it must be taken specially into account. This suspension has been brought to great perfection in the instruments constructed by M. Weber.

The other method of suspension is by means of a single wire which is connected to one extremity of the coil. The other extremity of the coil is connected to another wire which is made to hang down, in the same vertical straight line with the first wire, into a cup of mercury, as is shewn in Fig. 57, Art. 729. In certain cases it is convenient to fasten the extremities of the two wires to pieces by which they may be tightly stretched, care being taken that the line of these wires passes through the centre of gravity of the coil. The apparatus in this form may be used when the axis is not vertical; see Fig. 53.

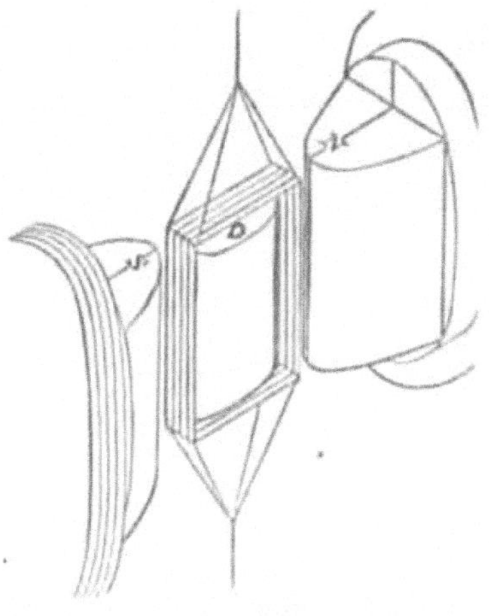

Fig. 53.

722.] The suspended coil may be used as an exceedingly sensitive galvanometer, for, by increasing the intensity of the magnetic force in the field in which it hangs, the force due to a feeble current in the coil may be greatly increased without adding to the mass of the coil. The magnetic force for this purpose may be produced by means of permanent magnets, or by electromagnets excited by an auxiliary current, and it may be powerfully concentrated on the suspended coil by means of soft iron armatures. Thus, in Sir W. Thomson's recording apparatus, Fig. 53, the coil is suspended between the opposite poles of the electromagnets *N* and *S*, and in order to concentrate the lines of magnetic force on the vertical sides of the coil, a piece of soft iron, *D*, is fixed between the poles of the magnets. This iron becoming magnetized by induction, produces a very powerful field of force, in the intervals between it and the two magnets, through which the vertical sides of the coil are free to move, so that the coil, even when the current through it is very feeble, is acted on by a considerable force tending to turn it about its vertical axis.

723.] Another application of the suspended coil is to determine, by comparison with a tangent galvanometer, the horizontal component of terrestrial magnetism.

The coil is suspended so that it is in stable equilibrium when its plane is parallel to the magnetic meridian. A current γ is passed through the coil and causes it to be deflected into a new position of equilibrium, making an angle θ with the magnetic meridian. If the suspension is bifilar, the moment of the couple which produces this deflexion is $F\sin\theta$, and this must be equal to $H\gamma g\cos\theta$, where H is the horizontal component of terrestrial magnetism, γ is the current in the coil, and g is the sum of the areas of all the windings of the coil. Hence

$$H\gamma = \frac{F}{g}\tan\theta.$$

If A is the moment of inertia of the coil about its axis of suspension, and T the time of a single vibration,

$$FT^2 = \pi^2 A,$$

and we obtain
$$H\gamma = \frac{\pi^2 A}{T^2 g}\tan\theta.$$

If the same current passes through the coil of a tangent galvanometer, and deflects the magnet through an angle ϕ,

$$\frac{\gamma}{H} = \frac{1}{G}\tan\phi,$$

where G is the principal constant of the tangent galvanometer, Art. 710.

From these two equations we obtain

$$H = \frac{\pi}{T}\sqrt{\frac{AG\tan\theta}{g\tan\phi}}, \qquad \gamma = \frac{\pi}{T}\sqrt{\frac{A\tan\theta\tan\phi}{Gg}}.$$

This method was given by F. Kohlrausch *.

* Pogg., *Ann.* cxxxviii, Feb. 1869.

724.] Sir William Thomson has constructed a single instrument by means of which the observations required to determine H and γ may be made simultaneously by the same observer.

The coil is suspended so as to be in equilibrium with its plane in the magnetic meridian, and is deflected from this position when the current flows through it. A very small magnet is suspended at the centre of the coil, and is deflected by the current in the direction opposite to that of the deflexion of the coil. Let the deflexion of the coil be θ, and that of the magnet ϕ, then the energy of the system is

$$H\gamma g \sin\theta + m\gamma G\sin(\theta-\phi) - Hm\cos\phi - F\cos\theta.$$

Differentiating with respect to θ and ϕ, we obtain the equations of equilibrium of the coil and of the magnet respectively,

$$H\gamma g\cos\theta + m\gamma G\cos(\theta-\phi) + F\sin\theta = 0,$$
$$-m\gamma G\cos(\theta-\phi) + Hm\sin\phi = 0.$$

From these equations we find, by eliminating H or γ, a quadratic equation from which γ or H may be found. If m, the magnetic moment of the suspended magnet, is very small, we obtain the following approximate values

$$H = \frac{\pi}{T}\sqrt{\frac{-AG\sin\theta\cos(\theta-\phi)}{g\cos\theta\sin\phi}} - \frac{1}{2}\frac{mG}{g}\frac{\cos(\theta-\phi)}{\cos\theta},$$

$$\gamma = \frac{\pi}{T}\sqrt{\frac{-A\sin\theta\sin\phi}{Gg\cos\theta\cos(\theta-\phi)}} - \frac{1}{2}\frac{m}{g}\frac{\sin\phi}{\cos\theta}.$$

In these expressions G and g are the principal electric constants of the coil, A its moment of inertia, T its time of vibration, m the magnetic moment of the magnet, H the intensity of the horizontal magnetic force, γ the strength of the current, θ the deflexion of the coil, and ϕ that of the magnet.

Since the deflexion of the coil is in the opposite direction to the deflexion of the magnet, these values of H and γ will always be real.

Weber's Electrodynamometer.

725.] In this instrument a small coil is suspended by two wires within a larger coil which is fixed. When a current is made to flow through both coils, the suspended coil tends to place itself parallel to the fixed coil. This tendency is counteracted by the moment of the forces arising from the bifilar suspension, and it is also affected by the action of terrestrial magnetism on the suspended coil.

In the ordinary use of the instrument the planes of the two coils are nearly at right angles to each other, so that the mutual action of the currents in the coils may be as great as possible, and the plane of the suspended coil is nearly at right angles to the magnetic meridian, so that the action of terrestrial magnetism may be as small as possible.

Let the magnetic azimuth of the plane of the fixed coil be a, and let the angle which the axis of the suspended coil makes with the plane of the fixed coil be $\theta+\beta$, where β is the value of this angle when the coil is in equilibrium and no current is flowing, and θ is the deflexion due to the current. The equation of equilibrium is

$$Gg\gamma_1\gamma_2\cos(\theta+\beta)-Hg\gamma_2\sin(\theta+\beta+a)-F\sin\theta=0.$$

Let us suppose that the instrument is adjusted so that a and β are both very small, and that $Hg\gamma_2$ is small compared with F. We have in this case, approximately,

$$\tan\theta=\frac{Gg\gamma_1\gamma_2\cos\beta}{F}-\frac{Hg\gamma_2\sin(a+\beta)}{F}-\frac{HGg^2\gamma_1\gamma_2^2}{F^2}-\frac{G^2g^2\gamma_1^2\gamma_2^2\sin\beta}{F^2}$$

If the deflexions when the signs of γ_1 and γ_2 are changed are as follows:

θ_1 when γ_1 is $+$ and γ_2 $+$,
θ_2 " $-$ " $-$,
θ_3 " $+$ " $-$,
θ_4 " $-$ " $+$;

then we find

$$\gamma_1\gamma_2=\tfrac{1}{4}\frac{F}{Gg\cos\beta}(\tan\theta_1+\tan\theta_2-\tan\theta_3-\tan\theta_4).$$

If it is the same current which flows through both coils we may put $\gamma_1\gamma_2=\gamma^2$, and thus obtain the value of γ.

When the currents are not very constant it is best to adopt this method, which is called the Method of Tangents.

If the currents are so constant that we can adjust β, the angle of the torsion-head of the instrument, we may get rid of the correction for terrestrial magnetism at once by the method of sines. In this method β is adjusted till the deflexion is zero, so that

$$\theta=-\beta.$$

If the signs of γ_1 and γ_2 are indicated by the suffixes of β as before,

$$F\sin\beta_1=-F\sin\beta_3=-Gg\gamma_1\gamma_2+Hg\gamma_2\sin a,$$
$$F\sin\beta_2=-F\sin\beta_4=-Gg\gamma_1\gamma_2-Hg\gamma_2\sin a,$$

and
$$\gamma_1\gamma_2=-\frac{F}{4Gg}(\sin\beta_1+\sin\beta_2-\sin\beta_3-\sin\beta_4).$$

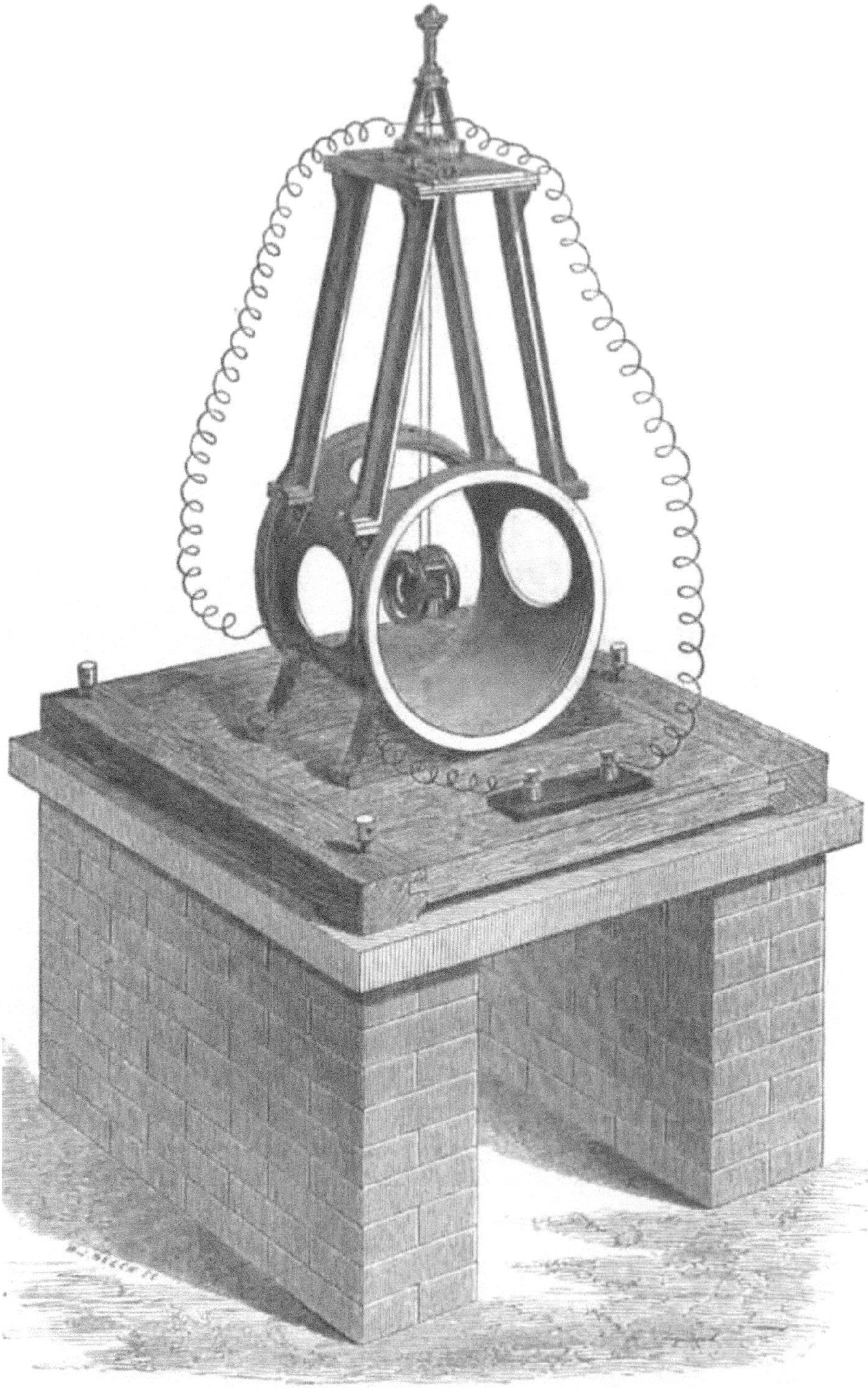

Fig. 54.

This is the method adopted by Mr. Latimer Clark in his use of the instrument constructed by the Electrical Committee of the British Association. We are indebted to Mr. Clark for the drawing of the electrodynamometer in Figure 54, in which Helmholtz's arrangement of two coils is adopted both for the fixed and for the suspended coil *. The torsion-head of the instrument, by which the bifilar suspension is adjusted, is represented in Fig. 55. The

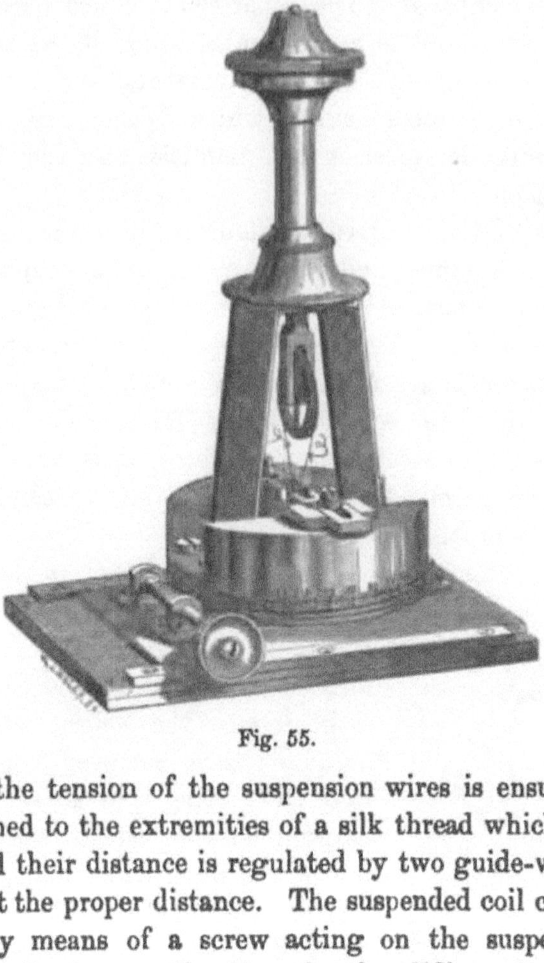

Fig. 55.

equality of the tension of the suspension wires is ensured by their being attached to the extremities of a silk thread which passes over a wheel, and their distance is regulated by two guide-wheels, which can be set at the proper distance. The suspended coil can be moved vertically by means of a screw acting on the suspension-wheel, and horizontally in two directions by the sliding pieces shewn at the bottom of Fig. 55. It is adjusted in azimuth by means of the torsion-screw, which turns the torsion-head round a vertical axis (see Art. 459). The azimuth of the suspended coil is ascertained by observing the reflexion of a scale in the mirror, shewn just beneath the axis of the suspended coil.

* In the actual instrument, the wires conveying the current to and from the coils are not spread out as displayed in the figure, but are kept as close together as possible, so as to neutralize each other's electromagnetic action.

The instrument originally constructed by Weber is described in his *Elektrodynamische Maasbestimmungen*. It was intended for the measurement of small currents, and therefore both the fixed and the suspended coils consisted of many windings, and the suspended coil occupied a larger part of the space within the fixed coil than in the instrument of the British Association, which was primarily intended as a standard instrument, with which more sensitive instruments might be compared. The experiments which he made with it furnish the most complete experimental proof of the accuracy of Ampère's formula as applied to closed currents, and form an important part of the researches by which Weber has raised the numerical determination of electrical quantities to a very high rank as regards precision.

Weber's form of the electrodynamometer, in which one coil is suspended within another, and is acted on by a couple tending to turn it about a vertical axis, is probably the best fitted for absolute measurements. A method of calculating the constants of such an arrangement is given in Art. 697.

726.] If, however, we wish, by means of a feeble current, to produce a considerable electromagnetic force, it is better to place the suspended coil parallel to the fixed coil, and to make it capable of motion to or from it.

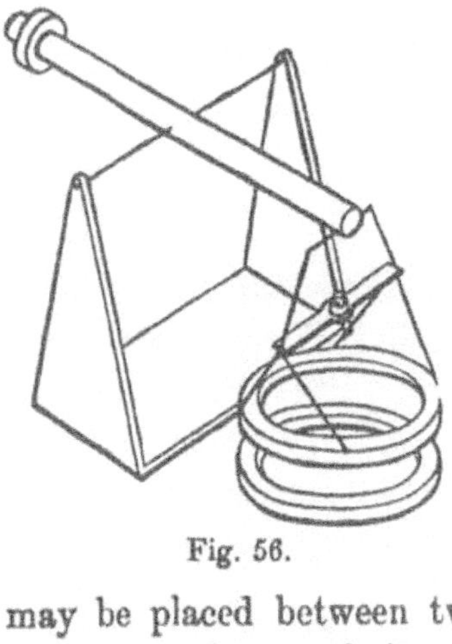

Fig. 56.

The suspended coil in Dr. Joule's current-weigher, Fig. 56, is horizontal, and capable of vertical motion, and the force between it and the fixed coil is estimated by the weight which must be added to or removed from the coil in order to bring it to the same relative position with respect to the fixed coil that it has when no current passes.

The suspended coil may also be fastened to the extremity of the horizontal arm of a torsion-balance, and may be placed between two fixed coils, one of which attracts it, while the other repels it, as in Fig. 57.

By arranging the coils as described in Art. 729, the force acting on the suspended coil may be made nearly uniform within a small distance of the position of equilibrium.

Another coil may be fixed to the other extremity of the arm of the torsion-balance and placed between two fixed coils. If the two suspended coils are similar, but with the current flowing in opposite directions, the effect of terrestrial magnetism on the

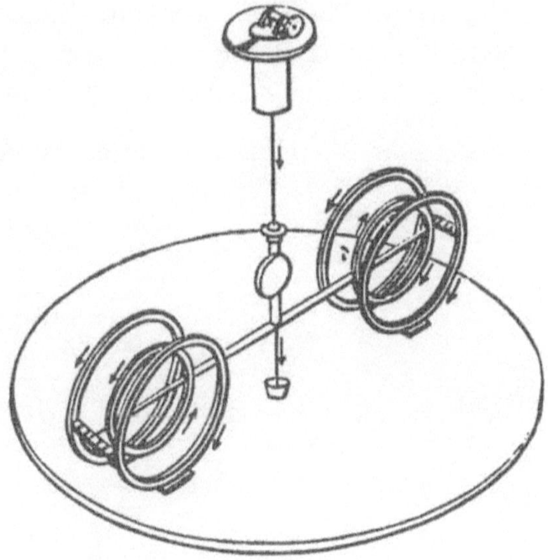

Fig. 57.

position of the arm of the torsion-balance will be completely eliminated.

727.] If the suspended coil is in the shape of a long solenoid, and is capable of moving parallel to its axis, so as to pass into the interior of a larger fixed solenoid having the same axis, then, if the current is in the same direction in both solenoids, the suspended solenoid will be sucked into the fixed one by a force which will be nearly uniform as long as none of the extremities of the solenoids are near one another.

728.] To produce a uniform longitudinal force on a small coil placed between two equal coils of much larger dimensions, we should make the ratio of the diameter of the large coils to the distance between their planes that of 2 to $\sqrt{3}$. If we send the same current through these coils in opposite directions, then, in the expression for ω, the terms involving odd powers of r disappear, and since $\sin^2 a = \frac{4}{7}$ and $\cos^2 a = \frac{3}{7}$, the term involving r^4 disappears also, and we have

$$\omega = \tfrac{8}{7}\sqrt{\tfrac{3}{7}}\,\pi n \gamma \left\{ 3\frac{r^2}{c^2} Q_2(\theta) + \tfrac{11}{7}\frac{r^6}{c^6} Q_6(\theta) + \&c. \right\},$$

which indicates a nearly uniform force on a small suspended coil. The arrangement of the coils in this case is that of the two outer coils in the galvanometer with three coils, described at Art. 715. See Fig. 51.

729.] If we wish to suspend a coil between two coils placed so near it that the distance between the mutually acting wires is small compared with the radius of the coils, the most uniform force is obtained by making the radius of either of the outer coils exceed that of the middle one by $\frac{1}{\sqrt{3}}$ of the distance between the planes of the middle and outer coils.

CHAPTER XVI.

ELECTROMAGNETIC OBSERVATIONS.

730.] So many of the measurements of electrical quantities depend on observations of the motion of a vibrating body that we shall devote some attention to the nature of this motion, and the best methods of observing it.

The small oscillations of a body about a position of stable equilibrium are, in general, similar to those of a point acted on by a force varying directly as the distance from a fixed point. In the case of the vibrating bodies in our experiments there is also a resistance to the motion, depending on a variety of causes, such as the viscosity of the air, and that of the suspension fibre. In many electrical instruments there is another cause of resistance, namely, the reflex action of currents induced in conducting circuits placed near vibrating magnets. These currents are induced by the motion of the magnet, and their action on the magnet is, by the law of Lenz, invariably opposed to its motion. This is in many cases the principal part of the resistance.

A metallic circuit, called a Damper, is sometimes placed near a magnet for the express purpose of damping or deadening its vibrations. We shall therefore speak of this kind of resistance as Damping.

In the case of slow vibrations, such as can be easily observed, the whole resistance, from whatever causes it may arise, appears to be proportional to the velocity. It is only when the velocity is much greater than in the ordinary vibrations of electromagnetic instruments that we have evidence of a resistance proportional to the square of the velocity.

We have therefore to investigate the motion of a body subject to an attraction varying as the distance, and to a resistance varying as the velocity.

731.] The following application, by Professor Tait*, of the principle of the Hodograph, enables us to investigate this kind of motion in a very simple manner by means of the equiangular spiral.

Let it be required to find the acceleration of a particle which describes a logarithmic or equiangular spiral with uniform angular velocity ω about the pole.

The property of this spiral is, that the tangent PT makes with the radius vector PS a constant angle a.

If v is the velocity at the point P, then

$$v \,.\, \sin a = \omega \,.\, SP.$$

Hence, if we draw SP' parallel to PT and equal to SP, the velocity at P will be given both in magnitude and direction by

$$v = \frac{\omega}{\sin a}\, SP'.$$

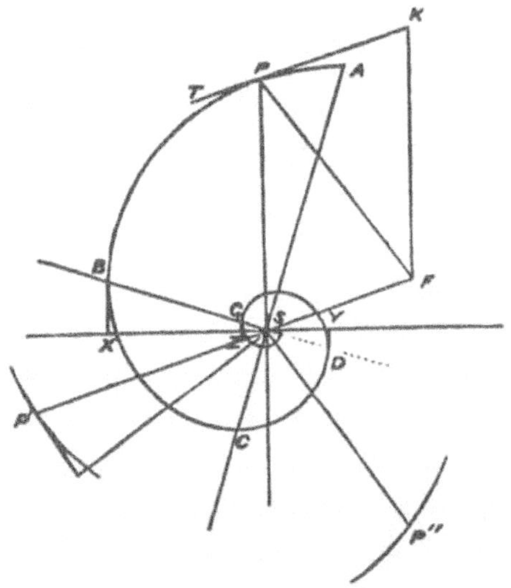

Fig. 58.

Hence P' will be a point in the hodograph. But SP' is SP turned through a constant angle $\pi - a$, so that the hodograph described by P' is the same as the original spiral turned about its pole through an angle $\pi - a$.

The acceleration of P is represented in magnitude and direction by the velocity of P' multiplied by the same factor, $\frac{\omega}{\sin a}$.

Hence, if we perform on SP' the same operation of turning it

* *Proc. R. S. Edin.*, Dec. 16, 1867.

through an angle $\pi - a$ into the position SP'', the acceleration of P will be equal in magnitude and direction to

$$\frac{\omega^2}{\sin^2 a} SP'',$$

where SP'' is equal to SP turned through an angle $2\pi - 2a$.

If we draw PF equal and parallel to SP'', the acceleration will be $\frac{\omega^2}{\sin^2 a} PF$, which we may resolve into

$$\frac{\omega^2}{\sin^2 a} PS \text{ and } \frac{\omega^2}{\sin^2 a} PK.$$

The first of these components is a central force towards S proportional to the distance.

The second is in a direction opposite to the velocity, and since

$$PK = 2\cos a\, P'S = -2\frac{\sin a \cos a}{\omega} v,$$

this force may be written

$$-2\frac{\omega \cos a}{\sin a} v.$$

The acceleration of the particle is therefore compounded of two parts, the first of which is an attractive force μr, directed towards S, and proportional to the distance, and the second is $-2kv$, a resistance to the motion proportional to the velocity, where

$$\mu = \frac{\omega^2}{\sin^2 a}, \text{ and } k = \omega\frac{\cos a}{\sin a}.$$

If in these expressions we make $a = \frac{\pi}{2}$, the orbit becomes a circle, and we have $\mu_0 = \omega_0{}^2$, and $k = 0$.

Hence, if the law of attraction remains the same, $\mu = \mu_0$, and

$$\omega = \omega_0 \sin a,$$

or the angular velocity in different spirals with the same law of attraction is proportional to the sine of the angle of the spiral.

732.] If we now consider the motion of a point which is the projection of the moving point P on the horizontal line XY, we shall find that its distance from S and its velocity are the horizontal components of those of P. Hence the acceleration of this point is also an attraction towards S, equal to μ times its distance from S, together with a retardation equal to k times its velocity.

We have therefore a complete construction for the rectilinear motion of a point, subject to an attraction proportional to the distance from a fixed point, and to a resistance proportional to the velocity. The motion of such a point is simply the horizontal part of the motion of another point which moves with uniform angular velocity in a logarithmic spiral.

733.] The equation of the spiral is

$$r = Ce^{-\phi \cot \alpha}.$$

To determine the horizontal motion, we put

$$\phi = \omega t, \qquad x = a + r \sin \phi,$$

where a is the value of x for the point of equilibrium.

If we draw BSD making an angle α with the vertical, then the tangents BX, DY, GZ, &c. will be vertical, and X, Y, Z, &c. will be the extremities of successive oscillations.

734.] The observations which are made on vibrating bodies are—

(1) The scale-reading at the stationary points. These are called Elongations.

(2) The time of passing a definite division of the scale in the positive or negative direction.

(3) The scale-reading at certain definite times. Observations of this kind are not often made except in the case of vibrations of long period *.

The quantities which we have to determine are—

(1) The scale-reading at the position of equilibrium.

(2) The logarithmic decrement of the vibrations.

(3) The time of vibration.

To determine the Reading at the Position of Equilibrium from Three Consecutive Elongations.

735.] Let x_1, x_2, x_3 be the observed scale-readings, corresponding to the elongations X, Y, Z, and let a be the reading at the position of equilibrium, S, and let r_1 be the value of SB,

$$x_1 - a = \quad r_1 \sin \alpha,$$

$$x_2 - a = -r_1 \sin \alpha e^{-\pi \cot \alpha},$$

$$x_3 - a = \quad r_1 \sin \alpha e^{-2\pi \cot \alpha}.$$

From these values we find

$$(x_1 - a)(x_3 - a) = (x_2 - a)^2,$$

whence $$a = \frac{x_1 x_3 - x_2^2}{x_1 + x_3 - 2x_2}.$$

When x_3 does not differ much from x_1 we may use as an approximate formula

$$a = \tfrac{1}{4}(x_1 + 2x_2 + x_3).$$

* See Gauss, *Resultate des Magnetischen Vereins*, 1836. II.

To determine the Logarithmic Decrement.

736.] The logarithm of the ratio of the amplitude of a vibration to that of the next following is called the Logarithmic Decrement. If we write ρ for this ratio

$$\rho = \frac{x_1 - x_2}{x_3 - x_2}, \qquad L = \log_{10}\rho, \qquad \lambda = \log_\epsilon \rho.$$

L is called the common logarithmic decrement, and λ the Napierian logarithmic decrement. It is manifest that

$$\lambda = L \log_\epsilon 10 = \pi \cot \alpha.$$

Hence

$$\alpha = \cot^{-1}\frac{\lambda}{\pi},$$

which determines the angle of the logarithmic spiral.

In making a special determination of λ we allow the body to perform a considerable number of vibrations. If c_1 is the amplitude of the first, and c_n that of the nth vibration,

$$\lambda = \frac{1}{n-1}\log_\epsilon\left(\frac{c_1}{c_n}\right).$$

If we suppose the accuracy of observation to be the same for small vibrations as for large ones, then, to obtain the best value of λ, we should allow the vibrations to subside till the ratio of c_1 to c_n becomes most nearly equal to ϵ, the base of the Napierian logarithms. This gives n the nearest whole number to $\frac{1}{\lambda} + 1$.

Since, however, in most cases time is valuable, it is best to take the second set of observations before the diminution of amplitude has proceeded so far.

737.] In certain cases we may have to determine the position of equilibrium from two consecutive elongations, the logarithmic decrement being known from a special experiment. We have then

$$a = \frac{x_1 + e^\lambda x_2}{1 + e^\lambda}.$$

Time of Vibration.

738.] Having determined the scale-reading of the point of equilibrium, a conspicuous mark is placed at that point of the scale, or as near it as possible, and the times of the passage of this mark are noted for several successive vibrations.

Let us suppose that the mark is at an unknown but very small distance x on the positive side of the point of equilibrium, and that

t_1 is the observed time of the first transit of the mark in the positive direction, and t_2, t_3, &c. the times of the following transits.

If T be the time of vibration, and P_1, P_2, P_3, &c. the times of transit of the true point of equilibrium,

$$t_1 = P_1 + \frac{x}{v_1}, \qquad t_2 = P_2 + \frac{x}{v_2},$$

where v_1, v_2, &c. are the successive velocities of transit, which we may suppose uniform for the very small distance x.

If ρ is the ratio of the amplitude of a vibration to the next in succession,
$$v_2 = -\frac{1}{\rho} v_1, \quad \text{and} \quad \frac{x}{v_2} = -\rho \frac{x}{v_1}.$$

If three transits are observed at times t_1, t_2, t_3, we find

$$\frac{x}{v_1} = \frac{t_1 - 2t_2 + t_3}{(\rho+1)^2}.$$

The period of vibration is therefore

$$T = \tfrac{1}{2}(t_3 - t_1) - \tfrac{1}{2}\frac{\rho - 1}{\rho + 1}(t_1 - 2t_2 + t_3).$$

The time of the second passage of the true point of equilibrium is

$$P_2 = \tfrac{1}{4}(t_1 + 2t_2 + t_3) - \tfrac{1}{4}\frac{(\rho-1)^2}{(\rho+1)^2}(t_1 - 2t_2 + t_3).$$

Three transits are sufficient to determine these three quantities, but any greater number may be combined by the method of least squares. Thus, for five transits,

$$T = \tfrac{1}{10}(2t_5 + t_4 - t_2 - 2t_1) - \tfrac{1}{10}(t_1 - 2t_2 + 2t_3 - 2t_4 + t_5)\frac{\rho - 1}{\rho + 1}\left(2 - \frac{\rho}{1 + \rho^2}\right).$$

The time of the third transit is,

$$P_3 = \tfrac{1}{8}(t_1 + 2t_2 + 2t_3 + 2t_4 + t_5) - \tfrac{1}{8}(t_1 - 2t_2 + 2t_3 - 2t_4 + t_5)\frac{(\rho-1)^2}{(\rho+1)^2}.$$

739.] The same method may be extended to a series of any number of vibrations. If the vibrations are so rapid that the time of every transit cannot be recorded, we may record the time of every third or every fifth transit, taking care that the directions of successive transits are opposite. If the vibrations continue regular for a long time, we need not observe during the whole time. We may begin by observing a sufficient number of transits to determine approximately the period of vibration, T, and the time of the middle transit, P, noting whether this transit is in the positive or the negative direction. We may then either go on counting the vibrations without recording the times of transit, or we may leave the apparatus unwatched. We then observe a second series of transits, and deduce the time of vibration T' and the time of middle transit P', noting the direction of this transit.

If T and T', the periods of vibration as deduced from the two sets of observations, are nearly equal, we may proceed to a more accurate determination of the period by combining the two series of observations.

Dividing $P'-P$ by T, the quotient ought to be very nearly an integer, even or odd according as the transits P and P' are in the same or in opposite directions. If this is not the case, the series of observations is worthless, but if the result is very nearly a whole number n, we divide $P'-P$ by n, and thus find the mean value of T for the whole time of swinging.

740.] The time of vibration T thus found is the actual mean time of vibration, and is subject to corrections if we wish to deduce from it the time of vibration in infinitely small arcs and without damping.

To reduce the observed time to the time in infinitely small arcs, we observe that the time of a vibration of amplitude a is in general of the form
$$T = T_1 (1 + \kappa c^2),$$
where κ is a coefficient, which, in the case of the ordinary pendulum, is $\frac{1}{64}$. Now the amplitudes of the successive vibrations are c, $c\rho^{-1}$, $c\rho^{-2}$, ... $c\rho^{1-n}$, so that the whole time of n vibrations is
$$nT = T_1 \left(n + \kappa \frac{\rho^2 c_1^{\,2} - c_n^{\,2}}{\rho^2 - 1}\right),$$
where T is the time deduced from the observations.

Hence, to find the time T_1 in infinitely small arcs, we have approximately,
$$T_1 = T\left\{1 - \frac{\kappa}{n}\,\frac{c_1^{\,2}\rho^2 - c_n^{\,2}}{\rho^2 - 1}\right\}.$$

To find the time T_0 when there is no damping, we have
$$T_0 = T_1 \sin a$$
$$= T_1 \frac{\pi}{\sqrt{\pi^2 + \lambda^2}}.$$

741.] The equation of the rectilinear motion of a body, attracted to a fixed point and resisted by a force varying as the velocity, is
$$\frac{d^2 x}{dt^2} + 2k\frac{dx}{dt} + \omega^2 (x - a) = 0, \qquad (1)$$
where x is the coordinate of the body at the time t, and a is the coordinate of the point of equilibrium.

To solve this equation, let

$$x-a = e^{-kt}y; \tag{2}$$

then

$$\frac{d^2y}{dt^2}+(\omega^2-k^2)y = 0; \tag{3}$$

the solution of which is

$$y = C\cos(\sqrt{\omega^2-k^2}\,t+a), \text{ when } k \text{ is less than } \omega; \tag{4}$$

$$y = A+Bt, \text{ when } k \text{ is equal to } \omega; \tag{5}$$

and $$y = C'\cosh(\sqrt{k^2-\omega^2}\,t+a'), \text{ when } k \text{ is greater than } \omega. \tag{6}$$

The value of x may be obtained from that of y by equation (2). When k is less than ω, the motion consists of an infinite series of oscillations, of constant periodic time, but of continually decreasing amplitude. As k increases, the periodic time becomes longer, and the diminution of amplitude becomes more rapid.

When k (half the coefficient of resistance) becomes equal to or greater than ω, (the square root of the acceleration at unit distance from the point of equilibrium,) the motion ceases to be oscillatory, and during the whole motion the body can only once pass through the point of equilibrium, after which it reaches a position of greatest elongation, and then returns towards the point of equilibrium, continually approaching, but never reaching it.

Galvanometers in which the resistance is so great that the motion is of this kind are called *dead beat* galvanometers. They are useful in many experiments, but especially in telegraphic signalling, in which the existence of free vibrations would quite disguise the movements which are meant to be observed.

Whatever be the values of k and ω, the value of a, the scale-reading at the point of equilibrium, may be deduced from five scale-readings, p, q, r, s, t, taken at equal intervals of time, by the formula

$$a = \frac{q(rs-qt)+r(pt-r^2)+s(qr-ps)}{(p-2q+r)(r-2s+t)-(q-2r+s)^2}.$$

On the Observation of the Galvanometer.

742.] To measure a constant current with the tangent galvanometer, the instrument is adjusted with the plane of its coils parallel to the magnetic meridian, and the zero reading is taken. The current is then made to pass through the coils, and the deflexion of the magnet corresponding to its new position of equilibrium is observed. Let this be denoted by ϕ.

Then, if H is the horizontal magnetic force, G the coefficient of the galvanometer, and γ the strength of the current,

$$\gamma = \frac{H}{G}\tan\phi. \tag{1}$$

If the coefficient of torsion of the suspension fibre is τMH (see Art. 452), we must use the corrected formula

$$\gamma = \frac{H}{G}(\tan\phi + \tau\phi\sec\phi). \qquad (2)$$

Best Value of the Deflexion.

743.] In some galvanometers the number of windings of the coil through which the current flows can be altered at pleasure. In others a known fraction of the current can be diverted from the galvanometer by a conductor called a Shunt. In either case the value of G, the effect of a unit-current on the magnet, is made to vary.

Let us determine the value of G, for which a given error in the observation of the deflexion corresponds to the smallest error of the deduced value of the strength of the current.

Differentiating equation (1), we find

$$\frac{d\gamma}{d\phi} = \frac{H}{G}\sec^2\phi. \qquad (3)$$

Eliminating G,

$$\frac{d\phi}{d\gamma} = \frac{1}{2\gamma}\sin 2\phi. \qquad (4)$$

This is a maximum for a given value of γ when the deflexion is 45°. The value of G should therefore be adjusted till $G\gamma$ is as nearly equal to H as is possible; so that for strong currents it is better not to use too sensitive a galvanometer.

On the Best Method of applying the Current.

744.] When the observer is able, by means of a key, to make or break the connexions of the circuit at any instant, it is advisable to operate with the key in such a way as to make the magnet arrive at its position of equilibrium with the least possible velocity. The following method was devised by Gauss for this purpose.

Suppose that the magnet is in its position of equilibrium, and that there is no current. The observer now makes contact for a short time, so that the magnet is set in motion towards its new position of equilibrium. He then breaks contact. The force is now towards the original position of equilibrium, and the motion is retarded. If this is so managed that the magnet comes to rest exactly at the new position of equilibrium, and if the observer again makes contact at that instant and maintains the contact, the magnet will remain at rest in its new position.

If we neglect the effect of the resistances and also the inequality of the total force acting in the new and the old positions, then, since we wish the new force to generate as much kinetic energy during the time of its first action as the original force destroys while the circuit is broken, we must prolong the first action of the current till the magnet has moved over half the distance from the first position to the second. Then if the original force acts while the magnet moves over the other half of its course, it will exactly stop it. Now the time required to pass from a point of greatest elongation to a point half way to the position of equilibrium is one-sixth of a complete period, or one-third of a single vibration.

The operator, therefore, having previously ascertained the time of a single vibration, makes contact for one-third of that time, breaks contact for another third of the same time, and then makes contact again during the continuance of the experiment. The magnet is then either at rest, or its vibrations are so small that observations may be taken at once, without waiting for the motion to die away. For this purpose a metronome may be adjusted so as to beat three times for each single vibration of the magnet.

The rule is somewhat more complicated when the resistance is of sufficient magnitude to be taken into account, but in this case the vibrations die away so fast that it is unnecessary to apply any corrections to the rule.

When the magnet is to be restored to its original position, the circuit is broken for one-third of a vibration, made again for an equal time, and finally broken. This leaves the magnet at rest in its former position.

If the reversed reading is to be taken immediately after the direct one, the circuit is broken for the time of a single vibration and then reversed. This brings the magnet to rest in the reversed position.

Measurement by the First Swing.

745.] When there is no time to make more than one observation, the current may be measured by the extreme elongation observed in the first swing of the magnet. If there is no resistance, the permanent deflexion ϕ is half the extreme elongation. If the resistance is such that the ratio of one vibration to the next is ρ, and if θ_0 is the zero reading, and θ_1 the extreme elongation in the first swing, the deflexion, ϕ, corresponding to the point of equilibrium is

$$\phi = \frac{\theta_0 + \rho\theta_1}{1+\rho}.$$

In this way the deflexion may be calculated without waiting for the magnet to come to rest in its position of equilibrium.

To make a Series of Observations.

746.] The best way of making a considerable number of measures of a constant current is by observing three elongations while the current is in the positive direction, then breaking contact for about the time of a single vibration, so as to let the magnet swing into the position of negative deflexion, then reversing the current and observing three successive elongations on the negative side, then breaking contact for the time of a single vibration and repeating the observations on the positive side, and so on till a sufficient number of observations have been obtained. In this way the errors which may arise from a change in the direction of the earth's magnetic force during the time of observation are eliminated. The operator, by carefully timing the making and breaking of contact, can easily regulate the extent of the vibrations, so as to make them sufficiently small without being indistinct. The motion of the magnet is graphically represented in Fig. 59, where the abscissa represents the time, and the ordinate the deflexion of the magnet. If $\theta_1 \dots \theta_6$ be the observed elongations, the deflexion is given by the equation

$$8\phi = \theta_1 + 2\theta_2 + \theta_3 - \theta_4 - 2\theta_5 - \theta_6.$$

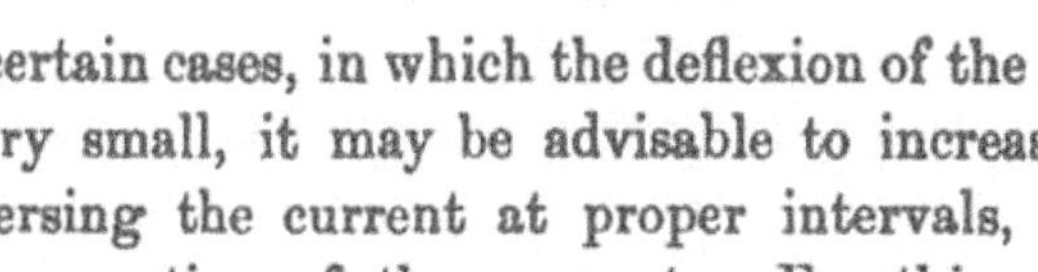

Fig. 59.

Method of Multiplication.

747.] In certain cases, in which the deflexion of the galvanometer magnet is very small, it may be advisable to increase the visible effect by reversing the current at proper intervals, so as to set up a swinging motion of the magnet. For this purpose, after ascertaining the time, T, of a single vibration of the magnet, the current is sent in the positive direction for a time T, then in the reversed direction for an equal time, and so on. When the motion of the magnet has become visible, we may make the reversal of the current at the observed times of greatest elongation.

Let the magnet be at the positive elongation θ_0, and let the current be sent through the coil in the negative direction. The point of equilibrium is then $-\phi$, and the magnet will swing to a negative elongation θ, such that

$$-\rho(\phi+\theta_1) = (\theta_0+\phi),$$

or $$-\rho\theta_1 = \theta_0+(\rho+1)\phi.$$

Similarly, if the current is now made positive while the magnet swings to θ_2, $$\rho\theta_2 = -\theta_1+(\rho+1)\phi,$$

or $$\rho^2\theta_2 = \theta_0+(\rho+1)^2\phi;$$

and if the current is reversed n times in succession, we find

$$(-1)^n\theta_n = \rho^{-n}\theta_0+\frac{\rho+1}{\rho-1}(1-\rho^{-n})\phi,$$

whence we may find ϕ in the form

$$\phi = (\theta_n-\rho^{-n}\theta_0)\frac{\rho-1}{\rho+1}\frac{1}{1-\rho^{-n}}.$$

If n is a number so great that ρ^{-n} may be neglected, the expression becomes

$$\phi = \theta_n\frac{\rho-1}{\rho+1}.$$

The application of this method to exact measurement requires an accurate knowledge of ρ, the ratio of one vibration of the magnet to the next under the influence of the resistances which it experiences. The uncertainties arising from the difficulty of avoiding irregularities in the value of ρ generally outweigh the advantages of the large angular elongation. It is only where we wish to establish the existence of a very small current by causing it to produce a visible movement of the needle that this method is really valuable.

On the Measurement of Transient Currents.

748.] When a current lasts only during a very small fraction of the time of vibration of the galvanometer-magnet, the whole quantity of electricity transmitted by the current may be measured by the angular velocity communicated to the magnet during the passage of the current, and this may be determined from the elongation of the first vibration of the magnet.

If we neglect the resistance which damps the vibrations of the magnet, the investigation becomes very simple.

Let γ be the intensity of the current at any instant, and Q the quantity of electricity which it transmits, then

$$Q = \int\gamma\,dt. \tag{1}$$

Let M be the magnetic moment, and A the moment of inertia of the magnet and suspended apparatus,

$$A\frac{d^2\theta}{dt^2}+MH\sin\theta = MG\gamma\cos\theta. \tag{2}$$

If the time of the passage of the current is very small, we may integrate with respect to t during this short time without regarding the change of θ, and we find

$$A\frac{d\theta}{dt} = MG\cos\theta_0\int\gamma\,dt+C = MGQ\cos\theta_0+C. \tag{3}$$

This shews that the passage of the quantity Q produces an angular momentum $MGQ\cos\theta_0$ in the magnet, where θ_0 is the value of θ at the instant of passage of the current. If the magnet is initially in equilibrium, we may make $\theta_0 = 0$.

The magnet then swings freely and reaches an elongation θ_1. If there is no resistance, the work done against the magnetic force during this swing is $MH(1-\cos\theta_1)$.

The energy communicated to the magnet by the current is

$$\tfrac{1}{2}A\overline{\frac{d\theta}{dt}}\Big|^2.$$

Equating these quantities, we find

$$\overline{\frac{d\theta}{dt}}\Big|^2 = 2\frac{MH}{A}(1-\cos\theta_1), \tag{4}$$

whence
$$\frac{d\theta}{dt} = 2\sqrt{\frac{MH}{A}}\sin\tfrac{1}{2}\theta_1$$
$$= \frac{MG}{A}Q \text{ by (3)}. \tag{5}$$

But if T be the time of a single vibration of the magnet,

$$T = \pi\sqrt{\frac{A}{MH}}, \tag{6}$$

and we find
$$Q = \frac{H}{G}\frac{T}{\pi}2\sin\tfrac{1}{2}\theta_1, \tag{7}$$

where H is the horizontal magnetic force, G the coefficient of the galvanometer, T the time of a single vibration, and θ_1 the first elongation of the magnet.

749.] In many actual experiments the elongation is a small angle, and it is then easy to take into account the effect of resistance, for we may treat the equation of motion as a linear equation.

Let the magnet be at rest at its position of equilibrium, let an angular velocity v be communicated to it instantaneously, and let its first elongation be θ_1.

The equation of motion is

$$\theta = C e^{-\omega_1 t \tan\beta} \sin \omega_1 t, \tag{8}$$

$$\frac{d\theta}{dt} = C\omega_1 \sec\beta\, e^{-\omega_1 t \tan\beta} \cos(\omega_1 t + \beta). \tag{9}$$

When $t = 0$, $\theta = 0$, and $\dfrac{d\theta}{dt} = C\omega_1 = v$.

When $\omega_1 t + \beta = \dfrac{\pi}{2}$,

$$\theta = C e^{-\left(\frac{\pi}{2} - \beta\right)\tan\beta} \cos\beta = \theta_1. \tag{10}$$

Hence
$$\theta_1 = \frac{v}{\omega_1} e^{-\left(\frac{\pi}{2} - \beta\right)\tan\beta} \cos\beta. \tag{11}$$

Now
$$\frac{MH}{A} = \omega^2 = \omega_1^2 \sec^2\beta, \tag{12}$$

$$\tan\beta = \frac{\lambda}{\pi}, \qquad \omega_1 = \frac{\pi}{T_1}, \tag{13}$$

$$v = \frac{MG}{A} Q. \tag{14}$$

Hence
$$\theta_1 = \frac{QG}{H} \frac{\sqrt{\pi^2 + \lambda^2}}{T_1} e^{-\frac{\lambda}{\pi}\tan^{-1}\frac{\pi}{\lambda}}, \tag{15}$$

and
$$Q = \frac{H}{G} \frac{T_1}{\sqrt{\pi^2 + \lambda^2}} e^{\frac{\lambda}{\pi}\tan^{-1}\frac{\pi}{\lambda}} \theta_1, \tag{16}$$

which gives the first elongation in terms of the quantity of electricity in the transient current, and conversely, where T_1 is the observed time of a single vibration as affected by the actual resistance of damping. When λ is small we may use the approximate formula

$$Q = \frac{H}{G} \frac{T}{\pi} (1 + \tfrac{1}{2}\lambda)\theta_1. \tag{17}$$

Method of Recoil.

750.] The method given above supposes the magnet to be at rest in its position of equilibrium when the transient current is passed through the coil. If we wish to repeat the experiment we must wait till the magnet is again at rest. In certain cases, however, in which we are able to produce transient currents of equal intensity, and to do so at any desired instant, the following method, described by Weber *, is the most convenient for making a continued series of observations.

* *Resultate des Magnetischen Vereins*, 1838, p. 98.

Suppose that we set the magnet swinging by means of a transient current whose value is Q_0. If, for brevity, we write

$$\frac{G}{H}\frac{\sqrt{\pi^2+\lambda^2}}{T_1}e^{-\frac{\lambda}{\pi}\tan^{-1}\frac{\pi}{\lambda}} = K, \tag{18}$$

then the first elongation

$$\theta_1 = KQ_0 = a_1 \text{ (say)}. \tag{19}$$

The velocity instantaneously communicated to the magnet at starting is

$$v_0 = \frac{MG}{A}Q_0. \tag{20}$$

When it returns through the point of equilibrium in a negative direction its velocity will be

$$v_1 = -ve^{-\lambda}. \tag{21}$$

The next negative elongation will be

$$\theta_2 = -\theta_1 e^{-\lambda} = b_1. \tag{22}$$

When the magnet returns to the point of equilibrium, its velocity will be

$$v_2 = v_0 e^{-2\lambda}. \tag{23}$$

Now let an instantaneous current, whose total quantity is $-Q$, be transmitted through the coil at the instant when the magnet is at the zero point. It will change the velocity v_2 into v_2-v, where

$$v = \frac{MG}{A}Q. \tag{24}$$

If Q is greater than $Q_0e^{-2\lambda}$, the new velocity will be negative and equal to

$$-\frac{MG}{A}(Q-Q_0e^{-2\lambda}).$$

The motion of the magnet will thus be reversed, and the next elongation will be negative,

$$\theta_3 = -K(Q-Q_0e^{-2\lambda}) = c_1 = -KQ+\theta_1e^{-2\lambda}. \tag{25}$$

The magnet is then allowed to come to its positive elongation

$$\theta_4 = -\theta_3e^{-\lambda} = d_1 = e^{-\lambda}(KQ-a_1e^{-2\lambda}), \tag{26}$$

and when it again reaches the point of equilibrium a positive current whose quantity is Q is transmitted. This throws the magnet back in the positive direction to the positive elongation

$$\theta_5 = KQ-\theta_3e^{-2\lambda}; \tag{27}$$

or, calling this the first elongation of a second series of four,

$$a_2 = KQ(1-e^{-2\lambda})+a_1e^{-4\lambda}. \tag{28}$$

Proceeding in this way, by observing two elongations + and −, then sending a positive current and observing two elongations

$-$ and $+$, then sending a positive current, and so on, we obtain a series consisting of sets of four elongations, in each of which

$$\frac{d-b}{a-c} = e^{-\lambda}, \tag{29}$$

and
$$KQ = \frac{(a-b)e^{-2\lambda} + d - c}{1+e^{-\lambda}}; \tag{30}$$

If n series of elongations have been observed, then we find the logarithmic decrement from the equation

$$\frac{\Sigma(d) - \Sigma(b)}{\Sigma(a) - \Sigma(c)} = e^{-\lambda}, \tag{31}$$

and Q from the equation

$$KQ(1+e^{-\lambda})(2n-1) = \Sigma_n(a-b-c+d)(1+e^{-2\lambda}) - (a_1-b_1) - (d_n-c_n)e^{-2\lambda}. \tag{32}$$

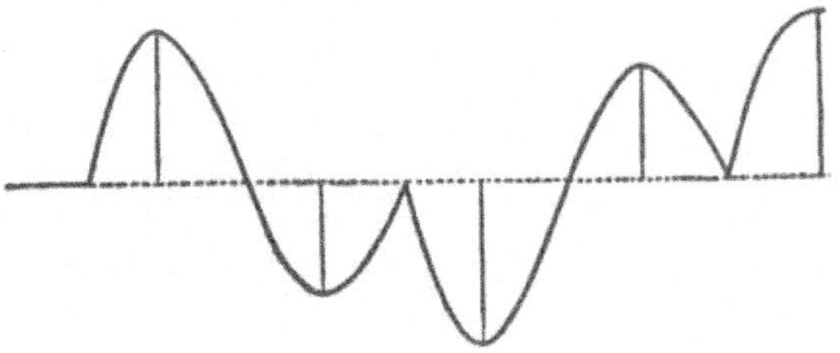

Fig. 60.

The motion of the magnet in the method of recoil is graphically represented in Fig. 60, where the abscissa represents the time, and the ordinate the deflexion of the magnet at that time. See Art. 760.

Method of Multiplication.

751.] If we make the transient current pass every time that the magnet passes through the zero point, and always so as to increase the velocity of the magnet, then, if θ_1, θ_2, &c. are the successive elongations,

$$\theta_2 = -KQ - e^{-\lambda}\theta_1, \tag{33}$$

$$\theta_3 = -KQ - e^{-\lambda}\theta_2. \tag{34}$$

The ultimate value to which the elongation tends after a great many vibrations is found by putting $\theta_n = -\theta_{n-1}$, whence we find

$$\theta = \pm \frac{1}{1-e^{-\lambda}} KQ. \tag{35}$$

If λ is small, the value of the ultimate elongation may be large, but since this involves a long continued experiment, and a careful determination of λ, and since a small error in λ introduces a large error in the determination of Q, this method is rarely useful for numerical determination, and should be reserved for obtaining evidence of the existence or non-existence of currents too small to be observed directly.

In all experiments in which transient currents are made to act on the moving magnet of the galvanometer, it is essential that the whole current should pass while the distance of the magnet from the zero point remains a small fraction of the total elongation. The time of vibration should therefore be large compared with the time required to produce the current, and the operator should have his eye on the motion of the magnet, so as to regulate the instant of passage of the current by the instant of passage of the magnet through its point of equilibrium.

To estimate the error introduced by a failure of the operator to produce the current at the proper instant, we observe that the effect of a force in increasing the elongation varies as

$$e^{\phi \tan \beta} \cos(\phi + \beta),$$

and that this is a maximum when $\phi = 0$. Hence the error arising from a mistiming of the current will always lead to an underestimation of its value, and the amount of the error may be estimated by comparing the cosine of the phase of the vibration at the time of the passage of the current with unity.

CHAPTER XVII.

COMPARISON OF COILS.

Experimental Determination of the Electrical Constants of a Coil.

752.] We have seen in Art. 717 that in a sensitive galvanometer the coils should be of small radius, and should contain many windings of the wire. It would be extremely difficult to determine the electrical constants of such a coil by direct measurement of its form and dimensions, even if we could obtain access to every winding of the wire in order to measure it. But in fact the greater number of the windings are not only completely hidden by the outer windings, but we are uncertain whether the pressure of the outer windings may not have altered the form of the inner ones after the coiling of the wire.

It is better therefore to determine the electrical constants of the coil by direct electrical comparison with a standard coil whose constants are known.

Since the dimensions of the standard coil must be determined by actual measurement, it must be made of considerable size, so that the unavoidable error of measurement of its diameter or circumference may be as small as possible compared with the quantity measured. The channel in which the coil is wound should be of rectangular section, and the dimensions of the section should be small compared with the radius of the coil. This is necessary, not so much in order to diminish the correction for the size of the section, as to prevent any uncertainty about the position of those windings of the coil which are hidden by the external windings*.

* Large tangent galvanometers are sometimes made with a single circular conducting ring of considerable thickness, which is sufficiently stiff to maintain its form without any support. This is not a good plan for a standard instrument. The distribution of the current within the conductor depends on the relative conductivity

The principal constants which we wish to determine are—

(1) The magnetic force at the centre of the coil due to a unit-current. This is the quantity denoted by G_1 in Art. 700.

(2) The magnetic moment of the coil due to a unit-current. This is the quantity g_1.

753.] *To determine* G_1. Since the coils of the working galvanometer are much smaller than the standard coil, we place the galvanometer within the standard coil, so that their centres coincide, the planes of both coils being vertical and parallel to the earth's magnetic force. We have thus obtained a differential galvanometer one of whose coils is the standard coil, for which the value of G_1 is known, while that of the other coil is G_1', the value of which we have to determine.

The magnet suspended in the centre of the galvanometer coil is acted on by the currents in both coils. If the strength of the current in the standard coil is γ, and that in the galvanometer coil γ', then, if these currents flowing in opposite directions produce a deflexion δ of the magnet,

$$H \tan \delta = G_1' \gamma' - G_1 \gamma, \tag{1}$$

where H is the horizontal magnetic force of the earth.

If the currents are so arranged as to produce no deflexion, we may find G_1' by the equation

$$G_1' = \frac{\gamma}{\gamma'} G_1. \tag{2}$$

We may determine the ratio of γ to γ' in several ways. Since the value of G_1 is in general greater for the galvanometer than for the standard coil, we may arrange the circuit so that the whole current γ flows through the standard coil, and is then divided so that γ' flows through the galvanometer and resistance coils, the combined resistance of which is R_1, while the remainder $\gamma - \gamma'$ flows through another set of resistance coils whose combined resistance is R_2.

of its various parts. Hence any concealed flaw in the continuity of the metal may cause the main stream of electricity to flow either close to the outside or close to the inside of the circular ring. Thus the true path of the current becomes uncertain. Besides this, when the current flows only once round the circle, especial care is necessary to avoid any action on the suspended magnet due to the current on its way to or from the circle, because the current in the electrodes is equal to that in the circle. In the construction of many instruments the action of this part of the current seems to have been altogether lost sight of.

The most perfect method is to make one of the electrodes in the form of a metal tube, and the other a wire covered with insulating material, and placed inside the tube and concentric with it. The external action of the electrodes when thus arranged is zero, by Art. 683.

We have then, by Art. 276,

$$\gamma' R_1 = (\gamma - \gamma') R_2, \tag{3}$$

or
$$\frac{\gamma}{\gamma'} = \frac{R_1 + R_2}{R_2}, \tag{4}$$

and
$$G_1' = \frac{R_1 + R_2}{R_2} G_1. \tag{5}$$

If there is any uncertainty about the actual resistance of the galvanometer coil (on account, say, of an uncertainty as to its temperature) we may add resistance coils to it, so that the resistance of the galvanometer itself forms but a small part of R_1, and thus introduces but little uncertainty into the final result.

754.] *To determine* g_1, the magnetic moment of a small coil due to a unit-current flowing through it, the magnet is still suspended at the centre of the standard coil, but the small coil is moved parallel to itself along the common axis of both coils, till the same current, flowing in opposite directions round the coils, no longer deflects the magnet. If the distance between the centres of the coils is r, we have now

$$G_1 = 2\frac{g_1}{r^3} + 3\frac{g_2}{r^4} + 4\frac{g_3}{r^5} + \&c. \tag{6}$$

By repeating the experiment with the small coil on the opposite side of the standard coil, and measuring the distance between the positions of the small coil, we eliminate the uncertain error in the determination of the position of the centres of the magnet and of the small coil, and we get rid of the terms in g_2, g_4, &c.

If the standard coil is so arranged that we can send the current through half the number of windings, so as to give a different value to G_1, we may determine a new value of r, and thus, as in Art. 454, we may eliminate the term involving g_3.

It is often possible, however, to determine g_3 by direct measurement of the small coil with sufficient accuracy to make it available in calculating the value of the correction to be applied to g_1 in the equation

$$g_1 = \frac{1}{2} G_1 r^3 - 2\frac{g_3}{r^2}. \tag{7}$$

where
$$g_3 = -\frac{1}{8}\pi a^2 (6a^2 + 3\xi^2 - 2\eta^2), \text{ by Art. 700.}$$

Comparison of Coefficients of Induction.

755.] It is only in a small number of cases that the direct calculation of the coefficients of induction from the form and position of the circuits can be easily performed. In order to attain a sufficient degree of accuracy, it is necessary that the distance between the circuits should be capable of exact measurement. But when the distance between the circuits is sufficient to prevent errors of measurement from introducing large errors into the result, the coefficient of induction itself is necessarily very much reduced in magnitude. Now for many experiments it is necessary to make the coefficient of induction large, and we can only do so by bringing the circuits close together, so that the method of direct measurement becomes impossible, and, in order to determine the coefficient of induction, we must compare it with that of a pair of coils arranged so that their coefficient may be obtained by direct measurement and calculation.

This may be done as follows :

Let A and a be the standard pair of coils, B and b the coils to be compared with them. Connect A and B in one circuit, and place the electrodes of the galvanometer, G, at P and Q, so that the resistance of PAQ is R, and that of QBP is S, K being the resistance of the galvanometer. Connect a and b in one circuit with the battery.

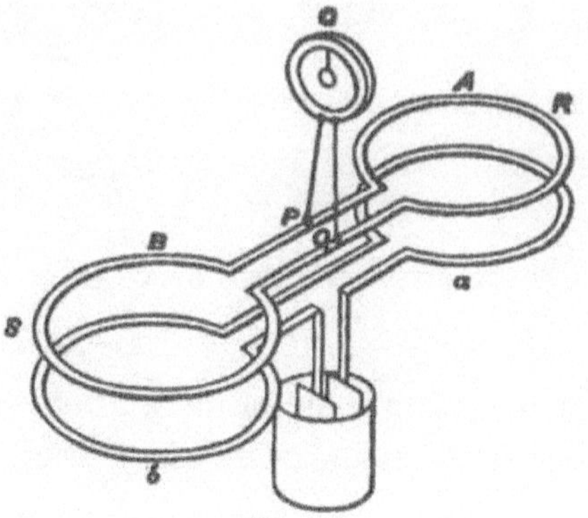

Fig. 61.

Let the current in A be $\dot{x}$, that in B, $\dot{y}$, and that in the galvanometer, $\dot{x}-\dot{y}$, that in the battery circuit being γ.

Then, if M_1 is the coefficient of induction between A and a, and M_2 that between B and b, the integral induction current through the galvanometer at breaking the battery circuit is

$$x-y = \gamma \frac{\frac{M_1}{R} - \frac{M_2}{S}}{1 + \frac{K}{R} + \frac{K}{S}}. \qquad (8)$$

By adjusting the resistances R and S till there is no current through the galvanometer at making or breaking the galvanometer circuit, the ratio of M_2 to M_1 may be determined by measuring that of S to R.

Comparison of a Coefficient of Self-induction with a Coefficient of Mutual Induction.

756.] In the branch AF of Wheatstone's Bridge let a coil be inserted, the coefficient of self-induction of which we wish to find. Let us call it L.

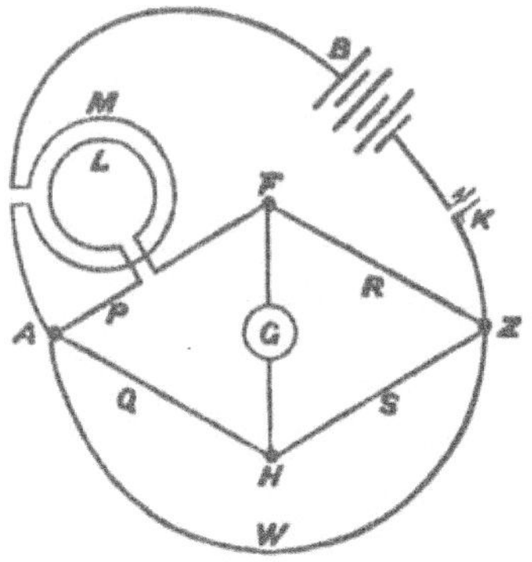

Fig. 62.

In the connecting wire between A and the battery another coil is inserted. The coefficient of mutual induction between this coil and the coil in AF is M. It may be measured by the method described in Art. 755.

If the current from A to F is x, and that from A to H is y, that from Z to A, through B, will be $x+y$. The external electromotive force from A to F is

$$A-F = Px + L\frac{dx}{dt} + M\left(\frac{dx}{dt} + \frac{dy}{dt}\right). \tag{9}$$

The external electromotive force along AH is

$$A-H = Qy. \tag{10}$$

If the galvanometer placed between F and H indicates no current, either transient or permanent, then by (9) and (10), since $H-F=0$,

$$Px = Qy; \tag{11}$$

and

$$L\frac{dx}{dt} + M\left(\frac{dx}{dt} + \frac{dy}{dt}\right) = 0, \tag{12}$$

whence

$$L = -\left(1 + \frac{P}{Q}\right)M. \tag{13}$$

Since L is always positive, M must be negative, and therefore the current must flow in opposite directions through the coils placed in P and in B. In making the experiment we may either begin by adjusting the resistances so that

$$PS = QR, \tag{14}$$

which is the condition that there may be no permanent current, and then adjust the distance between the coils till the galvanometer ceases to indicate a transient current on making and breaking the battery connexion; or, if this distance is not capable of adjustment, we may get rid of the transient current by altering the resistances Q and S in such a way that the ratio of Q to S remains constant.

If this double adjustment is found too troublesome, we may adopt a third method. Beginning with an arrangement in which the transient current due to self-induction is slightly in excess of that due to mutual induction, we may get rid of the inequality by inserting a conductor whose resistance is W between A and Z. The condition of no permanent current through the galvanometer is not affected by the introduction of W. We may therefore get rid of the transient current by adjusting the resistance of W alone. When this is done the value of L is

$$L = -\left(1 + \frac{P}{Q} + \frac{P+R}{W}\right) M. \tag{15}$$

Comparison of the Coefficients of Self-induction of Two Coils.

757.] Insert the coils in two adjacent branches of Wheatstone's Bridge. Let L and N be the coefficients of self-induction of the coils inserted in P and in R respectively, then the condition of no galvanometer current is

$$\left(Px + L\frac{dx}{dt}\right) Sy = Qy\left(Rx + N\frac{dx}{dt}\right), \tag{16}$$

whence $PS = QR$, for no permanent current, (17)

and $\dfrac{L}{P} = \dfrac{N}{R}$, for no transient current. (18)

Hence, by a proper adjustment of the resistances, both the permanent and the transient current can be got rid of, and then the ratio of L to N can be determined by a comparison of the resistances.

CHAPTER XVIII.

ELECTROMAGNETIC UNIT OF RESISTANCE.

On the Determination of the Resistance of a Coil in Electromagnetic Measure.

758.] THE resistance of a conductor is defined as the ratio of the numerical value of the electromotive force to that of the current which it produces in the conductor. The determination of the value of the current in electromagnetic measure can be made by means of a standard galvanometer, when we know the value of the earth's magnetic force. The determination of the value of the electromotive force is more difficult, as the only case in which we can directly calculate its value is when it arises from the relative motion of the circuit with respect to a known magnetic system.

759.] The first determination of the resistance of a wire in electromagnetic measure was made by Kirchhoff*. He employed two coils of known form, A_1 and A_2, and calculated their coefficient of mutual induction from the geometrical data of their form and position. These coils were placed in circuit with a galvanometer, G, and a battery, B, and two points of the circuit, P, between the coils, and Q, between the battery and galvanometer, were joined by the wire whose resistance, R, was to be measured.

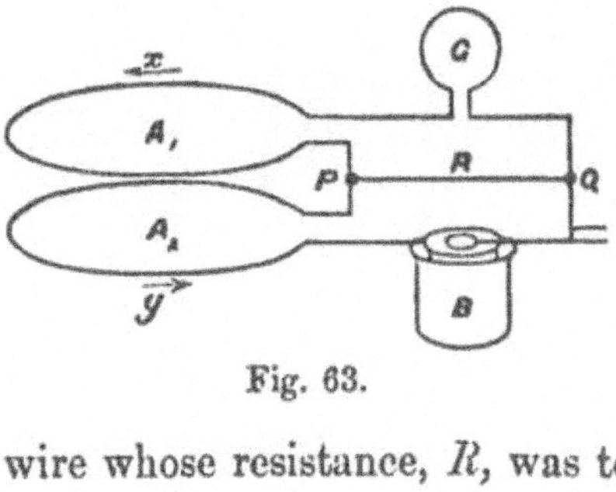

Fig. 63.

When the current is steady it is divided between the wire and the galvanometer circuit, and produces a certain permanent deflexion of the galvanometer. If the coil A_1 is now removed quickly

* 'Bestimmung der Constanten von welcher die Intensität inducirter elektrischer Ströme abhängt.' *Pogg. Ann.*, lxxvi (April 1849).

from A_2 and placed in a position in which the coefficient of mutual induction between A_1 and A_2 is zero (Art. 538), a current of induction is produced in both circuits, and the galvanometer needle receives an impulse which produces a certain transient deflexion.

The resistance of the wire, R, is deduced from a comparison between the permanent deflexion, due to the steady current, and the transient deflexion, due to the current of induction.

Let the resistance of QGA_1P be K, of PA_2BQ, B, and of PQ, R.

Let L, M and N be the coefficients of induction of A_1 and A_2.

Let $\dot{x}$ be the current in G, and $\dot{y}$ that in B, then the current from P to Q is $\dot{x}-\dot{y}$.

Let E be the electromotive force of the battery, then

$$(K+R)\,\dot{x}-R\dot{y}+\frac{d}{dt}\,(L\dot{x}+M\dot{y}) = 0, \tag{1}$$

$$R\dot{x}+(B+R)\,\dot{y}+\frac{d}{dt}\,(M\dot{x}+N\dot{y}) = E. \tag{2}$$

When the currents are constant, and everything at rest,

$$(K+R)\,\dot{x}-R\dot{y} = 0. \tag{3}$$

If M now suddenly becomes zero on account of the separation of A_1 from A_2, then, integrating with respect to t,

$$(K+R)\,x-Ry-M\dot{y} = 0, \tag{4}$$

$$-Rx+(B+R)\,y-M\dot{x} = \int E\,dt = 0. \tag{5}$$

whence
$$x = M\frac{(B+R)\,\dot{y}+R\dot{x}}{(B+R)\,(K+R)-R^2}. \tag{6}$$

Substituting the value of $\dot{y}$ in terms of $\dot{x}$ from (3), we find

$$\frac{x}{\dot{x}} = \frac{M}{R}\,\frac{(B+R)\,(K+R)+R^2}{(B+R)\,(K+R)-R^2} \tag{7}$$

$$= \frac{M}{R}\left\{1+\frac{2R^2}{(B+R)\,(K+R)}+\&c.\right\}. \tag{8}$$

When, as in Kirchhoff's experiment, both B and K are large compared with R, this equation is reduced to

$$\frac{x}{\dot{x}} = \frac{M}{R}. \tag{9}$$

Of these quantities, x is found from the throw of the galvanometer due to the induction current. See Art. 768. The permanent current, $\dot{x}$, is found from the permanent deflexion due to the steady current; see Art. 746. M is found either by direct calculation from the geometrical data, or by a comparison with a pair of coils, for which this calculation has been made; see Art. 755. From these three quantities R can be determined in electromagnetic measure.

These methods involve the determination of the period of vibration of the galvanometer magnet, and of the logarithmic decrement of its oscillations.

*Weber's Method by Transient Currents**.

760.] A coil of considerable size is mounted on an axle, so as to be capable of revolving about a vertical diameter. The wire of this coil is connected with that of a tangent galvanometer so as to form a single circuit. Let the resistance of this circuit be R. Let the large coil be placed with its positive face perpendicular to the magnetic meridian, and let it be quickly turned round half a revolution. There will be an induced current due to the earth's magnetic force, and the total quantity of electricity in this current in electromagnetic measure will be

$$Q = \frac{2 g_1 H}{R}, \tag{1}$$

where g_1 is the magnetic moment of the coil for unit current, which in the case of a large coil may be determined directly, by measuring the dimensions of the coil, and calculating the sum of the areas of its windings. H is the horizontal component of terrestrial magnetism, and R is the resistance of the circuit formed by the coil and galvanometer together. This current sets the magnet of the galvanometer in motion.

If the magnet is originally at rest, and if the motion of the coil occupies but a small fraction of the time of a vibration of the magnet, then, if we neglect the resistance to the motion of the magnet, we have, by Art. 748,

$$Q = \frac{H}{G}\frac{T}{\pi} 2 \sin \tfrac{1}{2}\theta, \tag{2}$$

where G is the constant of the galvanometer, T is the time of vibration of the magnet, and θ is the observed elongation. From these equations we obtain

$$R = \pi G g \frac{1}{T \sin \frac{1}{2}\theta}. \tag{3}$$

The value of H does not appear in this result, provided it is the same at the position of the coil and at that of the galvanometer. This should not be assumed to be the case, but should be tested by comparing the time of vibration of the same magnet, first at one of these places and then at the other.

* *Elekt. Maasb.*; or Pogg., *Ann.* lxxxii, 337 (1851).

761.] To make a series of observations Weber began with the coil parallel to the magnetic meridian. He then turned it with its positive face north, and observed the first elongation due to the negative current. He then observed the second elongation of the freely swinging magnet, and on the return of the magnet through the point of equilibrium he turned the coil with its positive face south. This caused the magnet to recoil to the positive side. The series was continued as in Art. 750, and the result corrected for resistance. In this way the value of the resistance of the combined circuit of the coil and galvanometer was ascertained.

In all such experiments it is necessary, in order to obtain sufficiently large deflexions, to make the wire of copper, a metal which, though it is the best conductor, has the disadvantage of altering considerably in resistance with alterations of temperature. It is also very difficult to ascertain the temperature of every part of the apparatus. Hence, in order to obtain a result of permanent value from such an experiment, the resistance of the experimental circuit should be compared with that of a carefully constructed resistance-coil, both before and after each experiment.

Weber's Method by observing the Decrement of the Oscillations of a Magnet.

762.] A magnet of considerable magnetic moment is suspended at the centre of a galvanometer coil. The period of vibration and the logarithmic decrement of the oscillations is observed, first with the circuit of the galvanometer open, and then with the circuit closed, and the conductivity of the galvanometer coil is deduced from the effect which the currents induced in it by the motion of the magnet have in resisting that motion.

If T is the observed time of a single vibration, and λ the Napierian logarithmic decrement for each single vibration, then, if we write

$$\omega = \frac{\pi}{T}, \tag{1}$$

and

$$a = \frac{\lambda}{T}, \tag{2}$$

the equation of motion of the magnet is of the form

$$\phi = Ce^{-at}\cos(\omega t + \beta). \tag{3}$$

This expresses the nature of the motion as determined by observation. We must compare this with the dynamical equation of motion.

Let M be the coefficient of induction between the galvanometer coil and the suspended magnet. It is of the form

$$M = G_1 g_1 Q_1(\theta) + G_2 g_2 Q_2(\theta) + \&c., \tag{4}$$

where G_1, G_2, &c. are coefficients belonging to the coil, g_1, g_2, &c. to the magnet, and $Q_1(\theta)$, $Q_2(\theta)$, &c., are zonal harmonics of the angle between the axes of the coil and the magnet. See Art. 700. By a proper arrangement of the coils of the galvanometer, and by building up the suspended magnet of several magnets placed side by side at proper distances, we may cause all the terms of M after the first to become insensible compared with the first. If we also put $\phi = \frac{\pi}{2} - \theta$, we may write

$$M = Gm \sin\phi, \tag{5}$$

where G is the principal coefficient of the galvanometer, m is the magnetic moment of the magnet, and ϕ is the angle between the axis of the magnet and the plane of the coil, which, in this experiment, is always a small angle.

If L is the coefficient of self-induction of the coil, and R its resistance, and γ the current in the coil,

$$\frac{d}{dt}(L\gamma + M) + R\gamma = 0, \tag{6}$$

or

$$L\frac{d\gamma}{dt} + R\gamma + Gm\cos\phi\frac{d\phi}{dt} = 0. \tag{7}$$

The moment of the force with which the current γ acts on the magnet is $\gamma\frac{dM}{d\phi}$, or $Gm\gamma\cos\phi$. The angle ϕ is in this experiment so small, that we may suppose $\cos\phi = 1$.

Let us suppose that the equation of motion of the magnet when the circuit is broken is

$$A\frac{d^2\phi}{dt^2} + B\frac{d\phi}{dt} + C\phi = 0, \tag{8}$$

where A is the moment of inertia of the suspended apparatus, $B\frac{d\phi}{dt}$ expresses the resistance arising from the viscosity of the air and of the suspension fibre, &c., and $C\phi$ expresses the moment of the force arising from the earth's magnetism, the torsion of the suspension apparatus, &c., tending to bring the magnet to its position of equilibrium.

The equation of motion, as affected by the current, will be

$$A\frac{d^2\phi}{dt^2} + B\frac{d\phi}{dt} + C\phi = Gm\gamma. \tag{9}$$

To determine the motion of the magnet, we have to combine this equation with (7) and eliminate γ. The result is

$$\left(R+L\frac{d}{dt}\right)\left(A\frac{d^2}{dt^2}+B\frac{d}{dt}+C\right)\phi+G^2m^2\frac{d\phi}{dt}=0, \qquad (10)$$

a linear differential equation of the third order.

We have no occasion, however, to solve this equation, because the data of the problem are the observed elements of the motion of the magnet, and from these we have to determine the value of R.

Let a_0 and ω_0 be the values of a and ω in equation (2) when the circuit is broken. In this case R is infinite, and the equation is reduced to the form (8). We thus find

$$B=2Aa_0, \qquad C=A(a_0^2+\omega_0^2). \qquad (11)$$

Solving equation (10) for R, and writing

$$\frac{d}{dt}=-(a+i\omega), \quad \text{where} \quad i=\sqrt{-1}, \qquad (12)$$

we find

$$R=\frac{G^2m^2}{A}\,\frac{a+i\omega}{a^2-\omega^2+2ia\omega-2a_0(a+i\omega)+a_0^2+\omega_0^2}+L(a+i\omega). \quad (13)$$

Since the value of ω is in general much greater than that of a, the best value of R is found by equating the terms in $i\omega$,

$$R=\frac{G^2m^2}{2A(a-a_0)}+\tfrac{1}{2}L\left(3a-a_0-\frac{\omega^2-\omega_0^2}{a-a_0}\right). \qquad (14)$$

We may also obtain a value of R by equating the terms not involving i, but as these terms are small, the equation is useful only as a means of testing the accuracy of the observations. From these equations we find the following testing equation,

$$G^2m^2\{a^2+\omega^2-a_0^2-\omega_0^2\}$$
$$=LA\{(a-a_0)^4+2(a-a_0)^2(\omega^2+\omega_0^2)+(\omega^2-\omega_0^2)^2\}. \quad (15)$$

Since $LA\omega^2$ is very small compared with G^2m^2, this equation gives

$$\omega^2-\omega_0^2=a_0^2-a^2; \qquad (16)$$

and equation (14) may be written

$$R=\frac{G^2m^2}{2A(a-a_0)}+2La. \qquad (17)$$

In this expression G may be determined either from the linear measurement of the galvanometer coil, or better, by comparison with a standard coil, according to the method of Art. 753. A is the moment of inertia of the magnet and its suspended apparatus, which is to be found by the proper dynamical method. ω, ω_0, a and a_0, are given by observation.

The determination of the value of m, the magnetic moment of the suspended magnet, is the most difficult part of the investigation, because it is affected by temperature, by the earth's magnetic force, and by mechanical violence, so that great care must be taken to measure this quantity when the magnet is in the very same circumstances as when it is vibrating.

The second term of R, that which involves L, is of less importance, as it is generally small compared with the first term. The value of L may be determined either by calculation from the known form of the coil, or by an experiment on the extra-current of induction. See Art. 756.

Thomson's Method by a Revolving Coil.

763.] This method was suggested by Thomson to the Committee of the British Association on Electrical Standards, and the experiment was made by M. M. Balfour Stewart, Fleeming Jenkin, and the author in 1863 *.

A circular coil is made to revolve with uniform velocity about a vertical axis. A small magnet is suspended by a silk fibre at the centre of the coil. An electric current is induced in the coil by the earth's magnetism, and also by the suspended magnet. This current is periodic, flowing in opposite directions through the wire of the coil during different parts of each revolution, but the effect of the current on the suspended magnet is to produce a deflexion from the magnetic meridian in the direction of the rotation of the coil.

764.] Let H be the horizontal component of the earth's magnetism.

Let γ be the strength of the current in the coil.

g the total area inclosed by all the windings of the wire.

G the magnetic force at the centre of the coil due to unit-current.

L the coefficient of self-induction of the coil.

M the magnetic moment of the suspended magnet.

θ the angle between the plane of the coil and the magnetic meridian.

ϕ the angle between the axis of the suspended magnet and the magnetic meridian

A the moment of inertia of the suspended magnet.

$MH\tau$ the coefficient of torsion of the suspension fibre.

α the azimuth of the magnet when there is no torsion.

R the resistance of the coil.

* See *Report of the British Association for* 1863.

The kinetic energy of the system is

$$T = \tfrac{1}{2} L\gamma^2 - Hg\gamma \sin\theta - MG\gamma \sin(\theta-\phi) + MH\cos\phi + \tfrac{1}{2} A\dot{\phi}^2. \quad (1)$$

The first term, $\frac{1}{2} L\gamma^2$, expresses the energy of the current as depending on the coil itself. The second term depends on the mutual action of the current and terrestrial magnetism, the third on that of the current and the magnetism of the suspended magnet, the fourth on that of the magnetism of the suspended magnet and terrestrial magnetism, and the last expresses the kinetic energy of the matter composing the magnet and the suspended apparatus which moves with it.

The potential energy of the suspended apparatus arising from the torsion of the fibre is

$$V = \frac{MH}{2} \tau (\phi^2 - 2\phi a). \quad (2)$$

The electromagnetic momentum of the current is

$$p = \frac{dT}{d\gamma} = L\gamma - Hg\sin\theta - MG\gamma\sin(\theta-\phi), \quad (3)$$

and if R is the resistance of the coil, the equation of the current is

$$R\gamma + \frac{d^2 T}{d\gamma\, dt} = 0, \quad (4)$$

or, since

$$\theta = \omega t, \quad (5)$$

$$\left(R + L\frac{d}{dt}\right)\gamma = Hg\omega\cos\theta + MG(\omega-\dot{\phi})\cos(\theta-\phi). \quad (6)$$

765.] It is the result alike of theory and observation that ϕ, the azimuth of the magnet, is subject to two kinds of periodic variations. One of these is a free oscillation, whose periodic time depends on the intensity of terrestrial magnetism, and is, in the experiment, several seconds. The other is a forced vibration whose period is half that of the revolving coil, and whose amplitude is, as we shall see, insensible. Hence, in determining γ, we may treat ϕ as sensibly constant.

We thus find

$$\gamma = \frac{Hg\omega}{R^2 + L^2\omega^2}(R\cos\theta + L\omega\sin\theta) \quad (7)$$

$$+ \frac{Mg(\omega-\dot{\phi})}{R^2 + L^2(\omega-\dot{\phi})^2}\left(R\cos(\theta-\phi) + L(\omega-\dot{\phi})\sin(\theta-\phi)\right), \quad (8)$$

$$+ Ce^{-\frac{R}{L}t}. \quad (9)$$

The last term of this expression soon dies away when the rotation is continued uniform.

The equation of motion of the suspended magnet is

$$\frac{d^2T}{d\dot{\phi}\,dt} - \frac{dT}{d\phi} + \frac{dV}{d\phi} = 0, \tag{10}$$

whence $A\ddot{\phi} - MG\gamma\cos(\theta-\phi) + MH(\sin\phi + \tau(\phi-a)) = 0.$ (11)

Substituting the value of γ, and arranging the terms according to the functions of multiples of θ, then we know from observation that

$$\phi = \phi_0 + be^{-lt}\cos nt + c\cos 2(\theta-\beta), \tag{12}$$

where ϕ_0 is the mean value of ϕ, and the second term expresses the free vibrations gradually decaying, and the third the forced vibrations arising from the variation of the deflecting current.

The value of n in equation (12) is $\frac{HM}{A}\sec\phi$. That of c, the amplitude of the forced vibrations, is $\frac{1}{4}\frac{n^2}{\omega^2}\sin\phi$. Hence, when the coil makes many revolutions during one free vibration of the magnet, the amplitude of the forced vibrations of the magnet is very small, and we may neglect the terms in (11) which involve c.

Beginning with the terms in (11) which do not involve θ, we find

$$\frac{MHGg\omega}{R^2+L^2\omega^2}(R\cos\phi_0 + L\omega\sin\phi_0) + \frac{M^2G^2(\omega-\dot{\phi})}{R^2+L^2(\omega-\dot{\phi})^2}R$$
$$= MH(\sin\phi_0 + \tau(\phi_0-a)). \tag{13}$$

Remembering that $\dot{\phi}$ is small, and that L is generally small compared with Gg, we find as a sufficiently approximate value of R,

$$R = \frac{Gg\omega}{2\tan\phi_0\left(1+\tau\frac{\phi-a}{\sin\phi}\right)}\left\{1 + \frac{GM}{gH}\sec\phi - \frac{2L}{gG}\left(\frac{2L}{Gg}-1\right)\tan^2\phi\right\}. \tag{14}$$

766.] The resistance is thus determined in electromagnetic measure in terms of the velocity ω and the deviation ϕ. It is not necessary to determine H, the horizontal terrestrial magnetic force, provided it remains constant during the experiment.

To determine $\frac{M}{H}$ we must make use of the suspended magnet to deflect the magnet of the magnetometer, as described in Art. 454. In this experiment M should be small, so that this correction becomes of secondary importance.

For the other corrections required in this experiment see the *Report of the British Association for* 1863, p. 168.

Joule's Calorimetric Method.

767.] The heat generated by a current γ in passing through a conductor whose resistance is R is, by Joule's law, Art. 242.

$$h = \frac{1}{J}\int R\gamma^2\,dt, \tag{1}$$

where J is the equivalent in dynamical measure of the unit of heat employed.

Hence, if R is constant during the experiment, its value is

$$R = \frac{Jh}{\int \gamma^2\,dt}. \tag{2}$$

This method of determining R involves the determination of h, the heat generated by the current in a given time, and of γ^2, the square of the strength of the current.

In Joule's experiments*, h was determined by the rise of temperature of the water in a vessel in which the conducting wire was immersed. It was corrected for the effects of radiation, &c. by alternate experiments in which no current was passed through the wire.

The strength of the current was measured by means of a tangent galvanometer. This method involves the determination of the intensity of terrestrial magnetism, which was done by the method described in Art. 457. These measurements were also tested by the current weigher, described in Art. 726, which measures γ^2 directly. The most direct method of measuring $\int \gamma^2\,dt$, however, is to pass the current through a self-acting electrodynamometer (Art. 725) with a scale which gives readings proportional to γ^2, and to make the observations at equal intervals of time, which may be done approximately by taking the reading at the extremities of every vibration of the instrument during the whole course of the experiment.

* *Report of the British Association for* 1867.

CHAPTER XIX.

COMPARISON OF THE ELECTROSTATIC WITH THE ELECTROMAGNETIC UNITS.

Determination of the Number of Electrostatic Units of Electricity in one Electromagnetic Unit.

768.] THE absolute magnitudes of the electrical units in both systems depend on the units of length, time, and mass which we adopt, and the mode in which they depend on these units is different in the two systems, so that the ratio of the electrical units will be expressed by a different number, according to the different units of length and time.

It appears from the table of dimensions, Art. 628, that the number of electrostatic units of electricity in one electromagnetic unit varies inversely as the magnitude of the unit of length, and directly as the magnitude of the unit of time which we adopt.

If, therefore, we determine a velocity which is represented numerically by this number, then, even if we adopt new units of length and of time, the number representing this velocity will still be the number of electrostatic units of electricity in one electromagnetic unit, according to the new system of measurement.

This velocity, therefore, which indicates the relation between electrostatic and electromagnetic phenomena, is a natural quantity of definite magnitude, and the measurement of this quantity is one of the most important researches in electricity.

To shew that the quantity we are in search of is really a velocity, we may observe that in the case of two parallel currents the attraction experienced by a length a of one of them is, by Art. 686,

$$F = 2CC'\frac{a}{b},$$

where C, C' are the numerical values of the currents in electromagnetic measure, and b the distance between them. If we make $b = 2a$, then

$$F = CC'.$$

Now the quantity of electricity transmitted by the current C in the time t is Ct in electromagnetic measure, or nCt in electrostatic measure, if n is the number of electrostatic units in one electromagnetic unit.

Let two small conductors be charged with the quantities of electricity transmitted by the two currents in the time t, and placed at a distance r from each other. The repulsion between them will be

$$F' = \frac{CC'n^2t^2}{r^2}.$$

Let the distance r be so chosen that this repulsion is equal to the attraction of the currents, then

$$\frac{CC'n^2t^2}{r^2} = CC'.$$

Hence

$$r = nt;$$

or the distance r must increase with the time t at the rate n. Hence n is a velocity, the absolute magnitude of which is the same, whatever units we assume.

769.] To obtain a physical conception of this velocity, let us imagine a plane surface charged with electricity to the electrostatic surface-density σ, and moving in its own plane with a velocity v. This moving electrified surface will be equivalent to an electric current-sheet, the strength of the current flowing through unit of breadth of the surface being σv in electrostatic measure, or $\frac{1}{n}\sigma v$ in electromagnetic measure, if n is the number of electrostatic units in one electromagnetic unit. If another plane surface, parallel to the first, is electrified to the surface-density σ', and moves in the same direction with the velocity v', it will be equivalent to a second current-sheet.

The electrostatic repulsion between the two electrified surfaces is, by Art. 124, $2\pi\sigma\sigma'$ for every unit of area of the opposed surfaces.

The electromagnetic attraction between the two current-sheets is, by Art. 653, $2\pi uu'$ for every unit of area, u and u' being the surface-densities of the currents in electromagnetic measure.

But $u = \frac{1}{n}\sigma v$, and $u' = \frac{1}{n}\sigma' v'$, so that the attraction is

$$2\pi\sigma\sigma'\frac{vv'}{n^2}.$$

The ratio of the attraction to the repulsion is equal to that of vv' to n^2. Hence, since the attraction and the repulsion are quantities of the same kind, n must be a quantity of the same kind as v, that is, a velocity. If we now suppose the velocity of each of the moving planes to be *equal* to n, the attraction will be equal to the repulsion, and there will be no mechanical action between them. Hence we may define the ratio of the electric units to be a velocity, such that two electrified surfaces, moving in the same direction with this velocity, have no mutual action. Since this velocity is about 288000 kilometres per second, it is impossible to make the experiment above described.

770.] If the electric surface-density and the velocity can be made so great that the magnetic force is a measurable quantity, we may at least verify our supposition that a moving electrified body is equivalent to an electric current.

It appears from Art. 57 that an electrified surface in air would begin to discharge itself by sparks when the electric force $2\pi\sigma$ reaches the value 130. The magnetic force due to the current-sheet is $2\pi\sigma\frac{v}{n}$. The horizontal magnetic force in Britain is about 0.175. Hence a surface electrified to the highest degree, and moving with a velocity of 100 metres per second, would act on a magnet with a force equal to about one-four-thousandth part of the earth's horizontal force, a quantity which can be measured. The electrified surface may be that of a non-conducting disk revolving in the plane of the magnetic meridian, and the magnet may be placed close to the ascending or descending portion of the disk, and protected from its electrostatic action by a screen of metal. I am not aware that this experiment has been hitherto attempted.

I. *Comparison of Units of Electricity.*

771.] Since the ratio of the electromagnetic to the electrostatic unit of electricity is represented by a velocity, we shall in future denote it by the symbol v. The first numerical determination of this velocity was made by Weber and Kohlrausch *.

Their method was founded on the measurement of the same quantity of electricity, first in electrostatic and then in electromagnetic measure.

The quantity of electricity measured was the charge of a Leyden jar. It was measured in electrostatic measure as the product of the

* *Elektrodynamische Maassbestimmungen:* and *Pogg. Ann.* xcix, (Aug. 10, 1856.)

capacity of the jar into the difference of potential of its coatings. The capacity of the jar was determined by comparison with that of a sphere suspended in an open space at a distance from other bodies. The capacity of such a sphere is expressed in electrostatic measure by its radius. Thus the capacity of the jar may be found and expressed as a certain length. See Art. 227.

The difference of the potentials of the coatings of the jar was measured by connecting the coatings with the electrodes of an electrometer, the constants of which were carefully determined, so that the difference of the potentials, E, became known in electrostatic measure.

By multiplying this by c, the capacity of the jar, the charge of the jar was expressed in electrostatic measure.

To determine the value of the charge in electromagnetic measure, the jar was discharged through the coil of a galvanometer. The effect of the transient current on the magnet of the galvanometer communicated to the magnet a certain angular velocity. The magnet then swung round to a certain deviation, at which its velocity was entirely destroyed by the opposing action of the earth's magnetism.

By observing the extreme deviation of the magnet the quantity of electricity in the current may be determined in electromagnetic measure, as in Art. 748, by the formula

$$Q = \frac{H}{G} \frac{T}{\pi} 2 \sin \tfrac{1}{2}\theta,$$

where Q is the quantity of electricity in electromagnetic measure. We have therefore to determine the following quantities:—

H, the intensity of the horizontal component of terrestrial magnetism; see Art. 456.

G, the principal constant of the galvanometer; see Art. 700.

T, the time of a single vibration of the magnet; and

θ, the deviation due to the transient current.

The value of v obtained by MM. Weber and Kohlrausch was

$$v = 310740000 \text{ metres per second.}$$

The property of solid dielectrics, to which the name of Electric Absorption has been given, renders it difficult to estimate correctly the capacity of a Leyden jar. The apparent capacity varies according to the time which elapses between the charging or discharging of the jar and the measurement of the potential, and the longer the time the greater is the value obtained for the capacity of the jar.

Hence, since the time occupied in obtaining a reading of the electrometer is large in comparison with the time during which the discharge through the galvanometer takes place, it is probable that the estimate of the discharge in electrostatic measure is too high, and the value of v, derived from it, is probably also too high.

II. *v expressed as a Resistance.*

772.] Two other methods for the determination of v lead to an expression of its value in terms of the resistance of a given conductor, which, in the electromagnetic system, is also expressed as a velocity.

In Sir William Thomson's form of the experiment, a constant current is made to flow through a wire of great resistance. The electromotive force which urges the current through the wire is measured electrostatically by connecting the extremities of the wire with the electrodes of an absolute electrometer, Arts. 217, 218. The strength of the current in the wire is measured in electromagnetic measure by the deflexion of the suspended coil of an electrodynamometer through which it passes, Art. 725. The resistance of the circuit is known in electromagnetic measure by comparison with a standard coil or Ohm. By multiplying the strength of the current by this resistance we obtain the electromotive force in electromagnetic measure, and from a comparison of this with the electrostatic measure the value of v is obtained.

This method requires the simultaneous determination of two forces, by means of the electrometer and electrodynamometer respectively, and it is only the ratio of these forces which appears in the result.

773.] Another method, in which these forces, instead of being separately measured, are directly opposed to each other, was employed by the present writer. The ends of the great resistance coil are connected with two parallel disks, one of which is moveable. The same difference of potentials which sends the current through the great resistance, also causes an attraction between these disks. At the same time, an electric current which, in the actual experiment, was distinct from the primary current, is sent through two coils, fastened, one to the back of the fixed disk, and the other to the back of the moveable disk. The current flows in opposite directions through these coils, so that they repel one another. By adjusting the distance of the two disks the attraction is exactly balanced by the repulsion, while at the same time another observer, by means of a differential galvanometer with shunts, determines the ratio of the primary to the secondary current.

In this experiment the only measurement which must be referred to a material standard is that of the great resistance, which must be determined in absolute measure by comparison with the Ohm. The other measurements are required only for the determination of ratios, and may therefore be determined in terms of any arbitrary unit.

Thus the ratio of the two forces is a ratio of equality.

The ratio of the two currents is found by a comparison of resistances when there is no deflexion of the differential galvanometer.

The attractive force depends on the square of the ratio of the diameter of the disks to their distance.

The repulsive force depends on the ratio of the diameter of the coils to their distance.

The value of v is therefore expressed directly in terms of the resistance of the great coil, which is itself compared with the Ohm.

The value of v, as found by Thomson's method, was 28.2 Ohms*; by Maxwell's, 28.8 Ohms†.

III. *Electrostatic Capacity in Electromagnetic Measure.*

774.] The capacity of a condenser may be ascertained in electromagnetic measure by a comparison of the electromotive force which produces the charge, and the quantity of electricity in the current of discharge. By means of a voltaic battery a current is maintained through a circuit containing a coil of great resistance. The condenser is charged by putting its electrodes in contact with those of the resistance coil. The current through the coil is measured by the deflexion which it produces in a galvanometer. Let ϕ be this deflexion, then the current is, by Art. 742,

$$\pi = \frac{H}{G}\tan\phi,$$

where H is the horizontal component of terrestrial magnetism, and G is the principal constant of the galvanometer.

If R is the resistance of the coil through which this current is made to flow, the difference of the potentials at the ends of the coil is

$$E = R\gamma,$$

* *Report of British Association*, 1869, p. 434.

† *Phil. Trans.*, 1868, p. 643; and *Report of British Association*, 1869, p. 436.

and the charge of electricity produced in the condenser, whose capacity in electromagnetic measure is C, will be

$$Q = EC.$$

Now let the electrodes of the condenser, and then those of the galvanometer, be disconnected from the circuit, and let the magnet of the galvanometer be brought to rest at its position of equilibrium. Then let the electrodes of the condenser be connected with those of the galvanometer. A transient current will flow through the galvanometer, and will cause the magnet to swing to an extreme deflexion θ. Then, by Art. 748, if the discharge is equal to the charge,

$$Q = \frac{H}{G}\frac{T}{\pi} 2 \sin \tfrac{1}{2}\theta.$$

We thus obtain as the value of the capacity of the condenser in electromagnetic measure

$$C = \frac{T}{\pi}\frac{1}{R}\frac{2 \sin \frac{1}{2}\theta}{\tan \phi}.$$

The capacity of the condenser is thus determined in terms of the following quantities:—

T, the time of vibration of the magnet of the galvanometer from rest to rest.

R, the resistance of the coil.

θ, the extreme limit of the swing produced by the discharge.

ϕ, the constant deflexion due to the current through the coil R.

This method was employed by Professor Fleeming Jenkin in determining the capacity of condensers in electromagnetic measure *.

If c be the capacity of the same condenser in electrostatic measure, as determined by comparison with a condenser whose capacity can be calculated from its geometrical data,

$$c = v^2 C.$$

Hence
$$v^2 = \pi R \frac{c}{T}\frac{\tan \phi}{2 \sin \frac{1}{2}\theta}.$$

The quantity v may therefore be found in this way. It depends on the determination of R in electromagnetic measure, but as it involves only the square root of R, an error in this determination will not affect the value of v so much as in the method of Arts. 772, 773.

Intermittent Current.

775.] If the wire of a battery-circuit be broken at any point, and

* *Report of British Association*, 1867.

the broken ends connected with the electrodes of a condenser, the current will flow into the condenser with a strength which diminishes as the difference of the potentials of the condenser increases, so that when the condenser has received the full charge corresponding to the electromotive force acting on the wire the current ceases entirely.

If the electrodes of the condenser are now disconnected from the ends of the wire, and then again connected with them in the reverse order, the condenser will discharge itself through the wire, and will then become recharged in the opposite way, so that a transient current will flow through the wire, the total quantity of which is equal to two charges of the condenser.

By means of a piece of mechanism (commonly called a Commutator, or *wippe*) the operation of reversing the connexions of the condenser can be repeated at regular intervals of time, each interval being equal to T. If this interval is sufficiently long to allow of the complete discharge of the condenser, the quantity of electricity transmitted by the wire in each interval will be $2\,EC$, where E is the electromotive force, and C is the capacity of the condenser.

If the magnet of a galvanometer included in the circuit is loaded, so as to swing so slowly that a great many discharges of the condenser occur in the time of one free vibration of the magnet, the succession of discharges will act on the magnet like a steady current whose strength is

$$\frac{2\,EC}{T}.$$

If the condenser is now removed, and a resistance coil substituted for it, and adjusted till the steady current through the galvanometer produces the same deflexion as the succession of discharges, and if R is the resistance of the whole circuit when this is the case,

$$\frac{E}{R} = \frac{2\,EC}{T}; \tag{1}$$

or

$$R = \frac{T}{2\,C}. \tag{2}$$

We may thus compare the condenser with its commutator in motion to a wire of a certain electrical resistance, and we may make use of the different methods of measuring resistance described in Arts. 345 to 357 in order to determine this resistance.

776.] For this purpose we may substitute for any one of the wires in the method of the Differential Galvanometer, Art. 346, or in that of Wheatstone's Bridge, Art. 347, a condenser with its commutator. Let us suppose that in either case a zero deflexion of the galvanometer has been obtained, first with the condenser and commutator, and then with a coil of resistance R_1 in its place, then the quantity $\frac{T}{2C}$ will be measured by the resistance of the circuit of which the coil R_1 forms part, and which is completed by the remainder of the conducting system including the battery. Hence the resistance, R, which we have to calculate, is equal to R_1, that of the resistance coil, together with R_2, the resistance of the remainder of the system (including the battery), the extremities of the resistance coil being taken as the electrodes of the system.

In the cases of the differential galvanometer and Wheatstone's Bridge it is not necessary to make a second experiment by substituting a resistance coil for the condenser. The value of the resistance required for this purpose may be found by calculation from the other known resistances in the system.

Using the notation of Art. 347, and supposing the condenser and commutator substituted for the conductor AC in Wheatstone's Bridge, and the galvanometer inserted in OA, and that the deflexion of the galvanometer is zero, then we know that the resistance of a coil, which placed in AC would give a zero deflexion, is

$$b = \frac{c\gamma}{\beta} = R_1. \tag{3}$$

The other part of the resistance, R_2, is that of the system of conductors AO, OC, AB, BC and OB, the points A and C being considered as the electrodes. Hence

$$R_2 = \frac{\beta(c+a)(\gamma+a)+ca(\gamma+a)+\gamma a(c+a)}{(c+a)(\gamma+a)+\beta(c+a+\gamma+a)}. \tag{4}$$

In this expression a denotes the internal resistance of the battery and its connexions, the value of which cannot be determined with certainty; but by making it small compared with the other resistances, this uncertainty will only slightly affect the value of R_2.

The value of the capacity of the condenser in electromagnetic measure is

$$C = \frac{t}{2(R_1+R_2)}. \tag{5}$$

777.] If the condenser has a large capacity, and the commutator is very rapid in its action, the condenser may not be fully discharged at each reversal. The equation of the electric current during the discharge is

$$Q+R_2C\frac{dQ}{dt}+EC = 0, \tag{6}$$

where Q is the charge, C the capacity of the condenser, R_2 the resistance of the rest of the system between the electrodes of the condenser, and E the electromotive force due to the connexions with the battery.

Hence
$$Q = (Q_0 + EC)e^{-\frac{t}{R_2 C}} - EC, \tag{7}$$
where Q_0 is the initial value of Q.

If τ is the time during which contact is maintained during each discharge, the quantity in each discharge is

$$Q = 2EC \frac{1 - e^{-\frac{\tau}{R_2 C}}}{1 + e^{-\frac{\tau}{R_2 C}}}. \tag{8}$$

By making c and γ in equation (4) large compared with β, a, or a, the time represented by $R_2 C$ may be made so small compared with τ, that in calculating the value of the exponential expression we may use the value of C in equation (5). We thus find

$$\frac{\tau}{R_2 C} = 2\frac{R_1 + R_2}{R_2}\frac{\tau}{T}, \tag{9}$$

where R_1 is the resistance which must be substituted for the condenser to produce an equivalent effect. R_2 is the resistance of the rest of the system, T is the interval between the beginning of a discharge and the beginning of the next discharge, and τ is the duration of contact for each discharge. We thus obtain for the corrected value of C in electromagnetic measure

$$C = \frac{1}{2}\frac{T}{R_1 + R_2}\frac{1 + e^{-2\frac{R_1 + R_2}{R_2}\frac{\tau}{T}}}{1 - e^{-2\frac{R_1 + R_2}{R_2}\frac{\tau}{T}}}. \tag{10}$$

IV. *Comparison of the Electrostatic Capacity of a Condenser with the Electromagnetic Capacity of Self-induction of a Coil.*

778.] If two points of a conducting circuit, between which the resistance is R, are connected with the electrodes of a condenser whose capacity is C, then, when an electromotive force acts on the circuit, part of the current, instead of passing through the resistance R, will be employed in charging the condenser. The current through R will therefore rise to its final value from zero in a gradual manner. It appears from the mathematical theory that the manner in which the current through

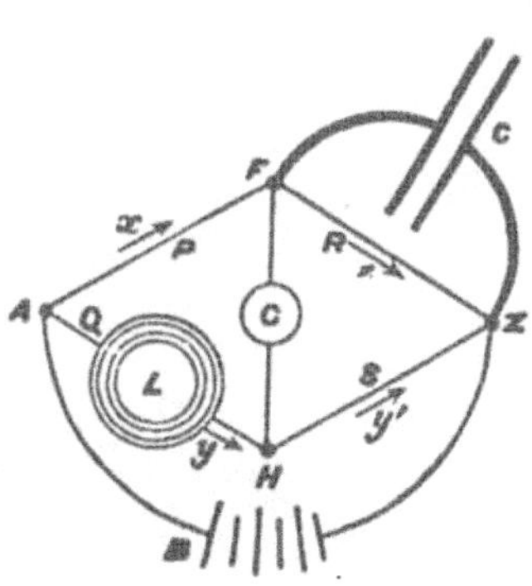

Fig. 64.

R rises from zero to its final value is expressed by a formula of exactly the same kind as that which expresses the value of a current urged by a constant electromotive force through the coil of an electromagnet. Hence we may place a condenser and an electromagnet on two opposite members of Wheatstone's Bridge in such a way that the current through the galvanometer is always zero, even at the instant of making or breaking the battery circuit.

In the figure, let P, Q, R, S be the resistances of the four members of Wheatstone's Bridge respectively. Let a coil, whose coefficient of self-induction is L, be made part of the member AH, whose resistance is Q, and let the electrodes of a condenser, whose capacity is C, be connected by pieces of small resistance with the points F and Z. For the sake of simplicity, we shall assume that there is no current in the galvanometer G, the electrodes of which are connected to F and H. We have therefore to determine the condition that the potential at F may be equal to that at H. It is only when we wish to estimate the degree of accuracy of the method that we require to calculate the current through the galvanometer when this condition is not fulfilled.

Let x be the total quantity of electricity which has passed through the member AF, and z that which has passed through FZ at the time t, then $x-z$ will be the charge of the condenser. The electromotive force acting between the electrodes of the condenser is, by Ohm's law, $R\frac{dz}{dt}$, so that if the capacity of the condenser is C,

$$x-z = RC\frac{dz}{dt}. \qquad (1)$$

Let y be the total quantity of electricity which has passed through the member AH, the electromotive force from A to H must be equal to that from A to F, or

$$Q\frac{dy}{dt} + L\frac{d^2y}{dt^2} = P\frac{dx}{dt}. \qquad (2)$$

Since there is no current through the galvanometer, the quantity which has passed through HZ must be also y, and we find

$$S\frac{dy}{dt} = R\frac{dz}{dt}. \qquad (3)$$

Substituting in (2) the value of x, derived from (1), and comparing with (3), we find as the condition of no current through the galvanometer

$$RQ\left(1+\frac{L}{Q}\frac{d}{dt}\right) = SP\left(1+RC\frac{d}{dt}\right). \qquad (4)$$

The condition of no final current is, as in the ordinary form of Wheatstone's Bridge,

$$QR = SP. \tag{5}$$

The condition of no current at making and breaking the battery connexion is

$$\frac{L}{Q} = RC. \tag{6}$$

Here $\frac{L}{Q}$ and RC are the time-constants of the members Q and R respectively, and if, by varying Q or R, we can adjust the members of Wheatstone's Bridge till the galvanometer indicates no current, either at making and breaking the circuit, or when the current is steady, then we know that the time-constant of the coil is equal to that of the condenser.

The coefficient of self-induction, L, can be determined in electromagnetic measure from a comparison with the coefficient of mutual induction of two circuits, whose geometrical data are known (Art. 756). It is a quantity of the dimensions of a line.

The capacity of the condenser can be determined in electrostatic measure by comparison with a condenser whose geometrical data are known (Art. 229). This quantity is also a length, c. The electromagnetic measure of the capacity is

$$C = \frac{c}{v^2}. \tag{7}$$

Substituting this value in equation (8), we obtain for the value of v^2

$$v^2 = \frac{c}{L} QR, \tag{8}$$

where c is the capacity of the condenser in electrostatic measure, L the coefficient of self-induction of the coil in electromagnetic measure, and Q and R the resistances in electromagnetic measure. The value of v, as determined by this method, depends on the determination of the unit of resistance, as in the second method, Arts. 772, 773.

V. *Combination of the Electrostatic Capacity of a Condenser with the Electromagnetic Capacity of Self-induction of a Coil.*

779.] Let C be the capacity of the condenser, the surfaces of which are connected by a wire of resistance R. In this wire let the coils L and L' be inserted, and let L denote the sum of their capacities of self-induction. The coil L' is hung by a bifilar suspension, and consists of two coils in vertical planes, between which passes a vertical axis which carries the magnet M, the axis of which revolves in a horizontal plane between the coils $L'L$. The coil L has a large coefficient of self-induction, and is fixed. The suspended coil L' is protected from the currents of air caused by the rotation of the magnet by enclosing the rotating parts in a hollow case.

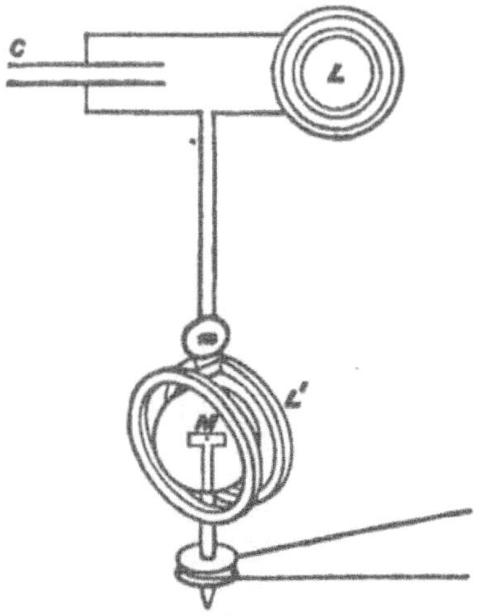

Fig. 65.

The motion of the magnet causes currents of induction in the coil, and these are acted on by the magnet, so that the plane of the suspended coil is deflected in the direction of the rotation of the magnet. Let us determine the strength of the induced currents, and the magnitude of the deflexion of the suspended coil.

Let x be the charge of electricity on the upper surface of the condenser C, then, if E is the electromotive force which produces this charge, we have, by the theory of the condenser,

$$x = CE. \tag{1}$$

We have also, by the theory of electric currents,

$$R\dot{x} + \frac{d}{dt}(L\dot{x} + M\cos\theta) + E = 0, \tag{2}$$

where M is the electromagnetic momentum of the circuit L', when the axis of the magnet is normal to the plane of the coil, and θ is the angle between the axis of the magnet and this normal.

The equation to determine x is therefore

$$CL\frac{d^2x}{dt^2} + CR\frac{dx}{dt} + x = CM\sin\theta\frac{d\theta}{dt}. \tag{3}$$

If the coil is in a position of equilibrium, and if the rotation of the magnet is uniform, the angular velocity being n,

$$\theta = nt. \tag{4}$$

The expression for the current consists of two parts, one of which is independent of the term on the right-hand of the equation, and diminishes according to an exponential function of the time. The other, which may be called the forced current, depends entirely on the term in θ, and may be written

$$x = A\sin\theta + B\cos\theta. \tag{5}$$

Finding the values of A and B by substitution in the equation (3), we obtain

$$x = MCn \frac{RCn\cos\theta - (1 - CLn^2)\sin\theta}{R^2C^2n^2 + (1 - CLn^2)^2}. \tag{6}$$

The moment of the force with which the magnet acts on the coil L', in which the current $\dot{x}$ is flowing, is

$$\Theta = \dot{x}\frac{d}{d\theta}(M\cos\theta) = M\sin\theta\frac{dx}{dt}. \tag{7}$$

Integrating this expression with respect to t, and dividing by t, we find, for the mean value of Θ,

$$\overline{\Theta} = \tfrac{1}{2}\frac{M^2RC^2n^3}{R^2C^2n^2 + (1 - CLn^2)^2}. \tag{8}$$

If the coil has a considerable moment of inertia, its forced vibrations will be very small, and its mean deflexion will be proportional to $\overline{\Theta}$.

Let D_1, D_2, D_3 be the observed deflexions corresponding to angular velocities n_1, n_2, n_3 of the magnet, then in general

$$P\frac{n}{D} = \left(\frac{1}{n} - CLn\right)^2 + R^2C^2, \tag{9}$$

where P is a constant.

Eliminating P and R from three equations of this form, we find

$$C^2L^2 = \frac{1}{n_1^2n_2^2n_3^2}\frac{\frac{n_1^3}{D_1}(n_2^2 - n_3^2) + \frac{n_2^3}{D_2}(n_3^2 - n_1^2) + \frac{n_3^3}{D_3}(n_1^2 - n_2^2)}{\frac{n_1}{D_1}(n_2^2 - n_3^2) + \frac{n_2}{D_2}(n_3^2 - n_1^2) + \frac{n_3}{D_3}(n_1^2 - n_2^2)}. \tag{10}$$

If n_2 is such that $CLn_2^2 = 1$, the value of $\frac{n}{D}$ will be a minimum for this value of n. The other values of n should be taken, one greater, and the other less, than n_2.

The value of CL, determined from this equation, is of the dimensions of the square of a time. Let us call it τ^2.

If C_s be the electrostatic measure of the capacity of the condenser, and L_m the electromagnetic measure of the self-induction of the coil, both C_s and L_m are lines, and the product

$$C_sL_m = v^2C_sL_s = v^2C_mL_m = v^2\tau^2; \tag{11}$$

and

$$v^2 = \frac{C_sL_m}{\tau^2}, \tag{12}$$

where τ^2 is the value of C^2L^2, determined by this experiment. The experiment here suggested as a method of determining v is of the same nature as one described by Sir W. R. Grove, *Phil. Mag.*,

March 1868, p. 184. See also remarks on that experiment, by the present writer, in the number for May 1868.

VI. *Electrostatic Measurement of Resistance.* (See Art. 355.)

780.] Let a condenser of capacity C be discharged through a conductor of resistance R, then, if x is the charge at any instant,

$$\frac{x}{C} + R\frac{dx}{dt} = 0. \tag{1}$$

Hence
$$x = x_0 e^{-\frac{t}{RC}}. \tag{2}$$

If, by any method, we can make contact for a short time, which is accurately known, so as to allow the current to flow through the conductor for the time t, then, if E_0 and E_1 are the readings of an electrometer put in connexion with the condenser before and after the operation,
$$RC(\log_e E_0 - \log_e E_1) = t. \tag{3}$$

If C is known in electrostatic measure as a linear quantity, R may be found from this equation in electrostatic measure as the reciprocal of a velocity.

If R_s is the numerical value of the resistance as thus determined, and R_m the numerical value of the resistance in electromagnetic measure,
$$v^2 = \frac{R_m}{R_s}. \tag{4}$$

Since it is necessary for this experiment that R should be very great, and since R must be small in the electromagnetic experiments of Arts. 763, &c., the experiments must be made on separate conductors, and the resistance of these conductors compared by the ordinary methods.

CHAPTER XX.

ELECTROMAGNETIC THEORY OF LIGHT.

781.] In several parts of this treatise an attempt has been made to explain electromagnetic phenomena by means of mechanical action transmitted from one body to another by means of a medium occupying the space between them. The undulatory theory of light also assumes the existence of a medium. We have now to shew that the properties of the electromagnetic medium are identical with those of the luminiferous medium.

To fill all space with a new medium whenever any new phenomenon is to be explained is by no means philosophical, but if the study of two different branches of science has independently suggested the idea of a medium, and if the properties which must be attributed to the medium in order to account for electromagnetic phenomena are of the same kind as those which we attribute to the luminiferous medium in order to account for the phenomena of light, the evidence for the physical existence of the medium will be considerably strengthened.

But the properties of bodies are capable of quantitative measurement. We therefore obtain the numerical value of some property of the medium, such as the velocity with which a disturbance is propagated through it, which can be calculated from electromagnetic experiments, and also observed directly in the case of light. If it should be found that the velocity of propagation of electromagnetic disturbances is the same as the velocity of light, and this not only in air, but in other transparent media, we shall have strong reasons for believing that light is an electromagnetic phenomenon, and the combination of the optical with the electrical evidence will produce a conviction of the reality of the medium similar to that which we obtain, in the case of other kinds of matter, from the combined evidence of the senses.

782.] When light is emitted, a certain amount of energy is expended by the luminous body, and if the light is absorbed by another body, this body becomes heated, shewing that it has received energy from without. During the interval of time after the light left the first body and before it reached the second, it must have existed as energy in the intervening space.

According to the theory of emission, the transmission of energy is effected by the actual transference of light-corpuscules from the luminous to the illuminated body, carrying with them their kinetic energy, together with any other kind of energy of which they may be the receptacles.

According to the theory of undulation, there is a material medium which fills the space between the two bodies, and it is by the action of contiguous parts of this medium that the energy is passed on, from one portion to the next, till it reaches the illuminated body.

The luminiferous medium is therefore, during the passage of light through it, a receptacle of energy. In the undulatory theory, as developed by Huygens, Fresnel, Young, Green, &c., this energy is supposed to be partly potential and partly kinetic. The potential energy is supposed to be due to the distortion of the elementary portions of the medium. We must therefore regard the medium as elastic. The kinetic energy is supposed to be due to the vibratory motion of the medium. We must therefore regard the medium as having a finite density.

In the theory of electricity and magnetism adopted in this treatise, two forms of energy are recognised, the electrostatic and the electrokinetic (see Arts. 630 and 636), and these are supposed to have their seat, not merely in the electrified or magnetized bodies, but in every part of the surrounding space, where electric or magnetic force is observed to act. Hence our theory agrees with the undulatory theory in assuming the existence of a medium which is capable of becoming a receptacle of two forms of energy *.

783.] Let us next determine the conditions of the propagation of an electromagnetic disturbance through a uniform medium, which we shall suppose to be at rest, that is, to have no motion except that which may be involved in electromagnetic disturbances.

* 'For my own part, considering the relation of a vacuum to the magnetic force, and the general character of magnetic phenomena external to the magnet, I am more inclined to the notion that in the transmission of the force there is such an action, external to the magnet, than that the effects are merely attraction and repulsion at a distance. Such an action may be a function of the æther; for it is not at all unlikely that, if there be an æther, it should have other uses than simply the conveyance of radiations.'—Faraday's *Experimental Researches*, 3075.

Let C be the specific conductivity of the medium, K its specific capacity for electrostatic induction, and μ its magnetic 'permeability.'

To obtain the general equations of electromagnetic disturbance, we shall express the true current $\mathfrak{C}$ in terms of the vector potential $\mathfrak{A}$ and the electric potential Ψ.

The true current $\mathfrak{C}$ is made up of the conduction current $\mathfrak{K}$ and the variation of the electric displacement $\dot{\mathfrak{D}}$, and since both of these depend on the electromotive force $\mathfrak{E}$, we find, as in Art. 611,

$$\mathfrak{C} = \left(C + \frac{1}{4\pi} K \frac{d}{dt}\right) \mathfrak{E}. \tag{1}$$

But since there is no motion of the medium, we may express the electromotive force, as in Art. 599,

$$\mathfrak{E} = -\dot{\mathfrak{A}} - \nabla\Psi. \tag{2}$$

Hence
$$\mathfrak{C} = -\left(C + \frac{1}{4\pi} K \frac{d}{dt}\right)\left(\frac{d\mathfrak{A}}{dt} + \nabla\Psi\right). \tag{3}$$

But we may determine a relation between $\mathfrak{C}$ and $\mathfrak{A}$ in a different way, as is shewn in Art. 616, the equations (4) of which may be written

$$4\pi\mu\mathfrak{C} = \nabla^2\mathfrak{A} + \nabla J, \tag{4}$$

where
$$J = \frac{dF}{dx} + \frac{dG}{dy} + \frac{dH}{dz}. \tag{5}$$

Combining equations (3) and (4), we obtain

$$\mu\left(4\pi C + K\frac{d}{dt}\right)\left(\frac{d\mathfrak{A}}{dt} + \nabla\Psi\right) + \nabla^2\mathfrak{A} + \nabla J = 0, \tag{6}$$

which we may express in the form of three equations as follows—

$$\left.\begin{aligned}
\mu\left(4\pi C + K\frac{d}{dt}\right)\left(\frac{dF}{dt} + \frac{d\Psi}{dx}\right) + \nabla^2 F + \frac{dJ}{dx} &= 0,\\
\mu\left(4\pi C + K\frac{d}{dt}\right)\left(\frac{dG}{dt} + \frac{d\Psi}{dy}\right) + \nabla^2 G + \frac{dJ}{dy} &= 0,\\
\mu\left(4\pi C + K\frac{d}{dt}\right)\left(\frac{dH}{dt} + \frac{d\Psi}{dz}\right) + \nabla^2 H + \frac{dJ}{dz} &= 0.
\end{aligned}\right\} \tag{7}$$

These are the general equations of electromagnetic disturbances.

If we differentiate these equations with respect to x, y, and z respectively, and add, we obtain

$$\mu\left(4\pi C + K\frac{d}{dt}\right)\left(\frac{dJ}{dt} - \nabla^2\Psi\right) = 0. \tag{8}$$

If the medium is a non-conductor, $C = 0$, and $\nabla^2\Psi$, which is proportional to the volume-density of free electricity, is independent of t. Hence J must be a linear function of t, or a constant, or zero, and we may therefore leave J and Ψ out of account in considering periodic disturbances.

Propagation of Undulations in a Non-conducting Medium.

784.] In this case $C = 0$, and the equations become

$$\left.\begin{aligned} K\mu\frac{d^2F}{dt^2} + \nabla^2 F &= 0, \\ K\mu\frac{d^2G}{dt^2} + \nabla^2 G &= 0, \\ K\mu\frac{d^2H}{dt^2} + \nabla^2 H &= 0. \end{aligned}\right\} \qquad (9)$$

The equations in this form are similar to those of the motion of an elastic solid, and when the initial conditions are given, the solution can be expressed in a form given by Poisson *, and applied by Stokes to the Theory of Diffraction †.

Let us write $$V = \frac{1}{\sqrt{K\mu}}. \qquad (10)$$

If the values of F, G, H, and of $\frac{dF}{dt}, \frac{dG}{dt}, \frac{dH}{dt}$ are given at every point of space at the epoch ($t = 0$), then we can determine their values at any subsequent time, t, as follows.

Let O be the point for which we wish to determine the value of F at the time t. With O as centre, and with radius Vt, describe a sphere. Find the initial value of F at every point of the spherical surface, and take the *mean*, $\bar{F}$, of all these values. Find also the initial values of $\frac{dF}{dt}$ at every point of the spherical surface, and let the mean of these values be $\overline{\frac{dF}{dt}}$.

Then the value of F at the point O, at the time t, is

$$F = \frac{d}{dt}(\bar{F}t) + t\,\overline{\frac{dF}{dt}}.$$

Similarly

$$\left.\begin{aligned} G &= \frac{d}{dt}(\bar{G}t) + t\,\overline{\frac{dG}{dt}}, \\ H &= \frac{d}{dt}(\bar{H}t) + t\,\overline{\frac{dH}{dt}}. \end{aligned}\right\} \qquad (11)$$

785.] It appears, therefore, that the condition of things at the point O at any instant depends on the condition of things at a distance Vt and at an interval of time t previously, so that any disturbance is propagated through the medium with the velocity V.

Let us suppose that when t is zero the quantities $\mathfrak{A}$ and $\dot{\mathfrak{A}}$ are

* *Mém. de l'Acad.*, tom. iii, p. 130.

† *Cambridge Transactions*, vol. ix, p. 10 (1850).

zero except within a certain space S. Then their values at O at the time t will be zero, unless the spherical surface described about O as centre with radius Vt lies in whole or in part within the space S. If O is outside the space S there will be no disturbance at O until Vt becomes equal to the shortest distance from O to the space S. The disturbance at O will then begin, and will go on till Vt is equal to the greatest distance from O to any part of S. The disturbance at O will then cease for ever.

786.] The quantity V, in Art. 793, which expresses the velocity of propagation of electromagnetic disturbances in a non-conducting medium is, by equation (9), equal to $\frac{1}{\sqrt{K\mu}}$.

If the medium is air, and if we adopt the electrostatic system of measurement, $K = 1$ and $\mu = \frac{1}{v^2}$, so that $V = v$, or the velocity of propagation is numerically equal to the number of electrostatic units of electricity in one electromagnetic unit. If we adopt the electromagnetic system, $K = \frac{1}{v^2}$ and $\mu = 1$, so that the equation $V = v$ is still true.

On the theory that light is an electromagnetic disturbance, propagated in the same medium through which other electromagnetic actions are transmitted, V must be the velocity of light, a quantity the value of which has been estimated by several methods. On the other hand, v is the number of electrostatic units of electricity in one electromagnetic unit, and the methods of determining this quantity have been described in the last chapter. They are quite independent of the methods of finding the velocity of light. Hence the agreement or disagreement of the values of V and of v furnishes a test of the electromagnetic theory of light.

787.] In the following table, the principal results of direct observation of the velocity of light, either through the air or through the planetary spaces, are compared with the principal results of the comparison of the electric units :—

Velocity of Light (mètres per second).		Ratio of Electric Units.	
Fizeau	314000000	Weber......	310740000
Aberration, &c., and Sun's Parallax } ...	308000000	Maxwell ...	288000000
Foucault	298360000	Thomson ...	282000000.

It is manifest that the velocity of light and the ratio of the units are quantities of the same order of magnitude. Neither of them can be said to be determined as yet with such a degree of accuracy as to enable us to assert that the one is greater or less than the other. It is to be hoped that, by further experiment, the relation between the magnitudes of the two quantities may be more accurately determined.

In the meantime our theory, which asserts that these two quantities are equal, and assigns a physical reason for this equality, is certainly not contradicted by the comparison of these results such as they are.

788.] In other media than air, the velocity V is inversely proportional to the square root of the product of the dielectric and the magnetic inductive capacities. According to the undulatory theory, the velocity of light in different media is inversely proportional to their indices of refraction.

There are no transparent media for which the magnetic capacity differs from that of air more than by a very small fraction. Hence the principal part of the difference between these media must depend on their dielectric capacity. According to our theory, therefore, the dielectric capacity of a transparent medium should be equal to the square of its index of refraction.

But the value of the index of refraction is different for light of different kinds, being greater for light of more rapid vibrations. We must therefore select the index of refraction which corresponds to waves of the longest periods, because these are the only waves whose motion can be compared with the slow processes by which we determine the capacity of the dielectric.

789.] The only dielectric of which the capacity has been hitherto determined with sufficient accuracy is paraffin, for which in the solid form M.M. Gibson and Barclay found *

$$K = 1.975. \qquad (12)$$

Dr. Gladstone has found the following values of the index of refraction of melted paraffin, sp. g. 0.779, for the lines A, D and H:—

Temperature	A	D	H
54°C	1.4306	1.4357	1.4499
57°C	1.4294	1.4343	1.4493;

from which I find that the index of refraction for waves of infinite length would be about 1.422.

The square root of K is 1.405.

The difference between these numbers is greater than can be ac-

* *Phil. Trans.* 1871, p. 573.

counted for by errors of observation, and shews that our theories of the structure of bodies must be much improved before we can deduce their optical from their electrical properties. At the same time, I think that the agreement of the numbers is such that if no greater discrepancy were found between the numbers derived from the optical and the electrical properties of a considerable number of substances, we should be warranted in concluding that the square root of K, though it may not be the complete expression for the index of refraction, is at least the most important term in it.

Plane Waves.

790.] Let us now confine our attention to plane waves, the front of which we shall suppose normal to the axis of z. All the quantities, the variation of which constitutes such waves, are functions of z and t only, and are independent of x and y. Hence the equations of magnetic induction, (A), Art. 591, are reduced to

$$a = -\frac{dG}{dz}, \qquad b = \frac{dF}{dz}, \qquad c = 0, \tag{13}$$

or the magnetic disturbance is in the plane of the wave. This agrees with what we know of that disturbance which constitutes light.

Putting $\mu\alpha$, $\mu\beta$ and $\mu\gamma$ for a, b and c respectively, the equations of electric currents, Art. 607, become

$$\left.\begin{aligned} 4\pi\mu u &= -\frac{db}{dz} = -\frac{d^2F}{dz^2}, \\ 4\pi\mu v &= \frac{da}{dz} = -\frac{d^2G}{dz^2}, \\ 4\pi\mu w &= 0. \end{aligned}\right\} \tag{14}$$

Hence the electric disturbance is also in the plane of the wave, and if the magnetic disturbance is confined to one direction, say that of x, the electric disturbance is confined to the perpendicular direction, or that of y.

But we may calculate the electric disturbance in another way, for if f, g, h are the components of electric displacement in a non-conducting medium

$$u = \frac{df}{dt}, \qquad v = \frac{dg}{dt}, \qquad w = \frac{dh}{dt}. \tag{15}$$

If P, Q, R are the components of the electromotive force

$$f = \frac{K}{4\pi}P, \qquad g = \frac{K}{4\pi}Q, \qquad w = \frac{K}{4\pi}R; \tag{16}$$

and since there is no motion of the medium, equations (B), Art. 598, become $P = -\frac{dF}{dt}, \quad Q = -\frac{dG}{dt}, \quad R = -\frac{dH}{dt}.$ (17)

Hence $u = -\frac{K}{4\pi}\frac{d^2F}{dt^2}, \quad v = -\frac{K}{4\pi}\frac{d^2G}{dt^2}, \quad w = -\frac{K}{4\pi}\frac{d^2F}{dt^2}.$ (18)

Comparing these values with those given in equation (14), we find

$$\left.\begin{aligned} \frac{d^2F}{dz^2} &= K\mu\frac{d^2F}{dt^2}, \\ \frac{d^2G}{dz^2} &= K\mu\frac{d^2G}{dt^2}, \\ 0 &= K\mu\frac{d^2H}{dt^2}. \end{aligned}\right\} \quad (19)$$

The first and second of these equations are the equations of propagation of a plane wave, and their solution is of the well-known form

$$\left.\begin{aligned} F &= f_1(z-Vt)+f_2(z+Vt), \\ G &= f_3(z-Vt)+f_4(z+Vt). \end{aligned}\right\} \quad (20)$$

The solution of the third equation is

$$K\mu H = A + Bt, \quad (21)$$

where A and B are functions of z. H is therefore either constant or varies directly with the time. In neither case can it take part in the propagation of waves.

791.] It appears from this that the directions, both of the magnetic and the electric disturbances, lie in the plane of the wave. The mathematical form of the disturbance therefore, agrees with that of the disturbance which constitutes light, in being transverse to the direction of propagation.

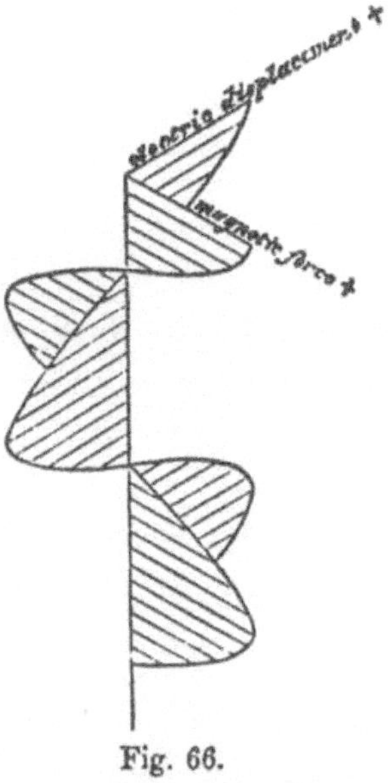

Fig. 66.

If we suppose $G = 0$, the disturbance will correspond to a plane-polarized ray of light.

The magnetic force is in this case parallel to the axis of y and equal to $\frac{1}{\mu}\frac{dF}{dz}$, and the electromotive force is parallel to the axis of x and equal to $-\frac{dF}{dt}$. The magnetic force is therefore in a plane perpendicular to that which contains the electric force.

The values of the magnetic force and of the electromotive force at a given instant at different points of the ray are represented in Fig. 66, for the case of a simple harmonic disturbance in one plane. This corresponds to a ray of plane-polarized light, but whether the plane of polarization corresponds to the plane of the magnetic disturbance, or to the plane of the electric disturbance, remains to be seen. See Art. 797.

Energy and Stress of Radiation.

792.] The electrostatic energy per unit of volume at any point of the wave in a non-conducting medium is

$$\tfrac{1}{2} f P = \frac{K}{8\pi} P^2 = \frac{K}{8\pi} \overline{\frac{dF}{dt}}\Big|^2. \tag{22}$$

The electrokinetic energy at the same point is

$$\frac{1}{8\pi} b\beta = \frac{1}{8\pi\mu} b^2 = \frac{1}{8\pi\mu} \overline{\frac{dF}{dz}}\Big|^2. \tag{23}$$

In virtue of equation (8) these two expressions are equal, so that at every point of the wave the intrinsic energy of the medium is half electrostatic and half electrokinetic.

Let p be the value of either of these quantities, that is, either the electrostatic or the electrokinetic energy per unit of volume, then, in virtue of the electrostatic state of the medium, there is a tension whose magnitude is p, in a direction parallel to x, combined with a pressure, also equal to p, parallel to y and z. See Art. 107.

In virtue of the electrokinetic state of the medium there is a tension equal to p in a direction parallel to y, combined with a pressure equal to p in directions parallel to x and z. See Art. 643.

Hence the combined effect of the electrostatic and the electrokinetic stresses is a *pressure* equal to $2p$ in the direction of the propagation of the wave. Now $2p$ also expresses the whole energy in unit of volume.

Hence in a medium in which waves are propagated there is a pressure in the direction normal to the waves, and numerically equal to the energy in unit of volume.

793.] Thus, if in strong sunlight the energy of the light which falls on one square foot is 83.4 foot pounds per second, the mean energy in one cubic foot of sunlight is about 0.0000000882 of a foot pound, and the mean pressure on a square foot is 0.0000000882 of a pound weight. A flat body exposed to sunlight would experience this pressure on its illuminated side only, and would therefore be repelled from the side on which the light falls. It is probable that a much greater energy of radiation might be obtained by means of the concentrated rays of the electric lamp. Such rays falling on a thin metallic disk, delicately suspended in a vacuum, might perhaps produce an observable mechanical effect. When a disturbance of any kind consists of terms involving sines or cosines of angles which vary with the time, the maximum energy is double of the mean energy. Hence, if P is the maximum electromotive force, and β the maximum magnetic force which are called into play during the propagation of light,

$$\frac{K}{8\pi}P^2 = \frac{\mu}{8\pi}\beta^2 = \text{mean energy in unit of volume.} \qquad (24)$$

With Pouillet's data for the energy of sunlight, as quoted by Thomson, *Trans. R. S. E.*, 1854, this gives in electromagnetic measure

$P = 60000000$, or about 600 Daniell's cells per mètre;

$\beta = 0.193$, or rather more than a tenth of the horizontal magnetic force in Britain.

Propagation of a Plane Wave in a Crystallized Medium.

794.] In calculating, from data furnished by ordinary electromagnetic experiments, the electrical phenomena which would result from periodic disturbances, millions of millions of which occur in a second, we have already put our theory to a very severe test, even when the medium is supposed to be air or vacuum. But if we attempt to extend our theory to the case of dense media, we become involved not only in all the ordinary difficulties of molecular theories, but in the deeper mystery of the relation of the molecules to the electromagnetic medium.

To evade these difficulties, we shall assume that in certain media the specific capacity for electrostatic induction is different in different directions, or in other words, the electric displacement, instead of being in the same direction as the electromotive force, and proportional to it, is related to it by a system of linear equations similar to those given in Art. 297. It may be shewn, as in Art. 436, that the system of coefficients must be symmetrical, so that, by a proper choice of axes, the equations become

$$f = \frac{1}{4\pi}K_1 P, \qquad g = \frac{1}{4\pi}K_2 Q, \qquad h = \frac{1}{4\pi}K_3 R, \qquad (1)$$

where K_1, K_2, and K_3 are the principal inductive capacities of the medium. The equations of propagation of disturbances are therefore

$$\left.\begin{aligned}\frac{d^2F}{dy^2}+\frac{d^2F}{dz^2}-\frac{d^2G}{dx\,dy}-\frac{d^2H}{dz\,dx}&=K_1\mu\left(\frac{d^2F}{dt^2}-\frac{d^2\Psi}{dx\,dt}\right),\\ \frac{d^2G}{dz^2}+\frac{d^2G}{dx^2}-\frac{d^2H}{dy\,dz}-\frac{d^2F}{dx\,dy}&=K_2\mu\left(\frac{d^2G}{dt^2}-\frac{d^2\Psi}{dy\,dt}\right),\\ \frac{d^2H}{dx^2}+\frac{d^2H}{dy^2}-\frac{d^2F}{dz\,dx}-\frac{d^2G}{dy\,dz}&=K_3\mu\left(\frac{d^2H}{dt^2}-\frac{d^2\Psi}{dz\,dt}\right).\end{aligned}\right\}\quad(2)$$

795.] If l, m, n are the direction-cosines of the normal to the wave-front, and V the velocity of the wave, and if

$$lx+my+nz-Vt=w, \quad (3)$$

and if we write F'', G'', H'', Ψ'' for the second differential coefficients of F, G, H, Ψ respectively with respect to w, and put

$$K_1\mu=\frac{1}{a^2},\qquad K_2\mu=\frac{1}{b^2},\qquad K_3\mu=\frac{1}{c^2}, \quad (4)$$

where a, b, c are the three principal velocities of propagation, the equations become

$$\left.\begin{aligned}\left(m^2+n^2-\frac{V^2}{a^2}\right)F''-lmG''-nlH''-V\Psi''\frac{l}{a^2}&=0,\\ -lmF''+\left(n^2+l^2-\frac{V^2}{b^2}\right)G''-mnH''-V\Psi''\frac{m}{b^2}&=0,\\ -nlF''-mnG''+\left(l^2+m^2-\frac{V^2}{c^2}\right)H''-V\Psi''\frac{n}{b^2}&=0.\end{aligned}\right\}\quad(5)$$

796.] If we write

$$\frac{l^2}{V^2-a^2}+\frac{m^2}{V^2-b^2}+\frac{n^2}{V^2-c^2}=U, \quad (6)$$

we obtain from these equations

$$\left.\begin{aligned}VU(VF''-l\Psi'')&=0,\\ VU(VG''-m\Psi'')&=0,\\ VU(VH''-n\Psi'')&=0.\end{aligned}\right\}\quad(7)$$

Hence, either $V=0$, in which case the wave is not propagated at all; or, $U=0$, which leads to the equation for V given by Fresnel; or the quantities within brackets vanish, in which case the vector whose components are F'', G'', H'' is normal to the wave-front and proportional to the electric volume-density. Since the medium is a non-conductor, the electric density at any given point is constant, and therefore the disturbance indicated by these equations is not periodic, and cannot constitute a wave. We may therefore consider $\Psi''=0$ in the investigation of the wave.

797.] The velocity of the propagation of the wave is therefore completely determined from the equation $U = 0$, or

$$\frac{l^2}{V^2-a^2}+\frac{m^2}{V^2-b^2}+\frac{n^2}{V^2-c^2}=0. \quad (8)$$

There are therefore two, and only two, values of V^2 corresponding to a given direction of wave-front.

If λ, μ, ν are the direction-cosines of the electric current whose components are u, v, w,

$$\lambda : \mu : \nu :: \frac{1}{a^2}F'' : \frac{1}{b^2}G'' : \frac{1}{c^2}H'', \quad (9)$$

then

$$l\lambda+m\mu+n\nu=0; \quad (10)$$

or the current is in the plane of the wave-front, and its direction in the wave-front is determined by the equation

$$\frac{l}{\lambda}(b^2-c^2)+\frac{m}{\mu}(c^2-a^2)+\frac{n}{\nu}(a^2-b^2)=0. \quad (11)$$

These equations are identical with those given by Fresnel if we define the plane of polarization as a plane through the ray perpendicular to the plane of the electric disturbance.

According to this electromagnetic theory of double refraction the wave of normal disturbance, which constitutes one of the chief difficulties of the ordinary theory, does not exist, and no new assumption is required in order to account for the fact that a ray polarized in a principal plane of the crystal is refracted in the ordinary manner *.

Relation between Electric Conductivity and Opacity.

798.] If the medium, instead of being a perfect insulator, is a conductor whose conductivity per unit of volume is C, the disturbance will consist not only of electric displacements but of currents of conduction, in which electric energy is transformed into heat, so that the undulation is absorbed by the medium.

If the disturbance is expressed by a circular function, we may write

$$F=e^{-pz}\cos(nt-qz), \quad (1)$$

for this will satisfy the equation

$$\frac{d^2F}{dz^2}=\mu K\frac{d^2F}{dt^2}+4\pi\mu C\frac{dF}{dt}, \quad (2)$$

provided

$$q^2-p^2=\mu Kn^2, \quad (3)$$

and

$$2pq=4\pi\mu Cn. \quad (4)$$

* See Stokes' 'Report on Double Refraction'; *Brit. Assoc. Reports*, 1862, p. 255.

The velocity of propagation is

$$V = \frac{n}{q}, \tag{5}$$

and the coefficient of absorption is

$$p = 2\pi\mu C V. \tag{6}$$

Let R be the resistance, in electromagnetic measure, of a plate whose length is l, breadth b, and thickness z,

$$R = \frac{l}{bzC}. \tag{7}$$

The proportion of the incident light which will be transmitted by this plate will be

$$e^{-2pz} = e^{-4\pi\mu\frac{l}{b}\frac{V}{R}}. \tag{8}$$

799.] Most transparent solid bodies are good insulators, and all good conductors are very opaque. There are, however, many exceptions to the law that the opacity of a body is the greater, the greater its conductivity.

Electrolytes allow an electric current to pass, and yet many of them are transparent. We may suppose, however, that in the case of the rapidly alternating forces which come into play during the propagation of light, the electromotive force acts for so short a time in one direction that it is unable to effect a complete separation between the combined molecules. When, during the other half of the vibration, the electromotive force acts in the opposite direction it simply reverses what it did during the first half. There is thus no true conduction through the electrolyte, no loss of electric energy, and consequently no absorption of light.

800.] Gold, silver, and platinum are good conductors, and yet, when formed into very thin plates, they allow light to pass through them. From experiments which I have made on a piece of gold leaf, the resistance of which was determined by Mr. Hockin, it appears that its transparency is very much greater than is consistent with our theory, unless we suppose that there is less loss of energy when the electromotive forces are reversed for every semi-vibration of light than when they act for sensible times, as in our ordinary experiments.

801.] Let us next consider the case of a medium in which the conductivity is large in proportion to the inductive capacity.

In this case we may leave out the term involving K in the equations of Art. 783, and they then become

$$\left.\begin{aligned} \nabla^2 F + 4\pi\mu C\frac{dF}{dt} &= 0, \\ \nabla^2 G + 4\pi\mu C\frac{dG}{dt} &= 0, \\ \nabla^2 H + 4\pi\mu C\frac{dH}{dt} &= 0. \end{aligned}\right\} \qquad (1)$$

Each of these equations is of the same form as the equation of the diffusion of heat given in Fourier's *Traité de Chaleur*.

802.] Taking the first as an example, the component F of the vector-potential will vary according to time and position in the same way as the temperature of a homogeneous solid varies according to time and position, the initial and the surface-conditions being made to correspond in the two cases, and the quantity $4\pi\mu C$ being numerically equal to the reciprocal of the thermometric conductivity of the substance, that is to say, the *number of units of volume of the substance which would be heated one degree by the heat which passes through a unit cube of the substance, two opposite faces of which differ by one degree of temperature, while the other faces are impermeable to heat* *.

The different problems in thermal conduction, of which Fourier has given the solution, may be transformed into problems in the diffusion of electromagnetic quantities, remembering that F, G, H are the components of a vector, whereas the temperature, in Fourier's problem, is a scalar quantity.

Let us take one of the cases of which Fourier has given a complete solution †, that of an infinite medium, the initial state of which is given.

The state of any point of the medium at the time t is found by taking the average of the state of every part of the medium, the weight assigned to each part in taking the average being

$$e^{-\frac{\pi\mu C r^2}{t}},$$

where r is the distance of that part from the point considered. This average, in the case of vector-quantities, is most conveniently taken by considering each component of the vector separately.

* See Maxwell's *Theory of Heat*, p. 235.

† *Traité de la Chaleur*, Art. 384. The equation which determines the temperature, v, at a point (x, y, z) after a time t, in terms of $f(\alpha, \beta, \gamma)$, the initial temperature at the point (α, β, γ), is

$$v = \iiint \frac{d\alpha\, d\beta\, d\gamma}{2^3\sqrt{k^3\pi^3 t^3}}\, e^{-\left(\frac{(\alpha-x)^2+(\beta-y)^2+(\gamma-z)^2}{4kt}\right)} f(\alpha, \beta, \gamma),$$

where k is the thermometric conductivity.

803.] We have to remark in the first place, that in this problem the thermal conductivity of Fourier's medium is to be taken inversely proportional to the electric conductivity of our medium, so that the time required in order to reach an assigned stage in the process of diffusion is greater the higher the electric conductivity. This statement will not appear paradoxical if we remember the result of Art. 655, that a medium of infinite conductivity forms a complete barrier to the process of diffusion of magnetic force.

In the next place, the time requisite for the production of an assigned stage in the process of diffusion is proportional to the square of the linear dimensions of the system.

There is no determinate velocity which can be defined as the velocity of diffusion. If we attempt to measure this velocity by ascertaining the time requisite for the production of a given amount of disturbance at a given distance from the origin of disturbance, we find that the smaller the selected value of the disturbance the greater the velocity will appear to be, for however great the distance, and however small the time, the value of the disturbance will differ mathematically from zero.

This peculiarity of diffusion distinguishes it from wave-propagation, which takes place with a definite velocity. No disturbance takes place at a given point till the wave reaches that point, and when the wave has passed, the disturbance ceases for ever.

804.] Let us now investigate the process which takes place when an electric current begins and continues to flow through a linear circuit, the medium surrounding the circuit being of finite electric conductivity. (Compare with Art. 660).

When the current begins, its first effect is to produce a current of induction in the parts of the medium close to the wire. The direction of this current is opposite to that of the original current, and in the first instant its total quantity is equal to that of the original current, so that the electromagnetic effect on more distant parts of the medium is initially zero, and only rises to its final value as the induction-current dies away on account of the electric resistance of the medium.

But as the induction-current close to the wire dies away, a new induction-current is generated in the medium beyond, so that the space occupied by the induction-current is continually becoming wider, while its intensity is continually diminishing.

This diffusion and decay of the induction-current is a phenomenon precisely analogous to the diffusion of heat from a part of the medium initially hotter or colder than the rest. We must remember, however, that since the current is a vector quantity, and since in a circuit the current is in opposite directions at opposite points of the circuit, we must, in calculating any given component of the induction-current, compare the problem with one in which equal quantities of heat and of cold are diffused from neighbouring places, in which case the effect on distant points will be of a smaller order of magnitude.

805.] If the current in the linear circuit is maintained constant, the induction currents, which depend on the initial change of state, will gradually be diffused and die away, leaving the medium in its permanent state, which is analogous to the permanent state of the flow of heat. In this state we have

$$\nabla^2 F = \nabla^2 G = \nabla^2 H = 0 \tag{2}$$

throughout the medium, except at the part occupied by the circuit, in which

$$\left.\begin{aligned} \nabla^2 F &= 4\pi u, \\ \nabla^2 G &= 4\pi v, \\ \nabla^2 H &= 4\pi w. \end{aligned}\right\} \tag{3}$$

These equations are sufficient to determine the values of F, G, H throughout the medium. They indicate that there are no currents except in the circuit, and that the magnetic forces are simply those due to the current in the circuit according to the ordinary theory. The rapidity with which this permanent state is established is so great that it could not be measured by our experimental methods, except perhaps in the case of a very large mass of a highly conducting medium such as copper.

NOTE.—In a paper published in Poggendorff's *Annalen*, June 1867, M. Lorenz has deduced from Kirchhoff's equations of electric currents (Pogg. *Ann.* cii. 1856), by the addition of certain terms which do not affect any experimental result, a new set of equations, indicating that the distribution of force in the electromagnetic field may be conceived as arising from the mutual action of contiguous elements, and that waves, consisting of transverse electric currents, may be propagated, with a velocity comparable to that of light, in non-conducting media. He therefore regards the disturbance which constitutes light as identical with these electric currents, and he shews that conducting media must be opaque to such radiations.

These conclusions are similar to those of this chapter, though obtained by an entirely different method. The theory given in this chapter was first published in the *Phil. Trans.* for 1865.

CHAPTER XXI.

MAGNETIC ACTION ON LIGHT.

806.] THE most important step in establishing a relation between electric and magnetic phenomena and those of light must be the discovery of some instance in which the one set of phenomena is affected by the other. In the search for such phenomena we must be guided by any knowledge we may have already obtained with respect to the mathematical or geometrical form of the quantities which we wish to compare. Thus, if we endeavour, as Mrs. Somerville did, to magnetize a needle by means of light, we must remember that the distinction between magnetic north and south is a mere matter of direction, and would be at once reversed if we reverse certain conventions about the use of mathematical signs. There is nothing in magnetism analogous to those phenomena of electrolysis which enable us to distinguish positive from negative electricity, by observing that oxygen appears at one pole of a cell and hydrogen at the other.

Hence we must not expect that if we make light fall on one end of a needle, that end will become a pole of a certain name, for the two poles do not differ as light does from darkness.

We might expect a better result if we caused circularly polarized light to fall on the needle, right-handed light falling on one end and left-handed on the other, for in some respects these kinds of light may be said to be related to each other in the same way as the poles of a magnet. The analogy, however, is faulty even here, for the two rays when combined do not neutralize each other, but produce a plane polarized ray.

Faraday, who was acquainted with the method of studying the strains produced in transparent solids by means of polarized light, made many experiments in hopes of detecting some action on polarized light while passing through a medium in which electrolytic conduction or dielectric induction exists *. He was not, however,

* *Experimental Researches*, 951–954 and 2216–2220.

able to detect any action of this kind, though the experiments were arranged in the way best adapted to discover effects of tension, the electric force or current being at right angles to the direction of the ray, and at an angle of forty-five degrees to the plane of polarization. Faraday varied these experiments in many ways without discovering any action on light due to electrolytic currents or to static electric induction.

He succeeded, however, in establishing a relation between light and magnetism, and the experiments by which he did so are described in the nineteenth series of his *Experimental Researches*. We shall take Faraday's discovery as our starting point for further investigation into the nature of magnetism, and we shall therefore describe the phenomenon which he observed.

807.] A ray of plane-polarized light is transmitted through a transparent diamagnetic medium, and the plane of its polarization, when it emerges from the medium, is ascertained by observing the position of an analyser when it cuts off the ray. A magnetic force is then made to act so that the direction of the force within the transparent medium coincides with the direction of the ray. The light at once reappears, but if the analyser is turned round through a certain angle, the light is again cut off. This shews that the effect of the magnetic force is to turn the plane of polarization, round the direction of the ray as an axis, through a certain angle, measured by the angle through which the analyser must be turned in order to cut off the light.

808.] The angle through which the plane of polarization is turned is proportional—

(1) To the distance which the ray travels within the medium. Hence the plane of polarization changes continuously from its position at incidence to its position at emergence.

(2) To the intensity of the resolved part of the magnetic force in the direction of the ray.

(3) The amount of the rotation depends on the nature of the medium. No rotation has yet been observed when the medium is air or any other gas.

These three statements are included in the more general one, that the angular rotation is numerically equal to the amount by which the magnetic potential increases, from the point at which the ray enters the medium to that at which it leaves it, multiplied by a coefficient, which, for diamagnetic media, is generally positive.

809.] In diamagnetic substances, the direction in which the plane of polarization is made to rotate is the same as the direction in which a positive current must circulate round the ray in order to produce a magnetic force in the same direction as that which actually exists in the medium.

Verdet, however, discovered that in certain ferromagnetic media, as, for instance, a strong solution of perchloride of iron in wood-spirit or ether, the rotation is in the opposite direction to the current which would produce the magnetic force.

This shews that the difference between ferromagnetic and diamagnetic substances does not arise merely from the 'magnetic permeability' being in the first case greater, and in the second less, than that of air, but that the properties of the two classes of bodies are really opposite.

The power acquired by a substance under the action of magnetic force of rotating the plane of polarization of light is not exactly proportional to its diamagnetic or ferromagnetic magnetizability. Indeed there are exceptions to the rule that the rotation is positive for diamagnetic and negative for ferromagnetic substances, for neutral chromate of potash is diamagnetic, but produces a negative rotation.

810.] There are other substances, which, independently of the application of magnetic force, cause the plane of polarization to turn to the right or to the left, as the ray travels through the substance. In some of these the property is related to an axis, as in the case of quartz. In others, the property is independent of the direction of the ray within the medium, as in turpentine, solution of sugar, &c. In all these substances, however, if the plane of polarization of any ray is twisted within the medium like a right-handed screw, it will still be twisted like a right-handed screw if the ray is transmitted through the medium in the opposite direction. The direction in which the observer has to turn his analyser in order to extinguish the ray after introducing the medium into its path, is the same with reference to the observer whether the ray comes to him from the north or from the south. The direction of the rotation in space is of course reversed when the direction of the ray is reversed. But when the rotation is produced by magnetic action, its direction in space is the same whether the ray be travelling north or south. The rotation is always in the same direction as that of the electric current which produces, or would produce, the actual magnetic state of the field, if the medium belongs to the positive class, or in the opposite direction if the medium belongs to the negative class.

It follows from this, that if the ray of light, after passing through the medium from north to south, is reflected by a mirror, so as to return through the medium from south to north, the rotation will be doubled when it results from magnetic action. When the rotation depends on the nature of the medium alone, as in turpentine, &c., the ray, when reflected back through the medium, emerges in the same plane as it entered, the rotation during the first passage through the medium having been exactly reversed during the second.

811.] The physical explanation of the phenomenon presents considerable difficulties, which can hardly be said to have been hitherto overcome, either for the magnetic rotation, or for that which certain media exhibit of themselves. We may, however, prepare the way for such an explanation by an analysis of the observed facts.

It is a well-known theorem in kinematics that two uniform circular vibrations, of the same amplitude, having the same periodic time, and in the same plane, but revolving in opposite directions, are equivalent, when compounded together, to a rectilinear vibration. The periodic time of this vibration is equal to that of the circular vibrations, its amplitude is double, and its direction is in the line joining the points at which two particles, describing the circular vibrations in opposite directions round the same circle, would meet. Hence if one of the circular vibrations has its phase accelerated, the direction of the rectilinear vibration will be turned, in the same direction as that of the circular vibration, through an angle equal to half the acceleration of phase.

It can also be proved by direct optical experiment that two rays of light, circularly-polarized in opposite directions, and of the same intensity, become, when united, a plane-polarized ray, and that if by any means the phase of one of the circularly-polarized rays is accelerated, the plane of polarization of the resultant ray is turned round half the angle of acceleration of the phase.

812.] We may therefore express the phenomenon of the rotation of the plane of polarization in the following manner:—A plane-polarized ray falls on the medium. This is equivalent to two circularly-polarized rays, one right-handed, the other left-handed (as regards the observer). After passing through the medium the ray is still plane-polarized, but the plane of polarization is turned, say, to the right (as regards the observer). Hence, of the two circularly-polarized rays, that which is right-handed must have had its phase accelerated with respect to the other during its passage through the medium.

In other words, the right-handed ray has performed a greater number of vibrations, and therefore has a smaller wave-length, within the medium, than the left-handed ray which has the same periodic time.

This mode of stating what takes place is quite independent of any theory of light, for though we use such terms as wave-length, circular-polarization, &c., which may be associated in our minds with a particular form of the undulatory theory, the reasoning is independent of this association, and depends only on facts proved by experiment.

813.] Let us next consider the configuration of one of these rays at a given instant. Any undulation, the motion of which at each point is circular, may be represented by a helix or screw. If the screw is made to revolve about its axis without any longitudinal motion, each particle will describe a circle, and at the same time the propagation of the undulation will be represented by the apparent longitudinal motion of the similarly situated parts of the thread of the screw. It is easy to see that if the screw is right-handed, and the observer is placed at that end towards which the undulation travels, the motion of the screw will appear to him left-handed, that is to say, in the opposite direction to that of the hands of a watch. Hence such a ray has been called, originally by French writers, but now by the whole scientific world, a left-handed circularly-polarized ray.

A right-handed circularly-polarized ray is represented in like manner by a left-handed helix. In Fig. 67 the right-handed helix *A*, on the right-hand of the figure, represents a left-handed ray, and the left-handed helix *B*, on the left-hand, represents a right-handed ray.

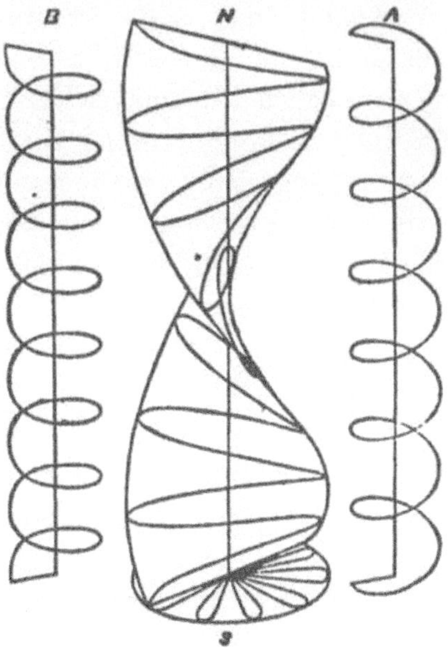

Fig. 67.

814.] Let us now consider two such rays which have the same wave-length within the medium. They are geometrically alike in all respects, except that one is the *perversion* of the other, like its image in a looking-glass. One of them, however, say A, has a shorter period of rotation than the other. If the motion is entirely due to the forces called into play by the displacement, this shews that greater forces are called into play by the same displacement when the configuration is like A than when it is like B. Hence in this case the left-handed ray will be accelerated with respect to the right-handed ray, and this will be the case whether the rays are travelling from N to S or from S to N.

This therefore is the explanation of the phenomenon as it is produced by turpentine, &c. In these media the displacement caused by a circularly-polarized ray calls into play greater forces of restitution when the configuration is like A than when it is like B. The forces thus depend on the configuration alone, not on the direction of the motion.

But in a diamagnetic medium acted on by magnetism in the direction SN, of the two screws A and B, that one always rotates with the greatest velocity whose motion, as seen by an eye looking from S to N, appears like that of a watch. Hence for rays from S to N the right-handed ray B will travel quickest, but for rays from N to S the left-handed ray A will travel quickest.

815.] Confining our attention to one ray only, the helix B has exactly the same configuration, whether it represents a ray from S to N or one from N to S. But in the first instance the ray travels faster, and therefore the helix rotates more rapidly. Hence greater forces are called into play when the helix is going round one way than when it is going round the other way. The forces, therefore, do not depend solely on the configuration of the ray, but also on the direction of the motion of its individual parts.

816.] The disturbance which constitutes light, whatever its physical nature may be, is of the nature of a vector, perpendicular to the direction of the ray. This is proved from the fact of the interference of two rays of light, which under certain conditions produces darkness, combined with the fact of the non-interference of two rays polarized in planes perpendicular to each other. For since the interference depends on the angular position of the planes of polarization, the disturbance must be a directed quantity or vector, and since the interference ceases when the planes of polarization are at right angles, the vector representing the disturbance must be perpendicular to the line of intersection of these planes, that is, to the direction of the ray.

817.] The disturbance, being a vector, can be resolved into components parallel to x and y, the axis of z being parallel to the direction of the ray. Let ξ and η be these components, then, in the case of a ray of homogeneous circularly-polarized light,

$$\xi = r\cos\theta, \qquad \eta = r\sin\theta, \tag{1}$$

where

$$\theta = nt - qz + a. \tag{2}$$

In these expressions, r denotes the magnitude of the vector, and θ the angle which it makes with the direction of the axis of x.

The periodic time, τ, of the disturbance is such that

$$n\tau = 2\pi. \tag{3}$$

The wave-length, λ, of the disturbance is such that

$$q\lambda = 2\pi. \tag{4}$$

The velocity of propagation is $\frac{n}{q}$.

The phase of the disturbance when t and z are both zero is a.

The circularly-polarized light is right-handed or left-handed according as q is negative or positive.

Its vibrations are in the positive or the negative direction of rotation in the plane of (x, y), according as n is positive or negative.

The light is propagated in the positive or the negative direction of the axis of z, according as n and q are of the same or of opposite signs.

In all media n varies when q varies, and $\frac{dn}{dq}$ is always of the same sign with $\frac{n}{q}$.

Hence, if for a given numerical value of n the value of $\frac{n}{q}$ is greater when n is positive than when n is negative, it follows that for a value of q, given both in magnitude and sign, the positive value of n will be greater than the negative value.

Now this is what is observed in a diamagnetic medium, acted on by a magnetic force, γ, in the direction of z. Of the two circularly-polarized rays of a given period, that is accelerated of which the direction of rotation in the plane of (x, y) is positive. Hence, of two circularly-polarized rays, both left-handed, whose wave-length within the medium is the same, that has the shortest period whose direction of rotation in the plane of xy is positive, that is, the ray which is propagated in the positive direction of z from south to north. We have therefore to account for the fact, that when in the equations of the system q and r are given, two values of n will satisfy the equations, one positive and the other negative, the positive value being numerically greater than the negative.

818.] We may obtain the equations of motion from a consideration of the potential and kinetic energies of the medium. The potential energy, V, of the system depends on its configuration, that is, on the relative position of its parts. In so far as it depends on the disturbance due to circularly-polarized light, it must be a function of r, the amplitude, and q, the coefficient of torsion, only. It may be different for positive and negative values of q of equal numerical value, and it probably is so in the case of media which of themselves rotate the plane of polarization.

The kinetic energy, T, of the system is a homogeneous function of the second degree of the velocities of the system, the coefficients of the different terms being functions of the coordinates.

819.] Let us consider the dynamical condition that the ray may be of constant intensity, that is, that r may be constant.

Lagrange's equation for the force in r becomes

$$\frac{d}{dt}\frac{dT}{d\dot{r}} - \frac{dT}{dr} + \frac{dV}{dr} = 0. \tag{5}$$

Since r is constant, the first term vanishes. We have therefore the equation

$$-\frac{dT}{dr} + \frac{dV}{dr} = 0, \tag{6}$$

in which q is supposed to be given, and we are to determine the value of the angular velocity $\dot{\theta}$, which we may denote by its actual value, n.

The kinetic energy, T, contains one term involving n^2; other terms may contain products of n with other velocities, and the rest of the terms are independent of n. The potential energy, V, is entirely independent of n. The equation is therefore of the form

$$An^2 + Bn + C = 0. \tag{7}$$

This being a quadratic equation, gives two values of n. It appears from experiment that both values are real, that one is positive and the other negative, and that the positive value is numerically the greater. Hence, if A is positive, both B and C are negative, for, if n_1 and n_2 are the roots of the equation,

$$A(n_1 + n_2) + B = 0. \tag{8}$$

The coefficient, B, therefore, is not zero, at least when magnetic force acts on the medium. We have therefore to consider the expression Bn, which is the part of the kinetic energy involving the first power of n, the angular velocity of the disturbance.

820.] Every term of T is of two dimensions as regards velocity. Hence the terms involving n must involve some other velocity. This velocity cannot be $\dot{r}$ or $\dot{q}$, because, in the case we consider, r and q are constant. Hence it is a velocity which exists in the medium independently of that motion which constitutes light. It must also be a velocity related to n in such a way that when it is multiplied by n the result is a scalar quantity, for only scalar quantities can occur as terms in the value of T, which is itself scalar. Hence this velocity must be in the same direction as n, or in the opposite direction, that is, it must be an *angular velocity* about the axis of z.

Again, this velocity cannot be independent of the magnetic force, for if it were related to a direction fixed in the medium, the phenomenon would be different if we turned the medium end for end, which is not the case.

We are therefore led to the conclusion that this velocity is an invariable accompaniment of the magnetic force in those media which exhibit the magnetic rotation of the plane of polarization.

821.] We have been hitherto obliged to use language which is perhaps too suggestive of the ordinary hypothesis of motion in the undulatory theory. It is easy, however, to state our result in a form free from this hypothesis.

Whatever light is, at each point of space there is something going on, whether displacement, or rotation, or something not yet imagined, but which is certainly of the nature of a vector or directed quantity, the direction of which is normal to the direction of the ray. This is completely proved by the phenomena of interference.

In the case of circularly-polarized light, the magnitude of this vector remains always the same, but its direction rotates round the direction of the ray so as to complete a revolution in the periodic time of the wave. The uncertainty which exists as to whether this vector is in the plane of polarization or perpendicular to it, does not extend to our knowledge of the direction in which it rotates in right-handed and in left-handed circularly-polarized light respectively. The direction and the angular velocity of this vector are perfectly known, though the physical nature of the vector and its absolute direction at a given instant are uncertain.

When a ray of circularly-polarized light falls on a medium under the action of magnetic force, its propagation within the medium is affected by the relation of the direction of rotation of the light to the direction of the magnetic force. From this we conclude, by the reasoning of Art. 821, that in the medium, when under the action of magnetic force, some rotatory motion is going on, the axis of rotation being in the direction of the magnetic forces; and that the rate of propagation of circularly-polarized light, when the direction of its vibratory rotation and the direction of the magnetic rotation of the medium are the same, is different from the rate of propagation when these directions are opposite.

The only resemblance which we can trace between a medium through which circularly-polarized light is propagated, and a medium through which lines of magnetic force pass, is that in both there is a motion of rotation about an axis. But here the resemblance stops, for the rotation in the optical phenomenon is that of the vector which represents the disturbance. This vector is always perpendicular to the direction of the ray, and rotates about it a known number of times in a second. In the magnetic phenomenon, that which rotates has no properties by which its sides can be distinguished, so that we cannot determine how many times it rotates in a second.

There is nothing, therefore, in the magnetic phenomenon which corresponds to the wave-length and the wave-propagation in the optical phenomenon. A medium in which a constant magnetic force is acting is not, in consequence of that force, filled with waves travelling in one direction, as when light is propagated through it. The only resemblance between the optical and the magnetic phenomenon is, that at each point of the medium something exists of the nature of an angular velocity about an axis in the direction of the magnetic force.

On the Hypothesis of Molecular Vortices.

822.] The consideration of the action of magnetism on polarized light leads, as we have seen, to the conclusion that in a medium under the action of magnetic force something belonging to the same mathematical class as an angular velocity, whose axis is in the direction of the magnetic force, forms a part of the phenomenon.

This angular velocity cannot be that of any portion of the medium of sensible dimensions rotating as a whole. We must therefore conceive the rotation to be that of very small portions of the medium, each rotating on its own axis. This is the hypothesis of molecular vortices.

The motion of these vortices, though, as we have shewn (Art. 575), it does not sensibly affect the visible motions of large bodies, may be such as to affect that vibratory motion on which the propagation of light, according to the undulatory theory, depends. The displacements of the medium, during the propagation of light, will produce a disturbance of the vortices, and the vortices when so disturbed may react on the medium so as to affect the mode of propagation of the ray.

823.] It is impossible, in our present state of ignorance as to the nature of the vortices, to assign the form of the law which connects the displacement of the medium with the variation of the vortices. We shall therefore assume that the variation of the vortices caused by the displacement of the medium is subject to the same conditions which Helmholtz, in his great memoir on Vortex-motion *, has shewn to regulate the variation of the vortices of a perfect liquid.

Helmholtz's law may be stated as follows:—Let P and Q be two neighbouring particles in the axis of a vortex, then, if in consequence of the motion of the fluid these particles arrive at the points $P'Q'$, the line $P'Q'$ will represent the new direction of the axis of the vortex, and its strength will be altered in the ratio of $P'Q'$ to PQ.

Hence if α, β, γ denote the components of the strength of a vortex, and if ξ, η, ζ denote the displacements of the medium, the value of α will become

$$\left.\begin{aligned} \alpha' &= \alpha + \alpha\frac{d\xi}{dx} + \beta\frac{d\xi}{dy} + \gamma\frac{d\xi}{dz}, \\ \beta' &= \beta + \alpha\frac{d\eta}{dx} + \beta\frac{d\eta}{dy} + \gamma\frac{d\eta}{dz}, \\ \gamma' &= \gamma + \alpha\frac{d\zeta}{dx} + \beta\frac{d\zeta}{dy} + \gamma\frac{d\zeta}{dz}. \end{aligned}\right\} \qquad (1)$$

We now assume that the same condition is satisfied during the small displacements of a medium in which α, β, γ represent, not the components of the strength of an ordinary vortex, but the components of magnetic force.

824.] The components of the angular velocity of an element of the medium are

$$\left.\begin{aligned} \omega_1 &= \tfrac{1}{2}\frac{d}{dt}\left(\frac{d\zeta}{dy} - \frac{d\eta}{dz}\right), \\ \omega_2 &= \tfrac{1}{2}\frac{d}{dt}\left(\frac{d\xi}{dz} - \frac{d\zeta}{dx}\right), \\ \omega_3 &= \tfrac{1}{2}\frac{d}{dt}\left(\frac{d\eta}{dx} - \frac{d\xi}{dy}\right). \end{aligned}\right\} \qquad (2)$$

* *Crelle's Journal*, vol. lv. (1858). Translated by Tait, *Phil. Mag.*, July, 1867.

The next step in our hypothesis is the assumption that the kinetic energy of the medium contains a term of the form

$$2C(\alpha\omega_1+\beta\omega_2+\gamma\omega_3). \tag{3}$$

This is equivalent to supposing that the angular velocity acquired by the element of the medium during the propagation of light is a quantity which may enter into combination with that motion by which magnetic phenomena are explained.

In order to form the equations of motion of the medium, we must express its kinetic energy in terms of the velocity of its parts, the components of which are $\dot{\xi}$, $\dot{\eta}$, $\dot{\zeta}$. We therefore integrate by parts, and find

$$2C\iiint(\alpha\omega_1+\beta\omega_2+\gamma\omega_3)\,dx\,dy\,dz$$

$$=C\iint(\gamma\dot{\eta}-\beta\dot{\zeta})\,dy\,dz+C\iint(\alpha\dot{\zeta}-\gamma\dot{\xi})\,dz\,dx+C\iint(\beta\dot{\xi}-\alpha\dot{\eta})\,dx\,dy$$

$$+C\iiint\left\{\dot{\xi}\left(\frac{d\gamma}{dy}-\frac{d\beta}{dz}\right)+\dot{\eta}\left(\frac{d\alpha}{dz}-\frac{d\gamma}{dy}\right)+\dot{\zeta}\left(\frac{d\beta}{dx}-\frac{d\alpha}{dy}\right)\right\}dx\,dy\,dz. \tag{4}$$

The double integrals refer to the bounding surface, which may be supposed at an infinite distance. We may, therefore, while investigating what takes place in the interior of the medium, confine our attention to the triple integral.

825.] The part of the kinetic energy in unit of volume, expressed by this triple integral, may be written

$$4\pi C(\dot{\xi}u+\dot{\eta}v+\dot{\zeta}w), \tag{5}$$

where u, v, w are the components of the electric current as given in equations (E), Art. 607.

It appears from this that our hypothesis is equivalent to the assumption that the velocity of a particle of the medium whose components are $\dot{\xi}$, $\dot{\eta}$, $\dot{\zeta}$ is a quantity which may enter into combination with the electric current whose components are u, v, w.

826.] Returning to the expression under the sign of triple integration in (4), substituting for the values of α, β, γ, those of α', β', γ', as given by equations (1), and writing

$$\frac{d}{dh} \text{ for } \alpha\frac{d}{dx}+\beta\frac{d}{dy}+\gamma\frac{d}{dz}; \tag{6}$$

the expression under the sign of integration becomes

$$C\left\{\dot{\xi}\frac{d}{dh}\left(\frac{d\zeta}{dy}-\frac{d\eta}{dz}\right)+\dot{\eta}\frac{d}{dh}\left(\frac{d\xi}{dz}-\frac{d\zeta}{dx}\right)+\dot{\zeta}\frac{d}{dh}\left(\frac{d\eta}{dx}-\frac{d\xi}{dy}\right)\right\}. \tag{7}$$

In the case of waves in planes normal to the axis of z the displacements are functions of z and t only, so that $\frac{d}{dh} = \gamma \frac{d}{dz}$, and this expression is reduced to

$$C\gamma \left(\frac{d^2\xi}{dz^2}\dot{\eta} - \frac{d^2\eta}{dz^2}\dot{\xi}\right). \tag{8}$$

The kinetic energy per unit of volume, so far as it depends on the velocities of displacement, may now be written

$$T = \tfrac{1}{2}\rho(\dot{\xi}^2 + \dot{\eta}^2 + \dot{\zeta}^2) + C\gamma\left(\frac{d^2\xi}{dz^2}\dot{\eta} - \frac{d^2\eta}{dz^2}\dot{\xi}\right), \tag{9}$$

where ρ is the density of the medium.

827.] The components, X and Y, of the impressed force, referred to unit of volume, may be deduced from this by Lagrange's equations, Art. 564.

$$X = \rho\frac{d^2\xi}{dt^2} - C\gamma\frac{d^3\eta}{dz^2\,dt}, \tag{10}$$

$$Y = \rho\frac{d^2\eta}{dt^2} + C\gamma\frac{d^3\xi}{dz^2\,dt}. \tag{11}$$

These forces arise from the action of the remainder of the medium on the element under consideration, and must in the case of an isotropic medium be of the form indicated by Cauchy,

$$X = A_0\frac{d^2\xi}{dz^2} + A_1\frac{d^4\xi}{dz^4} + \&c., \tag{12}$$

$$Y = A_0\frac{d^2\eta}{dz^2} + A_1\frac{d^4\eta}{dz^4} + \&c. \tag{13}$$

828.] If we now take the case of a circularly-polarized ray for which

$$\xi = r\cos(nt - qz), \qquad \eta = r\sin(nt - qz), \tag{14}$$

we find for the kinetic energy in unit of volume

$$T = \tfrac{1}{2}\rho r^2 n^2 - C\gamma r^2 q^2 n; \tag{15}$$

and for the potential energy in unit of volume

$$V = r^2(A_0 q^2 - A_1 q^4 + \&c.)$$
$$= r^2 Q, \tag{16}$$

where Q is a function of q^2.

The condition of free propagation of the ray given in Art. 820, equation (6), is

$$\frac{dT}{dr} = \frac{dV}{dr}, \tag{17}$$

which gives

$$\rho n^2 - 2C\gamma q^2 n = Q, \tag{18}$$

whence the value of n may be found in terms of q.

But in the case of a ray of given wave-period, acted on by

magnetic force, what we want to determine is the value of $\frac{dq}{d\gamma}$, when n is constant, in terms of $\frac{dq}{dn}$, when γ is constant. Differentiating (18)

$$(2\rho n - 2C\gamma q^2)\,dn - \left(\frac{dQ}{dq} + 4C\gamma q n\right)dq - 2Cq^2 n\,d\gamma = 0. \tag{19}$$

We thus find

$$\frac{dq}{d\gamma} = -\frac{Cq^2 n}{\rho n - C\gamma q^2}\frac{dq}{dn}. \tag{20}$$

829.] If λ is the wave-length in air, and i the corresponding index of refraction in the medium,

$$q\lambda = 2\pi i, \qquad n\lambda = 2\pi v. \tag{21}$$

The change in the value of q, due to magnetic action, is in every case an exceedingly small fraction of its own value, so that we may write

$$q = q_0 + \frac{dq}{d\gamma}\gamma, \tag{22}$$

where q_0 is the value of q when the magnetic force is zero. The angle, θ, through which the plane of polarization is turned in passing through a thickness c of the medium, is half the sum of the positive and negative values of qc, the sign of the result being changed, because the sign of q is negative in equations (14). We thus obtain

$$\theta = -c\gamma\frac{dq}{d\gamma} \tag{23}$$

$$= \frac{4\pi C}{v\rho}c\gamma\frac{i^2}{\lambda^2}\left(i - \lambda\frac{di}{d\lambda}\right)\frac{1}{1 - 2\pi C\gamma\frac{i^2}{v\rho\lambda}}. \tag{24}$$

The second term of the denominator of this fraction is approximately equal to the angle of rotation of the plane of polarization during its passage through a thickness of the medium equal to half a wave-length. It is therefore in all actual cases a quantity which we may neglect in comparison with unity.

Writing

$$\frac{4\pi C}{v\rho} = m, \tag{25}$$

we may call m the coefficient of magnetic rotation for the medium, a quantity whose value must be determined by observation. It is found to be positive for most diamagnetic, and negative for some paramagnetic media. We have therefore as the final result of our theory

$$\theta = mc\gamma\frac{i^2}{\lambda^2}\left(i - \lambda\frac{di}{d\lambda}\right), \tag{26}$$

where θ is the angular rotation of the plane of polarization, m a constant determined by observation of the medium, γ the intensity of the magnetic force resolved in the direction of the ray, c the length of the ray within the medium, λ the wave-length of the light in air, and i its index of refraction in the medium.

830.] The only test to which this theory has hitherto been subjected, is that of comparing the values of θ for different kinds of light passing through the same medium and acted on by the same magnetic force.

This has been done for a considerable number of media by M. Verdet *, who has arrived at the following results :—

(1) The magnetic rotations of the planes of polarization of the rays of different colours follow approximately the law of the inverse square of the wave-length.

(2) The exact law of the phenomena is always such that the product of the rotation by the square of the wave-length increases from the least refrangible to the most refrangible end of the spectrum.

(3) The substances for which this increase is most sensible are also those which have the greatest dispersive power.

He also found that in the solution of tartaric acid, which of itself produces a rotation of the plane of polarization, the magnetic rotation is by no means proportional to the natural rotation.

In an addition to the same memoir † Verdet has given the results of very careful experiments on bisulphide of carbon and on creosote, two substances in which the departure from the law of the inverse square of the wave-length was very apparent. He has also compared these results with the numbers given by three different formulæ,

$$\text{(I)} \qquad \theta = mc\gamma \frac{i^2}{\lambda^2}\left(i - \lambda \frac{di}{d\lambda}\right);$$

$$\text{(II)} \qquad \theta = mc\gamma \frac{1}{\lambda^2}\left(i - \lambda \frac{di}{d\lambda}\right);$$

$$\text{(III)} \qquad \theta = mc\gamma \quad \left(i - \lambda \frac{di}{d\lambda}\right).$$

The first of these formulæ, (I), is that which we have already obtained in Art. 829, equation (26). The second, (II), is that which results from substituting in the equations of motion, Art. 826, equations (10), (11), terms of the form $\frac{d^3\eta}{dt^3}$ and $-\frac{d^3\xi}{dt^3}$, instead of $\frac{d^3\eta}{dz^2dt}$

* Recherches sur les propriétés optiques développées dans les corps transparents par l'action du magnétisme, 4me partie. *Comptes Rendus*, t. lvi. p. 630 (6 April, 1863).

† *Comptes Rendus*, lvii. p. 670 (19 Oct., 1863).

and $-\frac{d^3\xi}{dz^2 dt}$. I am not aware that this form of the equations has been suggested by any physical theory. The third formula, (III), results from the physical theory of M. C. Neumann *, in which the equations of motion contain terms of the form $\frac{d\eta}{dt}$ and $-\frac{d\xi}{dt}$ †.

It is evident that the values of θ given by the formula (III) are not even approximately proportional to the inverse square of the wave-length. Those given by the formulæ (I) and (II) satisfy this condition, and give values of θ which agree tolerably well with the observed values for media of moderate dispersive power. For bisulphide of carbon and creosote, however, the values given by (II) differ very much from those observed. Those given by (I) agree better with observation, but, though the agreement is somewhat close for bisulphide of carbon, the numbers for creosote still differ by quantities much greater than can be accounted for by any errors of observation.

Magnetic Rotation of the Plane of Polarization (from Verdet).

Bisulphide of Carbon at 24°.9 C.

Lines of the spectrum	*C*	*D*	*E*	*F*	*G*
Observed rotation	592	768	1000	1234	1704
Calculated by I.	589	760	1000	1234	1713
,, II.	606	772	1000	1216	1640
,, III.	943	967	1000	1034	1091

Rotation of the ray $E = 25°.28'$.

Creosote at 24°.3 C.

Lines of the spectrum	*C*	*D*	*E*	*F*	*G*
Observed rotation	573	758	1000	1241	1723
Calculated by I.	617	780	1000	1210	1603
,, II.	623	789	1000	1200	1565
,, III.	976	993	1000	1017	1041

Rotation of the ray $E = 21°.58'$.

We are so little acquainted with the details of the molecular

* 'Explicare tentatur quomodo fiat ut lucis planum polarizationis per vires electricas vel magneticas declinetur.' *Halis Saxonum*, 1858.

† These three forms of the equations of motion were first suggested by Sir G. B. Airy (*Phil. Mag.*, June 1846) as a means of analysing the phenomenon then recently discovered by Faraday. Mac Cullagh had previously suggested equations containing terms of the form $\frac{d^3}{dz^3}$ in order to represent mathematically the phenomena of quartz. These equations were offered by Mac Cullagh and Airy, 'not as giving a mechanical explanation of the phenomena, but as shewing that the phenomena may be explained by equations, which equations appear to be such as might possibly be deduced from some plausible mechanical assumption, although no such assumption has yet been made.'

constitution of bodies, that it is not probable that any satisfactory theory can be formed relating to a particular phenomenon, such as that of the magnetic action on light, until, by an induction founded on a number of different cases in which visible phenomena are found to depend upon actions in which the molecules are concerned, we learn something more definite about the properties which must be attributed to a molecule in order to satisfy the conditions of observed facts.

The theory proposed in the preceding pages is evidently of a provisional kind, resting as it does on unproved hypotheses relating to the nature of molecular vortices, and the mode in which they are affected by the displacement of the medium. We must therefore regard any coincidence with observed facts as of much less scientific value in the theory of the magnetic rotation of the plane of polarization than in the electromagnetic theory of light, which, though it involves hypotheses about the electric properties of media, does not speculate as to the constitution of their molecules.

831.] NOTE.—The whole of this chapter may be regarded as an expansion of the exceedingly important remark of Sir William Thomson in the *Proceedings of the Royal Society*, June 1856:—'The magnetic influence on light discovered by Faraday depends on the direction of motion of moving particles. For instance, in a medium possessing it, particles in a straight line parallel to the lines of magnetic force, displaced to a helix round this line as axis, and then projected tangentially with such velocities as to describe circles, will have different velocities according as their motions are round in one direction (the same as the nominal direction of the galvanic current in the magnetizing coil), or in the contrary direction. But the elastic reaction of the medium must be the same for the same displacements, whatever be the velocities and directions of the particles; that is to say, the forces which are balanced by centrifugal force of the circular motions are equal, while the luminiferous motions are unequal. The absolute circular motions being therefore either equal or such as to transmit equal centrifugal forces to the particles initially considered, it follows that the luminiferous motions are only components of the whole motion; and that a less luminiferous component in one direction, compounded with a motion existing in the medium when transmitting no light, gives an equal resultant to that of a greater luminiferous motion in the contrary direction compounded with the same non-luminous motion. I think it is not only impossible to conceive any other than this dynamical explanation of the fact that circularly-polarized light transmitted through magnetized glass parallel to the lines of magnetizing force, with the same quality, right-handed always, or left-handed always, is propagated at different rates according as its course is in the direction or is contrary to the direction in which a north magnetic pole is drawn; but I believe it can be demonstrated that no other explanation of that fact is possible. Hence it appears that Faraday's optical discovery affords a demonstration of the reality of Ampère's explanation of the ultimate nature of magnetism; and gives a definition of magnetization in the dynamical theory of heat. The introduction of the principle of moments of momenta ("the conservation of areas") into the mechanical treatment of Mr. Rankine's hypothesis of "molecular vortices," appears to indicate a line perpendicular to the plane of resultant rotatory momentum ("the invariable plane") of the thermal motions as the magnetic axis of a magnetized body, and suggests the resultant moment of momenta of these motions as the definite measure of the "magnetic moment." The explanation of all phenomena of electromagnetic attraction or repulsion, and of electromagnetic induction, is to be looked for simply in the inertia and pressure of the matter of which the motions constitute heat. Whether this matter is or is not electricity, whether it is a continuous fluid interpermeating the spaces between molecular nuclei, or is itself molecularly grouped; or whether all matter is continuous, and molecular heterogeneousness consists in finite vortical or other relative motions of contiguous parts of a body; it is impossible to decide, and perhaps in vain to speculate, in the present state of science.'

A theory of molecular vortices, which I worked out at considerable length, was published in the *Phil. Mag.* for March, April, and May, 1861, Jan. and Feb. 1862.

I think we have good evidence for the opinion that some phenomenon of rotation is going on in the magnetic field, that this rotation is performed by a great number of very small portions of matter, each rotating on its own axis, this axis being parallel to the direction of the magnetic force, and that the rotations of these different vortices are made to depend on one another by means of some kind of mechanism connecting them.

The attempt which I then made to imagine a working model of this mechanism must be taken for no more than it really is, a demonstration that mechanism may be imagined capable of producing a connexion mechanically equivalent to the actual connexion of the parts of the electromagnetic field. The problem of determining the mechanism required to establish a given species of connexion between the motions of the parts of a system always admits of an infinite number of solutions. Of these, some may be more clumsy or more complex than others, but all must satisfy the conditions of mechanism in general.

The following results of the theory, however, are of higher value :—

(1) Magnetic force is the effect of the centrifugal force of the vortices.

(2) Electromagnetic induction of currents is the effect of the forces called into play when the velocity of the vortices is changing.

(3) Electromotive force arises from the stress on the connecting mechanism.

(4) Electric displacement arises from the elastic yielding of the connecting mechanism.

CHAPTER XXII.

FERROMAGNETISM AND DIAMAGNETISM EXPLAINED BY MOLECULAR CURRENTS.

On Electromagnetic Theories of Magnetism.

832.] We have seen (Art. 380) that the action of magnets on one another can be accurately represented by the attractions and repulsions of an imaginary substance called 'magnetic matter.' We have shewn the reasons why we must not suppose this magnetic matter to move from one part of a magnet to another through a sensible distance, as at first sight it appears to do when we magnetize a bar, and we were led to Poisson's hypothesis that the magnetic matter is strictly confined to single molecules of the magnetic substance, so that a magnetized molecule is one in which the opposite kinds of magnetic matter are more or less separated towards opposite poles of the molecule, but so that no part of either can ever be actually separated from the molecule (Art. 430).

These arguments completely establish the fact, that magnetization is a phenomenon, not of large masses of iron, but of molecules, that is to say, of portions of the substance so small that we cannot by any mechanical method cut one of them in two, so as to obtain a north pole separate from a south pole. But the nature of a magnetic molecule is by no means determined without further investigation. We have seen (Art. 442) that there are strong reasons for believing that the act of magnetizing iron or steel does not consist in imparting magnetization to the molecules of which it is composed, but that these molecules are already magnetic, even in unmagnetized iron, but with their axes placed indifferently in all directions, and that the act of magnetization consists in turning the molecules so that their axes are either rendered all parallel to one direction, or at least are deflected towards that direction.

833.] Still, however, we have arrived at no explanation of the nature of a magnetic molecule, that is, we have not recognized its likeness to any other thing of which we know more. We have therefore to consider the hypothesis of Ampère, that the magnetism of the molecule is due to an electric current constantly circulating in some closed path within it.

It is possible to produce an exact imitation of the action of any magnet on points external to it, by means of a sheet of electric currents properly distributed on its outer surface. But the action of the magnet on points in the interior is quite different from the action of the electric currents on corresponding points. Hence Ampère concluded that if magnetism is to be explained by means of electric currents, these currents must circulate within the molecules of the magnet, and must not flow from one molecule to another. As we cannot experimentally measure the magnetic action at a point in the interior of a molecule, this hypothesis cannot be disproved in the same way that we can disprove the hypothesis of currents of sensible extent within the magnet.

Besides this, we know that an electric current, in passing from one part of a conductor to another, meets with resistance and generates heat; so that if there were currents of the ordinary kind round portions of the magnet of sensible size, there would be a constant expenditure of energy required to maintain them, and a magnet would be a perpetual source of heat. By confining the circuits to the molecules, within which nothing is known about resistance, we may assert, without fear of contradiction, that the current, in circulating within the molecule, meets with no resistance.

According to Ampère's theory, therefore, all the phenomena of magnetism are due to electric currents, and if we could make observations of the magnetic force in the interior of a magnetic molecule, we should find that it obeyed exactly the same laws as the force in a region surrounded by any other electric circuit.

834.] In treating of the force in the interior of magnets, we have supposed the measurements to be made in a small crevasse hollowed out of the substance of the magnet, Art. 395. We were thus led to consider two different quantities, the magnetic force and the magnetic induction, both of which are supposed to be observed in a space from which the magnetic matter is removed. We were not supposed to be able to penetrate into the interior of a magnetic molecule and to observe the force within it.

If we adopt Ampère's theory, we consider a magnet, not as a

continuous substance, the magnetization of which varies from point to point according to some easily conceived law, but as a multitude of molecules, within each of which circulates a system of electric currents, giving rise to a distribution of magnetic force of extreme complexity, the direction of the force in the interior of a molecule being generally the reverse of that of the average force in its neighbourhood, and the magnetic potential, where it exists at all, being a function of as many degrees of multiplicity as there are molecules in the magnet.

835.] But we shall find, that, in spite of this apparent complexity, which, however, arises merely from the coexistence of a multitude of simpler parts, the mathematical theory of magnetism is greatly simplified by the adoption of Ampère's theory, and by extending our mathematical vision into the interior of the molecules.

In the first place, the two definitions of magnetic force are reduced to one, both becoming the same as that for the space outside the magnet. In the next place, the components of the magnetic force everywhere satisfy the condition to which those of induction are subject, namely,

$$\frac{d\alpha}{dx}+\frac{d\beta}{dy}+\frac{d\gamma}{dz}=0. \tag{1}$$

In other words, the distribution of magnetic force is of the same nature as that of the velocity of an incompressible fluid, or, as we have expressed it in Art. 25, the magnetic force has no convergence.

Finally, the three vector functions—the electromagnetic momentum, the magnetic force, and the electric current—become more simply related to each other. They are all vector functions of no convergence, and they are derived one from the other in order, by the same process of taking the space-variation, which is denoted by Hamilton by the symbol ∇.

836.] But we are now considering magnetism from a physical point of view, and we must enquire into the physical properties of the molecular currents. We assume that a current is circulating in a molecule, and that it meets with no resistance. If L is the coefficient of self-induction of the molecular circuit, and M the coefficient of mutual induction between this circuit and some other circuit, then if γ is the current in the molecule, and γ' that in the other circuit, the equation of the current γ is

$$\frac{d}{dt}(L\gamma+M\gamma')=-R\gamma; \tag{2}$$

and since by the hypothesis there is no resistance, $R = 0$, and we get by integration

$$L\gamma + M\gamma' = \text{constant}, = L\gamma_0, \text{ say.} \qquad (3)$$

Let us suppose that the area of the projection of the molecular circuit on a plane perpendicular to the axis of the molecule is A, this axis being defined as the normal to the plane on which the projection is greatest. If the action of other currents produces a magnetic force, X, in a direction whose inclination to the axis of the molecule is θ, the quantity $M\gamma'$ becomes $XA\cos\theta$, and we have as the equation of the current

$$L\gamma + XA\cos\theta = L\gamma_0, \qquad (4)$$

where γ_0 is the value of γ when $X = 0$.

It appears, therefore, that the strength of the molecular current depends entirely on its primitive value γ_0, and on the intensity of the magnetic force due to other currents.

837.] If we suppose that there is no primitive current, but that the current is entirely due to induction, then

$$\gamma = -\frac{XA}{L}\cos\theta. \qquad (5)$$

The negative sign shews that the direction of the induced current is opposite to that of the inducing current, and its magnetic action is such that in the interior of the circuit it acts in the opposite direction to the magnetic force. In other words, the molecular current acts like a small magnet whose poles are turned towards the poles of the same name of the inducing magnet.

Now this is an action the reverse of that of the molecules of iron under magnetic action. The molecular currents in iron, therefore, are not excited by induction. But in diamagnetic substances an action of this kind is observed, and in fact this is the explanation of diamagnetic polarity which was first given by Weber.

Weber's Theory of Diamagnetism.

838.] According to Weber's theory, there exist in the molecules of diamagnetic substances certain channels round which an electric current can circulate without resistance. It is manifest that if we suppose these channels to traverse the molecule in every direction, this amounts to making the molecule a perfect conductor.

Beginning with the assumption of a linear circuit within the molecule, we have the strength of the current given by equation (5).

The magnetic moment of the current is the product of its strength by the area of the circuit, or γA, and the resolved part of this in the direction of the magnetizing force is $\gamma A \cos\theta$, or, by (5),

$$-\frac{XA^2}{L}\cos^2\theta. \tag{6}$$

If there are n such molecules in unit of volume, and if their axes are distributed indifferently in all directions, then the average value of $\cos^2\theta$ will be $\frac{1}{3}$, and the intensity of magnetization of the substance will be

$$-\tfrac{1}{3}\frac{nXA^2}{L}. \tag{7}$$

Neumann's coefficient of magnetization is therefore

$$\kappa = -\tfrac{1}{3}\frac{nA^2}{L}. \tag{8}$$

The magnetization of the substance is therefore in the opposite direction to the magnetizing force, or, in other words, the substance is diamagnetic. It is also exactly proportional to the magnetizing force, and does not tend to a finite limit, as in the case of ordinary magnetic induction. See Arts. 442, &c.

839.] If the directions of the axes of the molecular channels are arranged, not indifferently in all directions, but with a preponderating number in certain directions, then the sum

$$\Sigma\frac{A^2}{L}\cos^2\theta$$

extended to all the molecules will have different values according to the direction of the line from which θ is measured, and the distribution of these values in different directions will be similar to the distribution of the values of moments of inertia about axes in different directions through the same point.

Such a distribution will explain the magnetic phenomena related to axes in the body, described by Plücker, which Faraday has called Magne-crystallic phenomena. See Art. 435.

840.] Let us now consider what would be the effect, if, instead of the electric current being confined to a certain channel within the molecule, the whole molecule were supposed a perfect conductor.

Let us begin with the case of a body the form of which is acyclic, that is to say, which is not in the form of a ring or perforated body, and let us suppose that this body is everywhere surrounded by a thin shell of perfectly conducting matter.

We have proved in Art. 654, that a closed sheet of perfectly conducting matter of any form, originally free from currents, becomes, when exposed to external magnetic force, a current-sheet, the action of which on every point of the interior is such as to make the magnetic force zero.

It may assist us in understanding this case if we observe that the distribution of magnetic force in the neighbourhood of such a body is similar to the distribution of velocity in an incompressible fluid in the neighbourhood of an impervious body of the same form.

It is obvious that if other conducting shells are placed within the first, since they are not exposed to magnetic force, no currents will be excited in them. Hence, in a solid of perfectly conducting material, the effect of magnetic force is to generate a system of currents which are entirely confined to the surface of the body.

841.] If the conducting body is in the form of a sphere of radius r, its magnetic moment is

$$-\tfrac{1}{2} r^3 X,$$

and if a number of such spheres are distributed in a medium, so that in unit of volume the volume of the conducting matter is k', then, by putting $\mu_1 = 1$, and $\mu_2 = 0$ in equation (17), Art. 314, we find the coefficient of magnetic permeability,

$$\mu = \frac{2 - 2k'}{2 + k'}, \tag{9}$$

whence we obtain for Poisson's magnetic coefficient

$$k = -\tfrac{1}{2} k', \tag{10}$$

and for Neumann's coefficient of magnetization by induction

$$\kappa = -\frac{3}{4\pi} \frac{k'}{2 + k'}. \tag{11}$$

Since the mathematical conception of perfectly conducting bodies leads to results exceedingly different from any phenomena which we can observe in ordinary conductors, let us pursue the subject somewhat further.

842.] Returning to the case of the conducting channel in the form of a closed curve of area A, as in Art. 836, we have, for the moment of the electromagnetic force tending to increase the angle θ,

$$\gamma\gamma' \frac{dM}{d\theta} = -\gamma X A \sin\theta \tag{12}$$

$$= \frac{X^2 A^2}{L} \sin\theta \cos\theta. \tag{13}$$

This force is positive or negative according as θ is less or greater than a right angle. Hence the effect of magnetic force on a perfectly conducting channel tends to turn it with its axis at right angles to the line of magnetic force, that is, so that the plane of the channel becomes parallel to the lines of force.

An effect of a similar kind may be observed by placing a penny or a copper ring between the poles of an electromagnet. At the instant that the magnet is excited the ring turns its plane towards the axial direction, but this force vanishes as soon as the currents are deadened by the resistance of the copper *.

843.] We have hitherto considered only the case in which the molecular currents are entirely excited by the external magnetic force. Let us next examine the bearing of Weber's theory of the magneto-electric induction of molecular currents on Ampère's theory of ordinary magnetism. According to Ampère and Weber, the molecular currents in magnetic substances are not excited by the external magnetic force, but are already there, and the molecule itself is acted on and deflected by the electromagnetic action of the magnetic force on the conducting circuit in which the current flows. When Ampère devised this hypothesis, the induction of electric currents was not known, and he made no hypothesis to account for the existence, or to determine the strength, of the molecular currents.

We are now, however, bound to apply to these currents the same laws that Weber applied to his currents in diamagnetic molecules. We have only to suppose that the primitive value of the current γ, when no magnetic force acts, is not zero but γ_0. The strength of the current when a magnetic force, X, acts on a molecular current of area A, whose axis is inclined θ to the line of magnetic force, is

$$\gamma = \gamma_0 - \frac{XA}{L}\cos\theta, \tag{14}$$

and the moment of the couple tending to turn the molecule so as to increase θ is

$$-\gamma_0 XA\sin\theta + \frac{X^2A^2}{2L}\sin 2\theta. \tag{15}$$

Hence, putting

$$A\gamma_0 = m, \qquad \frac{A}{L\gamma_0} = B, \tag{16}$$

in the investigation in Art. 443, the equation of equilibrium becomes

$$X\sin\theta - BX^2\sin\theta\cos\theta = D\sin(a-\theta). \tag{17}$$

The resolved part of the magnetic moment of the current in the direction of X is

$$\gamma A\cos\theta = \gamma_0 A\cos\theta - \frac{XA^2}{L}\cos^2\theta, \tag{18}$$

$$= m\cos\theta(1 - BX\cos\theta). \tag{19}$$

* See Faraday, *Exp. Res.*, 2310, &c.

844.] These conditions differ from those in Weber's theory of magnetic induction by the terms involving the coefficient B. If BX is small compared with unity, the results will approximate to those of Weber's theory of magnetism. If BX is large compared with unity, the results will approximate to those of Weber's theory of diamagnetism.

Now the greater γ_0, the primitive value of the molecular current, the smaller will B become, and if L is also large, this will also diminish B. Now if the current flows in a ring channel, the value of L depends on $\log \frac{R}{r}$, where R is the radius of the mean line of the channel, and r that of its section. The smaller therefore the section of the channel compared with its area, the greater will be L, the coefficient of self-induction, and the more nearly will the phenomena agree with Weber's original theory. There will be this difference, however, that as X, the magnetizing force, increases, the temporary magnetic moment will not only reach a maximum, but will afterwards diminish as X increases.

If it should ever be experimentally proved that the temporary magnetization of any substance first increases, and then diminishes as the magnetizing force is continually increased, the evidence of the existence of these molecular currents would, I think, be raised almost to the rank of a demonstration.

845.] If the molecular currents in diamagnetic substances are confined to definite channels, and if the molecules are capable of being deflected like those of magnetic substances, then, as the magnetizing force increases, the diamagnetic polarity will always increase, but, when the force is great, not quite so fast as the magnetizing force. The small absolute value of the diamagnetic coefficient shews, however, that the deflecting force on each molecule must be small compared with that exerted on a magnetic molecule, so that any result due to this deflexion is not likely to be perceptible.

If, on the other hand, the molecular currents in diamagnetic bodies are free to flow through the whole substance of the molecules, the diamagnetic polarity will be strictly proportional to the magnetizing force, and its amount will lead to a determination of the whole space occupied by the perfectly conducting masses, and, if we know the number of the molecules, to the determination of the size of each.

CHAPTER XXIII.

THEORIES OF ACTION AT A DISTANCE.

On the Explanation of Ampère's Formula given by Gauss and Weber.

846.] The attraction between the elements ds and ds' of two circuits, carrying electric currents of intensity i and i', is, by Ampère's formula,

$$-\frac{ii'\,ds\,ds'}{r^2}\left(2\cos\epsilon+3\frac{dr}{ds}\frac{dr}{ds'}\right); \tag{1}$$

or

$$\frac{ii'\,ds\,ds'}{r^2}\left(2r\frac{d^2r}{ds\,ds'}-\frac{dr}{ds}\frac{dr}{ds'}\right); \tag{2}$$

the currents being estimated in electromagnetic units. See Art. 526.

The quantities, whose meaning as they appear in these expressions we have now to interpret, are

$$\cos\epsilon,\quad \frac{dr}{ds}\frac{dr}{ds'},\quad \text{and}\quad \frac{d^2r}{ds\,ds'};$$

and the most obvious phenomenon in which to seek for an interpretation founded on a direct relation between the currents is the relative velocity of the electricity in the two elements.

847.] Let us therefore consider the relative motion of two particles, moving with constant velocities v and v' along the elements ds and ds' respectively. The square of the relative velocity of these particles is

$$u^2=v^2-2vv'\cos\epsilon+v'^2; \tag{3}$$

and if we denote by r the distance between the particles,

$$\frac{\partial r}{\partial t}=v\frac{dr}{ds}+v'\frac{dr}{ds'}, \tag{4}$$

$$\left(\frac{\partial r}{\partial t}\right)^2=v^2\left(\frac{dr}{ds}\right)^2+2vv'\frac{dr}{ds}\frac{dr}{ds'}+v'^2\left(\frac{dr}{ds'}\right)^2, \tag{5}$$

$$\frac{\partial^2 r}{\partial t^2}=v^2\frac{d^2r}{ds^2}+2vv'\frac{d^2r}{ds\,ds'}+v'^2\frac{d^2r}{ds'^2}, \tag{6}$$

where the symbol ∂ indicates that, in the quantity differentiated, the coordinates of the particles are to be expressed in terms of the time.

It appears, therefore, that the terms involving the product vv' in the equations (3), (5), and (6) contain the quantities occurring in (1) and (2) which we have to interpret. We therefore endeavour to express (1) and (2) in terms of u^2, $\overline{\frac{\partial r}{\partial t}}\Big|^2$, and $\frac{\partial^2 r}{\partial t^2}\cdot$ But in order to do so we must get rid of the first and third terms of each of these expressions, for they involve quantities which do not appear in the formula of Ampère. Hence we cannot explain the electric current as a transfer of electricity in one direction only, but we must combine two opposite streams in each current, so that the combined effect of the terms involving v^2 and v'^2 may be zero.

848.] Let us therefore suppose that in the first element, ds, we have one electric particle, e, moving with velocity v, and another, e_1, moving with velocity v_1, and in the same way two particles, e' and e'_1, in ds', moving with velocities v' and v'_1 respectively.

The term involving v^2 for the combined action of these particles is

$$\Sigma(v^2ee') = (v^2e + v_1{}^2e_1)(e' + e'_1). \tag{7}$$

Similarly

$$\Sigma(v'^2ee') = (v'^2e' + v'_1{}^2e'_1)(e + e_1); \tag{8}$$

and

$$\Sigma(vv'ee') = (ve + v_1e_1)(v'e' + v'_1e'_1). \tag{9}$$

In order that $\Sigma(v^2ee')$ may be zero, we must have either

$$e' + e'_1 = 0, \quad \text{or} \quad v^2e + v_1{}^2e_1 = 0. \tag{10}$$

According to Fechner's hypothesis, the electric current consists of a current of positive electricity in the positive direction, combined with a current of negative electricity in the negative direction, the two currents being exactly equal in numerical magnitude, both as respects the quantity of electricity in motion and the velocity with which it is moving. Hence both the conditions of (10) are satisfied by Fechner's hypothesis.

But it is sufficient for our purpose to assume, either—

That the quantity of positive electricity in each element is numerically equal to the quantity of negative electricity; or—

That the quantities of the two kinds of electricity are inversely as the squares of their velocities.

Now we know that by charging the second conducting wire as a whole, we can make $e' + e'_1$ either positive or negative. Such a charged wire, even without a current, according to this formula, would act on the first wire carrying a current in which $v^2e + v_1{}^2e_1$

has a value differing from zero. Such an action has never been observed.

Therefore, since the quantity $e'+e'_1$ may be shewn experimentally not to be always zero, and since the quantity $v^2e+v^2{}_1e_1$ is not capable of being experimentally tested, it is better for these speculations to assume that it is the latter quantity which invariably vanishes.

849.] Whatever hypothesis we adopt, there can be no doubt that the total transfer of electricity, reckoned algebraically, along the first circuit, is represented by

$$ve+v_1e_1 = ci\,ds;$$

where c is the number of units of statical electricity which are transmitted by the unit electric current in the unit of time, so that we may write equation (9)

$$\Sigma(vv'ee') = c^2ii'ds\,ds'. \tag{11}$$

Hence the sums of the four values of (3), (5), and (6) become

$$\Sigma(ee'u^2) = -2c^2ii'ds\,ds'\cos\epsilon; \tag{12}$$

$$\Sigma\left(ee'\left(\frac{\partial r}{\partial t}\right)^2\right) = 2c^2ii'ds\,ds'\frac{dr}{ds}\frac{dr}{ds'}, \tag{13}$$

$$\Sigma\left(ee'r\frac{\partial^2 r}{\partial t^2}\right) = 2c^2ii'ds\,ds'r\frac{d^2r}{ds\,ds'}, \tag{14}$$

and we may write the two expressions (1) and (2) for the attraction between ds and ds'

$$-\frac{1}{c^2}\Sigma\left[\frac{ee'}{r^2}\left(u^2-\tfrac{3}{2}\left(\frac{\partial r}{\partial t}\right)^2\right)\right], \tag{15}$$

and

$$\frac{1}{c^2}\Sigma\left[\frac{ee'}{r^2}\left(r\frac{\partial^2 r}{\partial t^2}-\tfrac{1}{2}\left(\frac{\partial r}{\partial t}\right)^2\right)\right]. \tag{16}$$

850.] The ordinary expression, in the theory of statical electricity, for the repulsion of two electrical particles e and e' is $\frac{ee'}{r^2}$, and

$$\Sigma\left(\frac{ee'}{r^2}\right) = \frac{(e+e_1)(e'+e'_1)}{r^2}, \tag{17}$$

which gives the electrostatic repulsion between the two elements if they are charged as wholes.

Hence, if we assume for the repulsion of the two particles either of the modified expressions

$$\frac{ee'}{r^2}\left[1+\frac{1}{c^2}\left(u^2-\tfrac{3}{2}\left(\frac{\partial r}{\partial t}\right)^2\right)\right], \tag{18}$$

or

$$\frac{ee'}{r^2}\left[1+\frac{1}{c^2}\left(r\frac{\partial^2 r}{\partial t^2}-\tfrac{1}{2}\left(\frac{\partial r}{\partial t}\right)^2\right)\right], \tag{19}$$

we may deduce from them both the ordinary electrostatic forces, and the forces acting between currents as determined by Ampère.

851.] The first of these expressions, (18), was discovered by Gauss* in July 1835, and interpreted by him as a fundamental law of electrical action, that 'Two elements of electricity in a state of relative motion attract or repel one another, but not in the same way as if they are in a state of relative rest.' This discovery was not, so far as I know, published in the lifetime of Gauss, so that the second expression, which was discovered independently by W. Weber, and published in the first part of his celebrated *Elektrodynamische Maasbestimmungen*†, was the first result of the kind made known to the scientific world.

852.] The two expressions lead to precisely the same result when they are applied to the determination of the mechanical force between two electric currents, and this result is identical with that of Ampère. But when they are considered as expressions of the physical law of the action between two electrical particles, we are led to enquire whether they are consistent with other known facts of nature.

Both of these expressions involve the relative velocity of the particles. Now, in establishing by mathematical reasoning the well-known principle of the conservation of energy, it is generally assumed that the force acting between two particles is a function of the distance only, and it is commonly stated that if it is a function of anything else, such as the time, or the velocity of the particles, the proof would not hold.

Hence a law of electrical action, involving the velocity of the particles, has sometimes been supposed to be inconsistent with the principle of the conservation of energy.

853.] The formula of Gauss is inconsistent with this principle, and must therefore be abandoned, as it leads to the conclusion that energy might be indefinitely generated in a finite system by physical means. This objection does not apply to the formula of Weber, for he has shewn‡ that if we assume as the potential energy of a system consisting of two electric particles,

$$\psi = \frac{ee'}{r}\left[1 - \frac{1}{2c^2}\left(\frac{\partial r}{\partial t}\right)^2\right], \tag{20}$$

the repulsion between them, which is found by differentiating this quantity with respect to r, and changing the sign, is that given by the formula (19).

* *Werke* (Göttingen edition, 1867), vol. v. p. 616.
† *Abh. Leibnizens Ges.*, Leipzig (1846).
‡ *Pogg. Ann.*, lxxiii. p. 229 (1848).

Hence the work done on a moving particle by the repulsion of a fixed particle is $\psi_0 - \psi_1$, where ψ_0 and ψ_1 are the values of ψ at the beginning and at the end of its path. Now ψ depends only on the distance, r, and on the velocity resolved in the direction of r. If, therefore, the particle describes any closed path, so that its position, velocity, and direction of motion are the same at the end as at the beginning, ψ_1 will be equal to ψ_0, and no work will be done on the whole during the cycle of operations.

Hence an indefinite amount of work cannot be generated by a particle moving in a periodic manner under the action of the force assumed by Weber.

854.] But Helmholtz, in his very powerful memoir on the 'Equations of Motion of Electricity in Conductors at Rest'*, while he shews that Weber's formula is not inconsistent with the principle of the conservation of energy, as regards only the work done during a complete cyclical operation, points out that it leads to the conclusion, that two electrified particles, which move according to Weber's law, may have at first finite velocities, and yet, while still at a finite distance from each other, they may acquire an infinite kinetic energy, and may perform an infinite amount of work.

To this Weber† replies, that the initial relative velocity of the particles in Helmholtz's example, though finite, is greater than the velocity of light; and that the distance at which the kinetic energy becomes infinite, though finite, is smaller than any magnitude which we can perceive, so that it may be physically impossible to bring two molecules so near together. The example, therefore, cannot be tested by any experimental method.

Helmholtz‡ has therefore stated a case in which the distances are not too small, nor the velocities too great, for experimental verification. A fixed non-conducting spherical surface, of radius a, is uniformly charged with electricity to the surface-density σ. A particle, of mass m and carrying a charge e of electricity, moves within the sphere with velocity v. The electrodynamic potential calculated from the formula (20) is

$$4\pi a\sigma e\left(1-\frac{v^2}{6c^2}\right), \tag{21}$$

and is independent of the position of the particle within the sphere. Adding to this V, the remainder of the potential energy arising

* *Crelle's Journal*, 72 (1870).

† *Elektr. Maasb. insbesondere über das Princip der Erhaltung der Energie.*

‡ *Berlin Monatsbericht*, April 1872; *Phil. Mag.*, Dec. 1872, *Supp.*

from the action of other forces, and $\frac{1}{2}mv^2$, the kinetic energy of the particle, we find as the equation of energy

$$\tfrac{1}{2}\left(m - \tfrac{4}{3}\frac{\pi a \sigma e}{c^2}\right)v^2 + 4\pi a \sigma e + V = \text{const.} \qquad (22)$$

Since the second term of the coefficient of v^2 may be increased indefinitely by increasing a, the radius of the sphere, while the surface-density σ remains constant, the coefficient of v^2 may be made negative. Acceleration of the motion of the particle would then correspond to diminution of its *vis viva*, and a body moving in a closed path and acted on by a force like friction, always opposite in direction to its motion, would continually increase in velocity, and that without limit. This impossible result is a necessary consequence of assuming any formula for the potential which introduces negative terms into the coefficient of v^2.

855.] But we have now to consider the application of Weber's theory to phenomena which can be realized. We have seen how it gives Ampère's expression for the force of attraction between two elements of electric currents. The potential of one of these elements on the other is found by taking the sum of the values of the potential ψ for the four combinations of the positive and negative currents in the two elements. The result is, by equation (20), taking the sum of the four values of $\left.\frac{dr}{dt}\right|^2$,

$$-ii'ds\,ds'\frac{1}{r}\frac{dr}{ds}\frac{dr}{ds'}, \qquad (23)$$

and the potential of one closed current on another is

$$-ii'\iint\frac{1}{r}\frac{dr}{ds}\frac{dr}{ds'}ds\,ds' = ii'M, \qquad (24)$$

where $M = \iint\frac{\cos e}{r}ds\,ds'$, as in Arts. 423, 524.

In the case of closed currents, this expression agrees with that which we have already (Art. 524) obtained*.

Weber's Theory of the Induction of Electric Currents.

856.] After deducing from Ampère's formula for the action between the elements of currents, his own formula for the action between moving electric particles, Weber proceeded to apply his formula to the explanation of the production of electric currents by

* In the whole of this investigation Weber adopts the electrodynamic system of units. In this treatise we always use the electromagnetic system. The electromagnetic unit of current is to the electrodynamic unit in the ratio of $\sqrt{2}$ to 1. Art. 526.

magneto-electric induction. In this he was eminently successful, and we shall indicate the method by which the laws of induced currents may be deduced from Weber's formula. But we must observe, that the circumstance that a law deduced from the phenomena discovered by Ampère is able also to account for the phenomena afterwards discovered by Faraday does not give so much additional weight to the evidence for the physical truth of the law as we might at first suppose.

For it has been shewn by Helmholtz and Thomson (see Art. 543), that if the phenomena of Ampère are true, and if the principle of the conservation of energy is admitted, then the phenomena of induction discovered by Faraday follow of necessity. Now Weber's law, with the various assumptions about the nature of electric currents which it involves, leads by mathematical transformations to the formula of Ampère. Weber's law is also consistent with the principle of the conservation of energy in so far that a potential exists, and this is all that is required for the application of the principle by Helmholtz and Thomson. Hence we may assert, even before making any calculations on the subject, that Weber's law will explain the induction of electric currents. The fact, therefore, that it is found by calculation to explain the induction of currents, leaves the evidence for the physical truth of the law exactly where it was.

On the other hand, the formula of Gauss, though it explains the phenomena of the attraction of currents, is inconsistent with the principle of the conservation of energy, and therefore we cannot assert that it will explain all the phenomena of induction. In fact, it fails to do so, as we shall see in Art. 859.

857.] We must now consider the electromotive force tending to produce a current in the element ds', due to the current in ds, when ds is in motion, and when the current in it is variable.

According to Weber, the action on the material of the conductor of which ds' is an element, is the *sum* of all the actions on the electricity which it carries. The electromotive force, on the other hand, on the electricity in ds', is the *difference* of the electric forces acting on the positive and the negative electricity within it. Since all these forces act in the line joining the elements, the electromotive force on ds' is also in this line, and in order to obtain the electromotive force in the direction of ds' we must resolve the force in that direction. To apply Weber's formula, we must calculate the various terms which occur in it, on the supposition that the element ds is in motion relatively to ds', and that the currents in both elements vary with the time. The expressions thus found will contain terms involving v^2, vv', v'^2, v, v', and terms not involving v or v', all of which are multiplied by ee'. Examining, as we did before, the four values of each term, and considering first the mechanical force which arises from the sum of the four values, we find that the only term which we must take into account is that involving the product $vv'ee'$.

If we then consider the force tending to produce a current in the second element, arising from the difference of the action of the first element on the positive and the negative electricity of the second element, we find that the only term which we have to examine is that which involves vee'. We may write the four terms included in $\Sigma(vee')$, thus

$$e'(ve+v_1 e_1) \quad \text{and} \quad e'_1(ve+v_1 e_1).$$

Since $e'+e'_1 = 0$, the mechanical force arising from these terms is zero, but the electromotive force acting on the positive electricity e' is $(ve+v_1 e_1)$, and that acting on the negative electricity e'_1 is equal and opposite to this.

858.] Let us now suppose that the first element ds is moving relatively to ds' with velocity V in a certain direction, and let us denote by $\widehat{Vds}$ and $\widehat{Vds'}$, the angle between the direction of V and that of ds and of ds' respectively, then the square of the relative velocity, u, of two electric particles is

$$u^2 = v^2+v'^2+V^2-2vv'\cos\epsilon+2Vv\cos\widehat{Vds}-2Vv'\cos\widehat{Vds'}. \quad (25)$$

The term in vv' is the same as in equation (3). That in v, on which the electromotive force depends, is

$$2Vv\cos\widehat{Vds}.$$

We have also for the value of the time-variation of r in this case

$$\frac{\partial r}{\partial t} = v\frac{dr}{ds}+v'\frac{dr}{ds'}+\frac{dr}{dt}, \quad (26)$$

where $\frac{\partial r}{\partial t}$ refers to the motion of the electric particles, and $\frac{dr}{dt}$ to that of the material conductor. If we form the square of this quantity, the term involving vv', on which the mechanical force depends, is the same as before, in equation (5), and that involving v, on which the electromotive force depends, is

$$2v\frac{dr}{ds}\frac{dr}{dt}.$$

Differentiating (26) with respect to t, we find

$$\frac{\partial^2 r}{\partial t^2} = v^2\frac{d^2 r}{ds^2} + 2vv'\frac{d^2 r}{ds\,ds'} + v'^2\frac{d^2 r}{ds'^2} + \frac{dv}{dt}\frac{dr}{ds} + \frac{dv'}{dt}\frac{dr}{ds'} \qquad (27)$$
$$+ v\frac{dv}{ds}\frac{dr}{ds} + v'\frac{dv'}{ds}\frac{dr}{ds'} + \frac{d^2 r}{dt^2}.$$

We find that the term involving vv' is the same as before in (6). The term whose sign alters with that of v is $\frac{dv}{dt}\frac{dr}{ds}$.

859.] If we now calculate by the formula of Gauss (equation (18)), the resultant electrical force in the direction of the second element ds', arising from the action of the first element ds, we obtain

$$\frac{1}{r^2}\,ds\,ds'\,iV(2\cos\widehat{V ds} - 3\cos\widehat{Vr}\cos\widehat{r\,ds})\cos\widehat{r\,ds'}. \qquad (28)$$

As in this expression there is no term involving the rate of variation of the current i, and since we know that the variation of the primary current produces an inductive action on the secondary circuit, we cannot accept the formula of Gauss as a true expression of the action between electric particles.

860.] If, however, we employ the formula of Weber, (19), we obtain

$$\frac{1}{r^2}\,ds\,ds'\left(r\frac{dr}{ds}\frac{di}{dt} - i\frac{dr}{ds}\frac{dr}{dt}\right)\frac{dr}{ds'}, \qquad (29)$$

or

$$\frac{dr}{ds}\frac{dr}{ds'}\frac{d}{dt}\left(\frac{i}{r}\right)ds\,ds'. \qquad (30)$$

If we integrate this expression with respect to s and s', we obtain for the electromotive force on the second circuit

$$\frac{d}{dt}\,i\iint\frac{1}{r}\frac{dr}{ds}\frac{dr}{ds'}\,ds\,ds'. \qquad (31)$$

Now, when the first circuit is closed,

$$\int\frac{d^2 r}{ds\,ds'}\,ds = 0.$$

Hence $$\int\frac{1}{r}\frac{dr}{ds}\frac{dr}{ds'}\,ds = \int\left(\frac{1}{r}\frac{dr}{ds}\frac{dr}{ds'} + \frac{d^2 r}{ds\,ds'}\right)ds = -\int\frac{\cos\epsilon}{r}\,ds. \qquad (32)$$

But $$\iint\frac{\cos\epsilon}{r}\,ds\,ds' = M, \text{ by Arts. 423, 524.} \qquad (33)$$

Hence we may write the electromotive force on the second circuit

$$-\frac{d}{dt}(iM), \qquad (34)$$

which agrees with what we have already established by experiment; Art. 539.

On Weber's Formula, considered as resulting from an Action transmitted from one Electric Particle to the other with a Constant Velocity.

861.] In a very interesting letter of Gauss to W. Weber * he refers to the electrodynamic speculations with which he had been occupied long before, and which he would have published if he could then have established that which he considered the real keystone of electrodynamics, namely, the deduction of the force acting between electric particles in motion from the consideration of an action between them, not instantaneous, but propagated in time, in a similar manner to that of light. He had not succeeded in making this deduction when he gave up his electrodynamic researches, and he had a subjective conviction that it would be necessary in the first place to form a consistent representation of the manner in which the propagation takes place.

Three eminent mathematicians have endeavoured to supply this keystone of electrodynamics.

862.] In a memoir presented to the Royal Society of Göttingen in 1858, but afterwards withdrawn, and only published in Poggendorff's *Annalen* in 1867, after the death of the author, Bernhard Riemann deduces the phenomena of the induction of electric currents from a modified form of Poisson's equation

$$\frac{d^2V}{dx^2}+\frac{d^2V}{dy^2}+\frac{d^2V}{dz^2}+4\pi\rho=\frac{1}{a^2}\frac{d^2V}{dt^2},$$

where V is the electrostatic potential, and a a velocity.

This equation is of the same form as those which express the propagation of waves and other disturbances in elastic media. The author, however, seems to avoid making explicit mention of any medium through which the propagation takes place.

The mathematical investigation given by Riemann has been examined by Clausius †, who does not admit the soundness of the mathematical processes, and shews that the hypothesis that potential is propagated like light does not lead either to the formula of Weber, or to the known laws of electrodynamics.

863.] Clausius has also examined a far more elaborate investigation by C. Neumann on the 'Principles of Electrodynamics' ‡. Neumann, however, has pointed out § that his theory of the transmission of potential from one electric particle to another is quite different from that proposed by Gauss, adopted by Riemann, and criticized

* March 19, 1845, *Werke*, bd. v. 629.

† Pogg., bd. cxxxv. 612.

‡ Tübingen, 1868.

§ *Mathematische Annalen*, i. 317.

by Clausius, in which the propagation is like that of light. There is, on the contrary, the greatest possible difference between the transmission of potential, according to Neumann, and the propagation of light.

A luminous body sends forth light in all directions, the intensity of which depends on the luminous body alone, and not on the presence of the body which is enlightened by it.

An electric particle, on the other hand, sends forth a potential, the value of which, $\frac{ee'}{r}$, depends not only on e, the emitting particle, but on e', the receiving particle, and on the distance r between the particles *at the instant of emission*.

In the case of light the intensity diminishes as the light is propagated further from the luminous body; the emitted potential flows to the body on which it acts without the slightest alteration of its original value.

The light received by the illuminated body is in general only a fraction of that which falls on it; the potential as received by the attracted body is identical with, or equal to, the potential which arrives at it.

Besides this, the velocity of transmission of the potential is not, like that of light, constant relative to the æther or to space, but rather like that of a projectile, constant relative to the velocity of the emitting particle at the instant of emission.

It appears, therefore, that in order to understand the theory of Neumann, we must form a very different representation of the process of the transmission of potential from that to which we have been accustomed in considering the propagation of light. Whether it can ever be accepted as the 'construirbar Vorstellung' of the process of transmission, which appeared necessary to Gauss, I cannot say, but I have not myself been able to construct a consistent mental representation of Neumann's theory.

864.] Professor Betti *, of Pisa, has treated the subject in a different way. He supposes the closed circuits in which the electric currents flow to consist of elements each of which is polarized periodically, that is, at equidistant intervals of time. These polarized elements act on one another as if they were little magnets whose axes are in the direction of the tangent to the circuits. The periodic time of this polarization is the same in all electric circuits. Betti supposes the action of one polarized element on an-

* *Nuovo Cimento*, xxvii (1868).

other at a distance to take place, not instantaneously, but after a time proportional to the distance between the elements. In this way he obtains expressions for the action of one electric circuit on another, which coincide with those which are known to be true. Clausius, however, has, in this case also, criticized some parts of the mathematical calculations into which we shall not here enter.

865.] There appears to be, in the minds of these eminent men, some prejudice, or *à priori* objection, against the hypothesis of a medium in which the phenomena of radiation of light and heat, and the electric actions at a distance take place. It is true that at one time those who speculated as to the causes of physical phenomena, were in the habit of accounting for each kind of action at a distance by means of a special æthereal fluid, whose function and property it was to produce these actions. They filled all space three and four times over with æthers of different kinds, the properties of which were invented merely to 'save appearances,' so that more rational enquirers were willing rather to accept not only Newton's definite law of attraction at a distance, but even the dogma of Cotes*, that action at a distance is one of the primary properties of matter, and that no explanation can be more intelligible than this fact. Hence the undulatory theory of light has met with much opposition, directed not against its failure to explain the phenomena, but against its assumption of the existence of a medium in which light is propagated.

866.] We have seen that the mathematical expressions for electrodynamic action led, in the mind of Gauss, to the conviction that a theory of the propagation of electric action in time would be found to be the very key-stone of electrodynamics. Now we are unable to conceive of propagation in time, except either as the flight of a material substance through space, or as the propagation of a condition of motion or stress in a medium already existing in space. In the theory of Neumann, the mathematical conception called Potential, which we are unable to conceive as a material substance, is supposed to be projected from one particle to another, in a manner which is quite independent of a medium, and which, as Neumann has himself pointed out, is extremely different from that of the propagation of light. In the theories of Riemann and Betti it would appear that the action is supposed to be propagated in a manner somewhat more similar to that of light.

But in all of these theories the question naturally occurs:—If

* Preface to Newton's *Principia*, 2nd edition.

something is transmitted from one particle to another at a distance, what is its condition after it has left the one particle and before it has reached the other? If this something is the potential energy of the two particles, as in Neumann's theory, how are we to conceive this energy as existing in a point of space, coinciding neither with the one particle nor with the other? In fact, whenever energy is transmitted from one body to another in time, there must be a medium or substance in which the energy exists after it leaves one body and before it reaches the other, for energy, as Torricelli* remarked, 'is a quintessence of so subtile a nature that it cannot be contained in any vessel except the inmost substance of material things.' Hence all these theories lead to the conception of a medium in which the propagation takes place, and if we admit this medium as an hypothesis, I think it ought to occupy a prominent place in our investigations, and that we ought to endeavour to construct a mental representation of all the details of its action, and this has been my constant aim in this treatise.

* *Lezioni Accademiche* (Firenze, 1715), p. 25.

INDEX.

The References are to the Articles.

ABERRATION of light, 78.
Absorption, electric, 53, 227, 329.
— of light, 798.
Accumulators or condensers, 50, 226–228.
Action at a distance, 105, 641–646, 846–866.
Acyclic region, 19, 113.
Æther, 782 *n.*
Airy, Sir G. B., 454, 830.
Ampère, André Marie, 482, 502–528, 638, 687, 833, 846.
Anion, 237.
Anode, 237.
Arago's disk, 668, 669.
Astatic balance, 504.
Atmospheric electricity, 221.
Attraction, electric, 27, 38, 103.
— explained by stress in a medium, 105.

Barclay and Gibson, 229, 789.
Battery, voltaic, 232.
Beetz, W., 255, 265, 442.
Betti, E., 173, 864.
Bifilar suspension, 459.
Bismuth, 425.
Borda, J. C., 3.
Bowl, spherical, 176–181.
Bridge, Wheatstone's*, 347, 756, 775, 778.
— electrostatic, 353.
Bright, Sir C., and Clark, 354, 367.
Brodie, Sir B. C., 359.
Broun, John Allan, 462.
Brush, 56.
Buff, Heinrich, 271, 368.

Capacity (electrostatic), 50, 226.
— of a condenser, 50, 87, 102, 196, 227–229, 771, 774–780.
Capacity, calculation of, 102, 196.
— measurement of, 227–229.
— in electromagnetic measure, 774, 775.
Capacity (electromagnetic) of a coil, 706, 756, 778, 779.
Cathode, 237.
Cation, 237.
Cauchy, A. L., 827.
Cavendish, Henry, 38.
Cayley, A., 553.
Centrobaric, 101.
Circuits, electric, 578–584.
Circular currents, 694–706.
— solid angle subtended by, 695.
Charge, electric, 31.
Clark, Latimer, 358, 629, 725.
Classification of electrical quantities, 620–629.
Clausius, R., 70, 256, 863.
Clifford, W. K., 138.
Coefficients of electrostatic capacity and induction, 87, 102.
— of potential, 87.
— of resistance and conductivity, 297, 298.
— of induced magnetization, 426.
— of electromagnetic induction, 755.
— of self-induction, 756, 757.
Coercive force, 424, 444.
Coils, resistance, 335–344.
— electromagnetic, 694–706.
— measurement of, 708.
— comparison of, 752–757.
Comparison of capacities, 229.
— of coils, 752–757.
— of electromotive forces, 358.
— of resistances, 345–358.
Concentration, 26, 77.
Condenser, 50, 226–228.

* Sir Charles Wheatstone, in his paper on 'New Instruments and Processes,' *Phil. Trans.*, 1843, brought this arrangement into public notice, with due acknowledgment of the original inventor, Mr. S. Hunter Christie, who had described it in his paper on 'Induced Currents,' *Phil. Trans.*, 1833, under the name of a Differential Arrangement. See the remarks of Mr. Latimer Clark in the *Society of Telegraph Engineers*, May 8, 1872.

Condenser, capacity of, 50, 87, 102, 196, 227-229, 771, 774-780.
Conduction, 29, 241-254.
Conduction, linear, 273-284.
— superficial, 294.
— in solids, 285-334.
— electrolytic, 255-265.
— in dielectrics, 325-334.
Conductivity, equations of, 298, 609.
— and opacity, 798.
Conductor, 29, 80, 86.
Conductors, systems of electrified, 84-94.
Confocal quadric surfaces, 147-154, 192.
Conjugate circuits, 538, 759.
— conductors, 282, 347.
— functions, 182-206.
— harmonics, 138.
Constants, principal, of a coil, 700, 753, 754.
Conservation of energy, 92, 242, 262, 543.
Contact force, 246.
Continuity in time and space, 7.
— equation of, 36, 295.
Convection, 55, 238, 248.
Convergence, 25.
Copper, 51, 360, 362, 761.
Cotes, Roger, 865.
Coulomb, C. A., 38, 74, 215, 223, 373.
Coulomb's law, 79, 80.
Crystal, conduction in, 297.
— magnetic properties of, 435, 436, 438.
— propagation of light in a, 794-797.
Cumming, James, 252.
Curl, 25.
Current, electric, 230.
— best method of applying, 744.
— -function, } 294, 647-681.
— -sheet, }
— induced, 582.
— steady, 232.
— thermoelectric, 249-254.
— transient, 232, 530, 536, 537, 582, 748, 758, 760, 771, 776.
Current-weigher, 726.
Cyclic region, 18, 113, 481.
Cylinder, electrification of, 189.
— magnetization of, 436, 438, 439.
— currents in, 682-690.
Cylindric coils, 676-681.

Damped vibrations, 732-742, 762.
Damper, 730.
Daniell's cell, 232, 272.
Dead beat galvanometer, 741.
Decrement, logarithmic, 736.
Deflexion, 453, 743.
Delambre, J. B. J., 3.
Dellmann, F., 221.
Density, electric, 64.
— of a current, 285.
— measurement of, 223.
Diamagnetism, 429, 440, 838.
Dielectric, 52, 109, 111, 229, 325-334, 366-370, 784.
Diffusion of magnetic force, 801.
Dip, 461.
Dipolar, 173, 381.
Dimensions, 2, 42, 87, 278, 620-629.
Directed quantities (or vectors), 10.
Directrix, 517.
Discharge, 55.
Discontinuity, 8.
Disk, 177.
— Arago's, 668, 669.
Displacement, electric, 60, 75, 76, 111, 328-334, 608, 783, 791.
Dygogram, 441.

Earnshaw, S., 116.
Earth, magnetism of, 465-474.
Electric brush, 56.
— charge, 31.
— conduction, 29.
— convection, 211, 238, 248, 255, 259.
— current, 230.
— discharge, 55-57.
— displacement, 60, 75, 76, 111, 328-324, 608, 783, 791.
— energy, 85.
— glow, 55.
— induction, 28.
— machine, 207.
— potential, 70.
— spark, 57.
— tension, 48, 59, 107, 108, 111.
— wind, 55.
Electrode, 237.
Electrodynamic system of measurement, 526.
Electrodynamometer, 725.
Electrolysis, 236, 255-272.
Electrolyte, 237, 255.
Electrolytic conduction, 255-272, 363, 799.
— polarization, 257, 264-272.
Electromagnetic force, 475, 580, 583.
— measurement, 495.
— momentum, 585.
— observations, 730-780.
— and electrostatic units compared, 768-780.
— rotation, 491.
Electromagnetism, dynamical theory of, 568-577.
Electrometers, 214-220.
Electromotive force, 49, 69, 111, 241, 246-254, 358, 569, 579.
Electrophorus, 208.
Electroscope, 33, 214.
Electrostatic measurements, 214-220.
— polarization, 59, 111.
— attraction, 103-111.
— system of units, 620, &c.
Electrotonic state, 540.
Elongation, 734.
Ellipsoid, 150, 302, 437, 439.
Elliptic integrals, 149, 437, 701.
Energy, 6, 85, 630-638, 782, 792.

Equations of conductivity, 298, 609.
— of continuity, 35.
— of electric currents, 607.
— of total currents, 610.
— of electromagnetic force, 603.
— of electromotive force, 598.
— of Laplace, 77.
— of magnetization, 400, 605.
— of magnetic induction, 591.
— of Poisson, 77.
— of resistance, 297.
Equilibrium, points of, 112-117.

False magnetic poles, 468.
Faraday, M., his discoveries, 52, 55, 236, 255, 530, 531, 534, 546, 668, 806.
— his experiments, 28, 429, 530, 668.
— his methods, 37, 82, 122, 493, 528, 529, 541, 592, 594, 604.
— his speculations, 54, 60, 83, 107, 109, 245, 429, 502, 540, 547, 569, 645, 782.
Farad, 629.
Fechner, G. T., 231, 274, 848.
Felici, R., 536-539, 669.
Ferromagnetic, 425, 429, 844.
Field, electric, 44.
— electromagnetic, 585-619.
— of uniform force, 672.
First swing, 745.
Fizeau, H. L., 787.
Fluid, electric, 36, 37.
— incompressible, 61, 111, 295, 329, 334.
— magnetic, 380.
Flux, 12.
Force, electromagnetic, 475, 580, 583.
— electromotive, 49, 69, 111, 233, 241, 246-254, 358, 569, 579, 595, 598.
— mechanical, 92, 93, 103-111, 174, 580, 602.
— measurement of, 6.
— acting at a distance, 105.
— lines of, 82, 117-123, 404.
Foucault, L., 787.
Fourier, J. B. J., 2*n*, 243, 332, 333, 801-805.

Galvanometer, 240, 707.
— differential, 346.
— sensitive, 717.
— standard, 708.
— observation of, 742-751.
Gases, electric discharge in, 55-77, 370.
— resistance of, 369.
Gassiot, J. P., 57.
Gaugain, J. M., 366, 712.
Gauge electrometer, 218.
Gauss, C. F., 18, 70, 131, 140, 144, 409, 421, 454, 459, 470, 706, 733, 744, 851.
Geometric mean distance, 691-693.
Geometry of position, 421.
Gibson and Barclay, 229, 789.
Gladstone, Dr. J. H., 789.
Glass, 51, 271, 368.
Glow, electric, 55.
Grassmann, H., 526, 687.
Grating, electric effect of, 203.
Green, George, 70, 89, 318, 439.
Green's function, 88, 101.
— theorem, 100.
Groove, electric effect of, 199.
Grove, Sir W. R., 272, 779.
Guard-ring, 201, 217, 228.
Gutta-percha, 51, 367.

Hamilton, Sir W. Rowan, 10, 561.
Hard iron, 424, 444.
Harris, Sir W. Snow, 38, 216.
Heat, conduction of, 801.
— generated by the current, 242, 283, 299.
— specific, of electricity, 253.
Helix, 813.
Helmholtz, H., 88, 100, 202, 421, 543, 713, 823, 854.
Heterostatic electrometers, 218.
Hockin, Charles, 352, 360, 800.
Holtz, W., electrical machine, 212.
Hornstein, Karl, 471 *n*.
Huygens, Christian, 782.
Hydraulic ram, 550.
Hyposine, 151.

Idiostatic electrometers, 218.
Images, electric, 119, 155-181, 189.
— magnetic, 318.
— moving, 662.
Imaginary magnetic matter, 380.
Induced currents, 528-552.
— in a plane sheet, 656-669.
— Weber's theory of, 856.
Induced magnetization, 424-448.
Induction, electrostatic, 28, 75, 76, 111.
— magnetic, 400.
Inertia, electric, 550.
— moments and products of, 565.
Insulators, 29.
Inversion, electric, 162-181, 188, 316.
Ion, 237, 255.
Iron, 424.
— perchloride of, 809.
Irreconcileable curves, 20, 421.

Jacobi, M. H., 336.
Jenkins, William, 546. See *Phil. Mag.*, 1834, pt. ii, p. 351.
Jenkin, Fleeming, 763, 774.
Jochmann, E., 669.
Joule, J. P., 242, 262, 448, 457, 463, 726, 767.

'Keystone of electrodynamics,' 861.
Kinetics, 553-565.
Kirchhoff, Gustav, 282, 316, 439, 758.
Kohlrausch, Rudolph, 265, 365, 723, 771.

Lagrange's (J. L.) dynamical equations, 553–565.
Lamé, G., 17, 147.
Lamellar magnet, 412.
Laplace, P. S., 70.
Laplace's coefficients, 128–146.
— equation, 26, 77, 144, 301.
— expansion, 140.
Leibnitz, G. W., 18, 424.
Lenz, E., 265, 530, 542.
Light, electromagnetic theory of, 781–805.
— and magnetism, 806–831.
Line-density, 64, 81.
— -integral, 16–20.
— of electric force, 69, 622.
— of magnetic force, 401, 481, 498, 499, 590, 606, 607, 622.
Lines of equilibrium, 112.
— of flow, 22, 293.
— of electric induction, 82, 117–123.
— of magnetic induction, 404, 489, 529, 541, 597, 702.
Linnæus, C., 23.
Liouville, J., 173, 176.
Listing, J. B., 18, 23, 421.
Lorenz, L., 805 *n*.
Loschmidt, J., 5.

Magnecrystallic phenomena, 425, 435, 839.
Magnet, its properties, 371.
— direction of axis, 372–390.
— magnetic moment of, 384, 390.
— centre and principal axes, 392.
— potential energy of, 389.
Magnetic action of light, 806.
— disturbances, 473.
— force, law of, 374.
— — direction of, 372, 452.
— — intensity of, 453.
— induction, 400.
Magnetic 'matter,' 380.
— measurements, 449–464.
— poles, 468.
— survey, 466.
— variations, 472.
Magnetism of ships, 441.
— terrestrial, 465–474.
Magnetization, components of, 384.
— induced, 424–430.
— Ampère's theory of, 638, 833.
— Poisson's theory of, 429.
— Weber's theory of, 442, 838.
Magnus' (G.) law, 251.
Mance's, Henry, method, 357.
Matthiessen, Aug., 352, 360.
Measurement, theory of, 1.
— of result of electric force, 38.
— of electrostatic capacity, 226–229.
— of electromotive force or potential, 216, 358.
— of resistance, 335–357.
— of constant currents, 746.
— of transient currents, 748.
Measurement of coils, 708, 752–757.
— magnetic, 449–464.
Medium, electromagnetic, 866.
— luminiferous, 806.
Mercury, resistance of, 361.
Metals, resistance of, 363.
Michell, John, 38.
Miller, W. H., 23.
Mirror method, 450.
Molecular charge of electricity, 259.
— currents, 833.
— standards, 5.
— vortices, 822.
Molecules, size of, 5.
— electric, 260.
— magnetic, 430, 832–845.
Moment, magnetic, 384.
— of inertia, 565.
Momentum, 6.
— electrokinetic, 578, 585.
Mossotti, O. F., 62.
Motion, equations of, 553–565.
Moving axes, 600.
— conductors, 602.
— images, 662.
Multiple conductors, 276, 344.
— functions, 9.
Multiplication, method of, 747, 751.

Neumann, F. E., coefficient of magnetization, 430.
— magnetization of ellipsoid, 439.
— theory of induced currents, 542.
Neumann, C. G., 190, 830, 863.
Nicholson's Revolving Doubler, 209.
Nickel, 425.
Null methods, 214, 346, 503.

Örsted, H. C., 239, 475.
Ohm, G. S., 241, 333.
Ohm's Law, 241.
Ohm, the, 338, 340, 629.
Opacity, 798.
Ovary ellipsoid, 152.

Paalzow, A., 364.
Paraboloids, confocal, 154.
Paramagnetic (same as Ferromagnetic), 425, 429, 844.
Peltier, A., 249.
Periodic functions, 9.
Periphractic region, 22, 113.
Permeability, magnetic, 428, 614.
Phillips, S. E., 342.
Plan of this Treatise, 59.
Plane current-sheet, 656–669.
Planetary ellipsoid, 151.
Platymeter, electro-, 229.
Plücker, Julius, 839.
Points of equilibrium, 112.
Poisson, S. D., 155, 431, 437, 674.
Poisson's equation, 77, 148.

Poisson's theory of magnetism, 427, 429, 431, 441, 832.
— theory of wave-propagation, 784.
Polar definition of magnetic force, 398.
Polarity, 381.
Polarization, electrostatic, 59, 111.
— electrolytic, 257, 264–272.
— magnetic, 381.
— of light, 381, 791.
— circular, 813,
Poles of a magnet, 373.
— magnetic of the earth, 468.
Positive and negative, conventions about, 23, 27, 36, 37, 63, 68–81, 231, 374, 394, 417, 489, 498.
Potential, 16.
— electric, 45, 70, 220.
— magnetic, 383, 391.
— of magnetization, 412.
— of two circuits, 423.
— of two circles, 698.
Potential, vector-, 405, 422, 590, 617, 657.
Principal axes, 299, 302.
Problems, electrostatic, 155–205.
— electrokinematic, 306–333.
— magnetic, 431–441.
— electromagnetic, 647–706.
Proof of the law of the Inverse Square, 74.
Proof plane, 223.

Quadrant electrometer, 219.
Quadric surfaces, 147–154.
Quantity, expression for a physical, 1.
Quantities, classification of electromagnetic, 620–629.
Quaternions, 11, 303, 490, 522, 618.
Quinke, G., 316 *n*.

Radiation, forces concerned in, 792.
Rankine, W. J. M., 115, 831.
Ray of electromagnetic disturbance, 791.
Reciprocal properties, electrostatic, 88.
— electrokinematic, 281, 348.
— magnetic, 421, 423.
— electromagnetic, 536.
— kinetic, 565.
Recoil, method of, 750.
Residual charge, 327–334.
— magnetization, 444.
Replenisher, 210.
Resistance of conductors, 51, 275.
— tables of, 362–365.
— equations of, 297.
— unit of, 758–767.
— electrostatic measure of, 355, 780.
Resultant electric force at a point, 68.
Riemann, Bernhard, 421, 862.
Right and left-handed systems of axes, 23, 498, 501.
— crcularly-polarized rays, 813.
Ritchie, W., 542.
Ritter's (J. W.) Secondary Pile, 271.
Rotation of plane of polarization, 806.
— magnetism, a phenomenon of, 821.
Rühlmann, R., 370.
Rule of electromagnetic direction, 477, 494, 496.

Scalar, 11.
Scale for mirror observations, 450.
Sectorial harmonic, 132, 138.
Seebeck, T. J., 250.
Selenium, 51, 362.
Self-induction, 7.
— measurement of, 756, 778, 779.
— coil of maximum, 706.
Sensitive galvanometer, 717.
Series of observations, 746, 750.
Shell, magnetic, 409, 484, 485, 606, 652, 670, 694, 696.
Siemens, C. W., 336, 361.
Sines, method of, 455, 710.
Singular points, 128.
Slope, 17.
Smee, A., 272.
Smith, Archibald, 441.
Smith, W. R., 123, 316.
Soap bubble, 125.
Solenoid, magnetic, 407.
— electromagnetic, 676–681, 727.
Solenoidal distribution, 21, 82, 407.
Solid angle, 409, 417–422, 485, 695.
Space-variation, 17, 71, 835.
Spark, 57, 370.
Specific inductive capacity, 52, 83, 94, 111, 229, 325, 334, 627, 788.
— conductivity, 278, 627.
— resistance, 277, 627.
— heat of electricity, 253.
Sphere, 125.
Spherical harmonics, 128–146, 391, 431.
Spiral, logarithmic, 731.
Standard electrometer, 217.
— galvanometer, 708.
Stokes, G. G., 24, 115, 784.
Stoney, G. J., 5.
Stratified conductors, 319.
Stress, electrostatic, 107, 111.
— electrokinetic, 641, 645, 646.
Strutt, Hon. J. W., 102, 306.
Surface-integral, 15, 21, 75, 402.
— -density, 64, 78, 223.
Surface, equipotential, 46.
— electrified, 78.
Suspended coil, 721–729.
Suspension, bifilar, 459.
— Joule's, 463.
— Thomson's, 721.
— unifilar, 449.

Tables of coefficients of a coil, 700.
— of dimensions, 621–629.
— of electromotive force, 358.
— of magnetic rotation, 830.

Tables for magnetization of a cylinder, 439.
— of resistance, 363–365.
— of velocity of light and of electromagnetic disturbance, 787.
— of temporary and residual magnetization, 445.
Tait, P. G., 25, 254, 387, 522, 687, 731.
Tangent galvanometer, 710.
Tangents, method of, 454, 710.
Telegraph cable, 332, 689.
Temporary magnetization, 444.
Tension, electrostatic, 48, 59, 107, 108.
— electromagnetic, 645, 646.
Terrestrial magnetism, 465–474.
Thalèn, Tobias Robert, 430.
Theorem, Green's, 100.
— Earnshaw's, 116.
— Coulomb's, 80.
— Thomson's, 98.
— Gauss', 144, 409.
Theory of one fluid, 37.
— of two fluids, 36.
— of magnetic matter, 380.
— of magnetic molecules, 430, 832–845.
— of molecular currents, 833.
— of molecular vortices, 822.
— of action at a distance, 105, 641–646, 846–866.
Thermo-electric currents, 249–254.
Thickness of galvanometer wire, 716, 719.
Thomson, Sir William,
— electric images, 43, 121, 155–181, 173.
— experiments, 51, 57, 248, 369, 772.
— instruments, 127, 201, 210, 211, 216–222, 272, 722, 724.
— magnetism, 318, 398, 400, 407–416, 428.
— resistance, 338, 351, 356, 763.
— thermo-electricity, 207, 242, 249, 252, 253.
— theorems, 98, 138, 263, 299, 304, 652.
— theory of electricity, 27, 37, 543, 831, 856.
— vortex motion, 20, 100, 487, 702.
Thomson and Tait's Natural Philosophy, 132, 141, 144, 162, 303, 553, 676.
Time, periodic of vibration, 456, 738.
Time-integral, 541, 558.
Torricelli, Evangelista, 866.
Torsion-balance, 38, 215, 373, 726.
Transient currents, 232, 530, 536, 537, 582, 748, 758, 760, 771, 776.

Units of physical quantities, 2.
— three fundamental, 3.
— derived, 6.
— electrostatic, 41, 625.
— magnetic, 374, 625.
— electrodynamic, 526.
— electromagnetic, 526, 620.
— classification of, 620–629.
— practical, 629.
— of resistance, 758–767.
— ratios of the two systems, 768–780.

Variation of magnetic elements, 472.
Varley, C. F., 210, 271, 332, 369.
Vector, 10.
Vector-potential, 405, 422, 590, 617, 657.
Velocity represented by the unit of resistance, 338, 628, 758.
— — by the ratio of electric units, 768–780.
— of electromagnetic disturbance, 784.
— of light, 787.
— of the electric current, 569.
Verdet, M. E., 809, 830.
Vibration, time of, 456, 738.
Volt, 629.
Volta, A., 246.
Voltameter, 237.
Vortices, molecular, 822–831.

Water, resistance of, 365.
Wave-propagation, 784, 785.
Weber, W., 231, 338, 346.
— electrodynamometer, 725.
— induced magnetism, 442–448, 838.
— unit of resistance, 760–762.
— ratio of electric units, 227, 771.
— electrodynamic formula, 846–861.
Wertheim, W., 447.
Wheatstone's Bridge, 347.
— electrostatic, 353, 756, 775, 778.
Whewell, W., 237.
Wiedemann, G., 236, 370, 446, 447.
Wind, electric, 55.
Wippe, 775.
Work, 6.

Zero reading, 735.
Zonal harmonic, 132.

PLATES

Fig. XIV.

Art. 388.

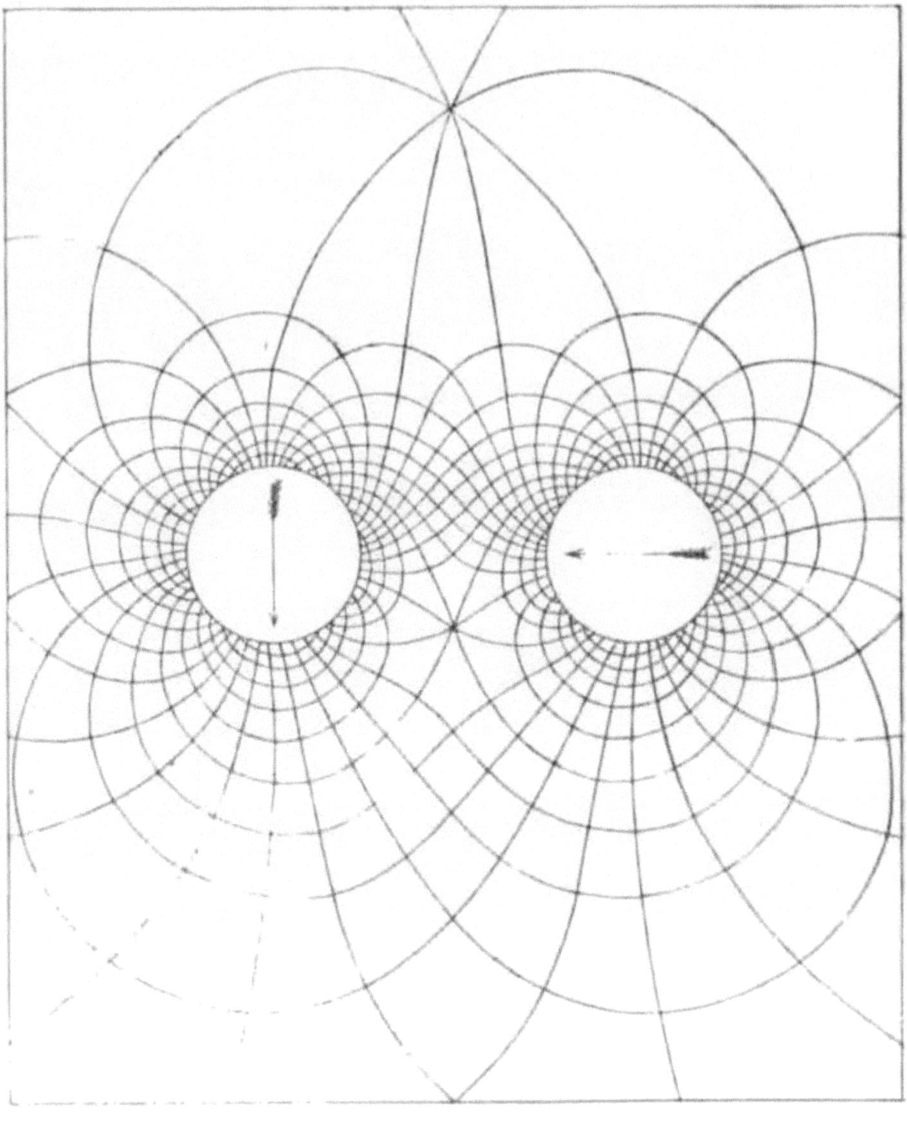

Two Cylinders magnetized transversely.

FIG XV.

Art. 434.

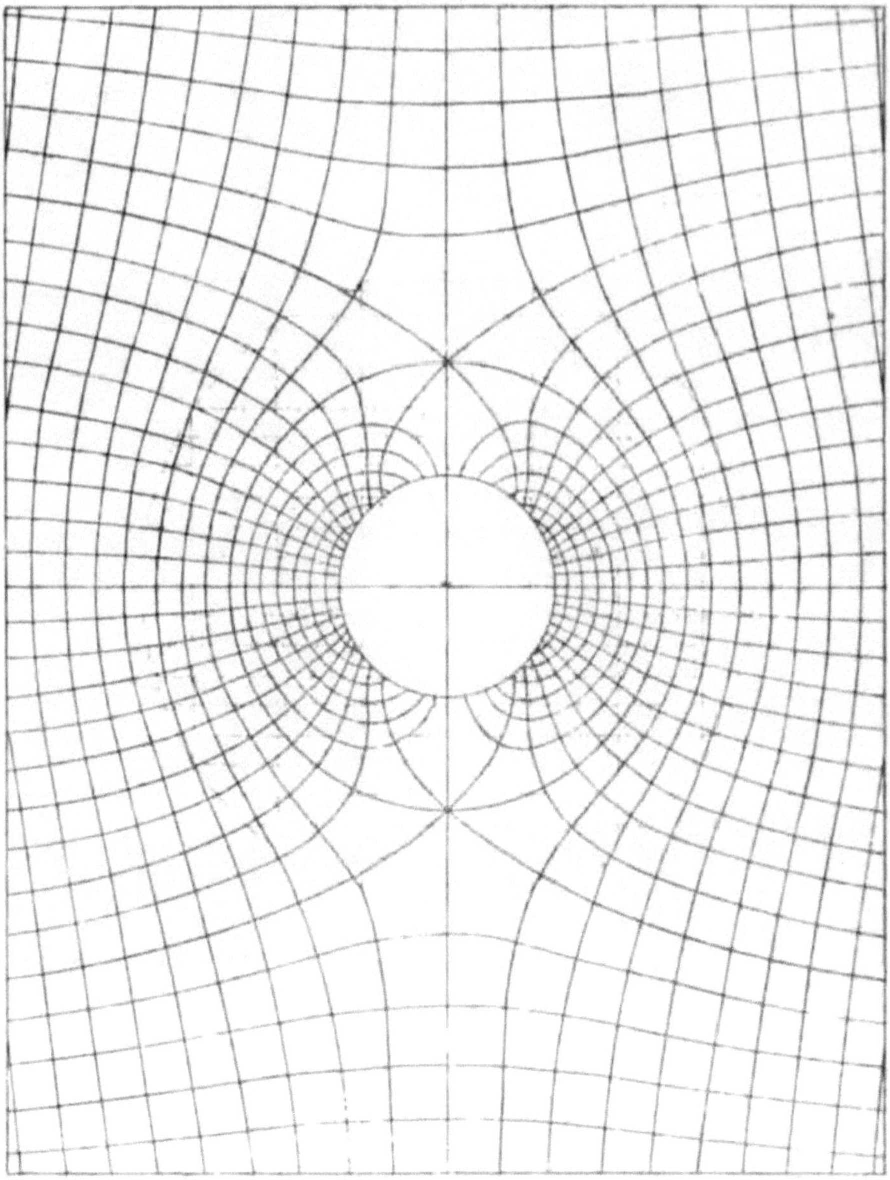

Cylinder magnetized transversely, placed North and South in a uniform magnetic field.

FIG. XVI

Art. 436.

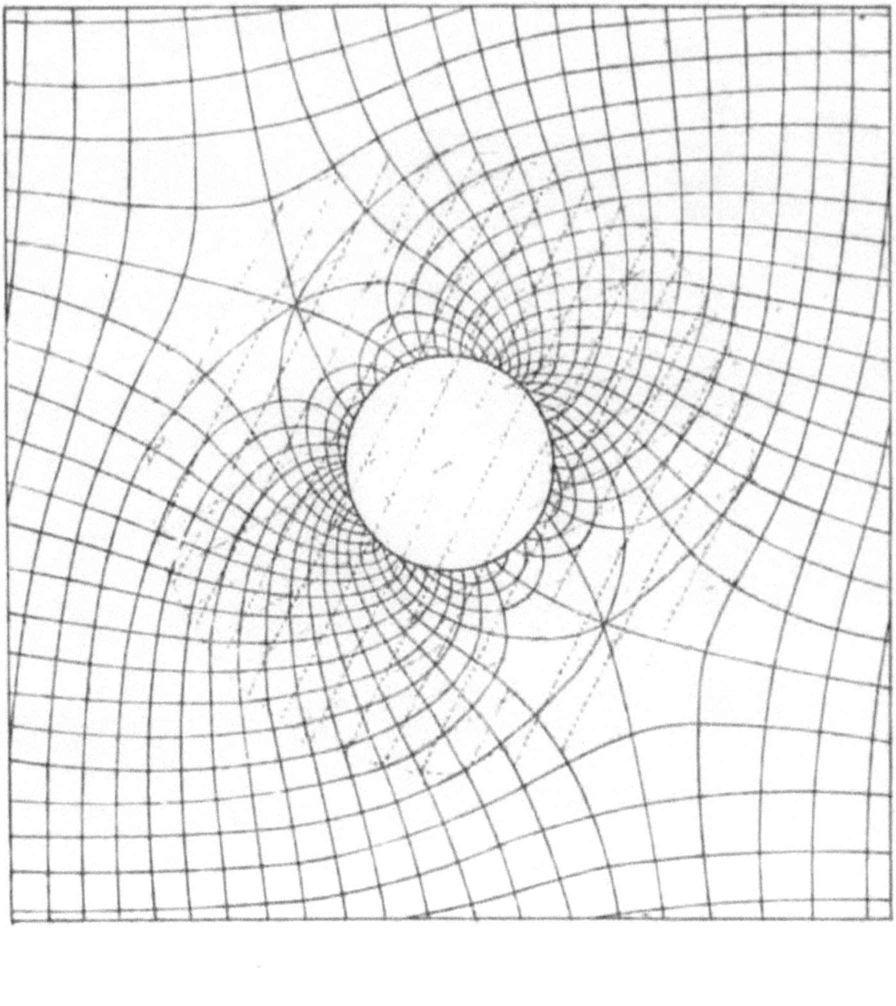

Cylinder magnetized transversely, placed East and West in a uniform magnetic field.

FIG. XVII.

Art. 496.

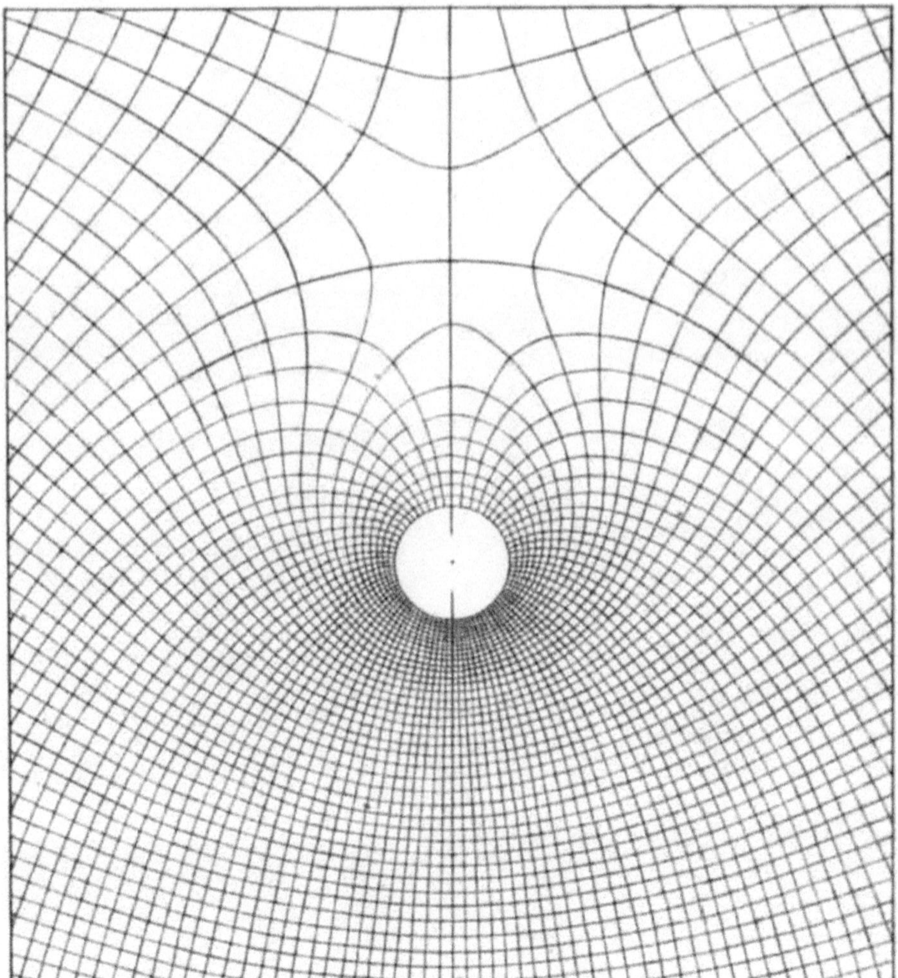

Uniform magnetic field disturbed by an Electric Current in a straight conductor.

Fig. XVIII.

Art. 487, 702.

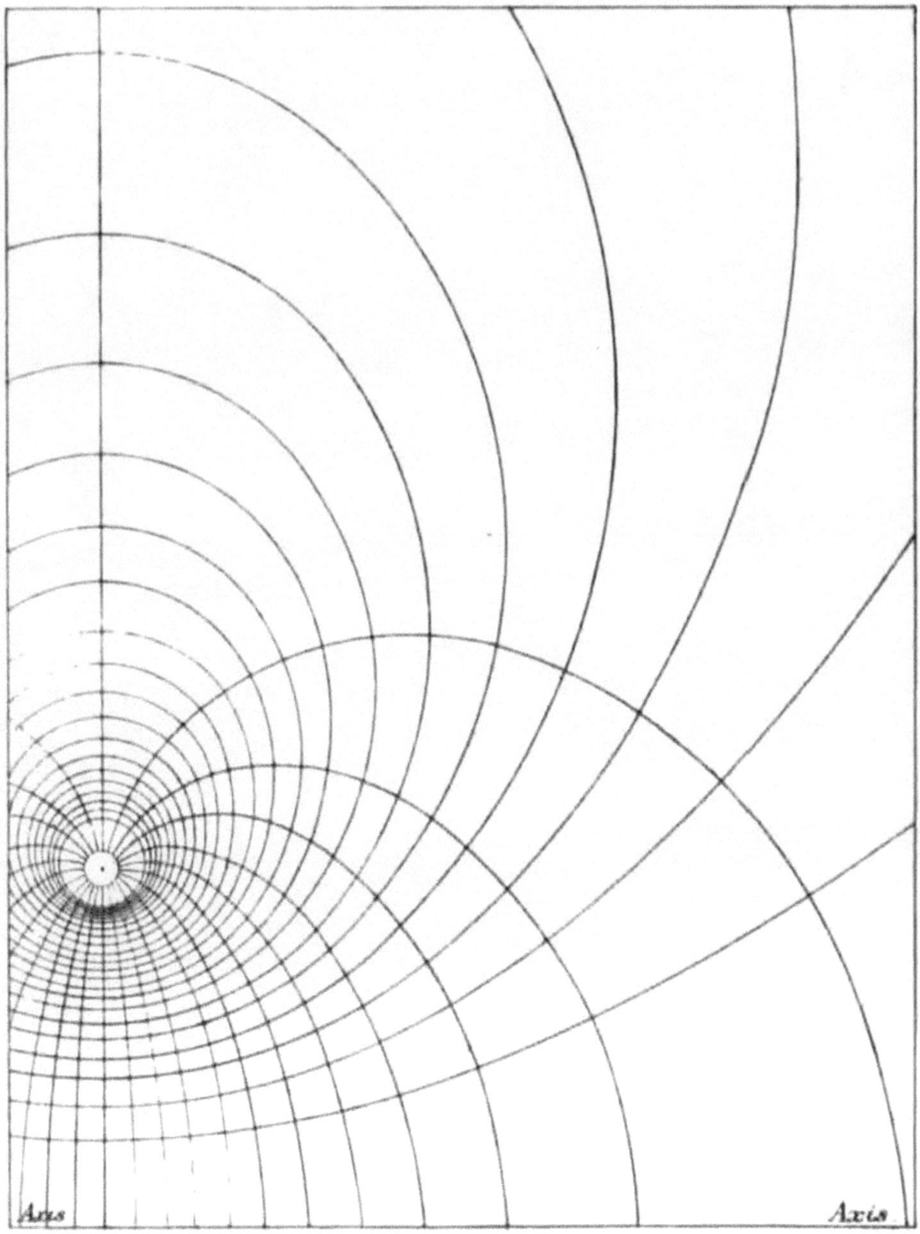

Circular Current.

FIG XIX.
Art 713

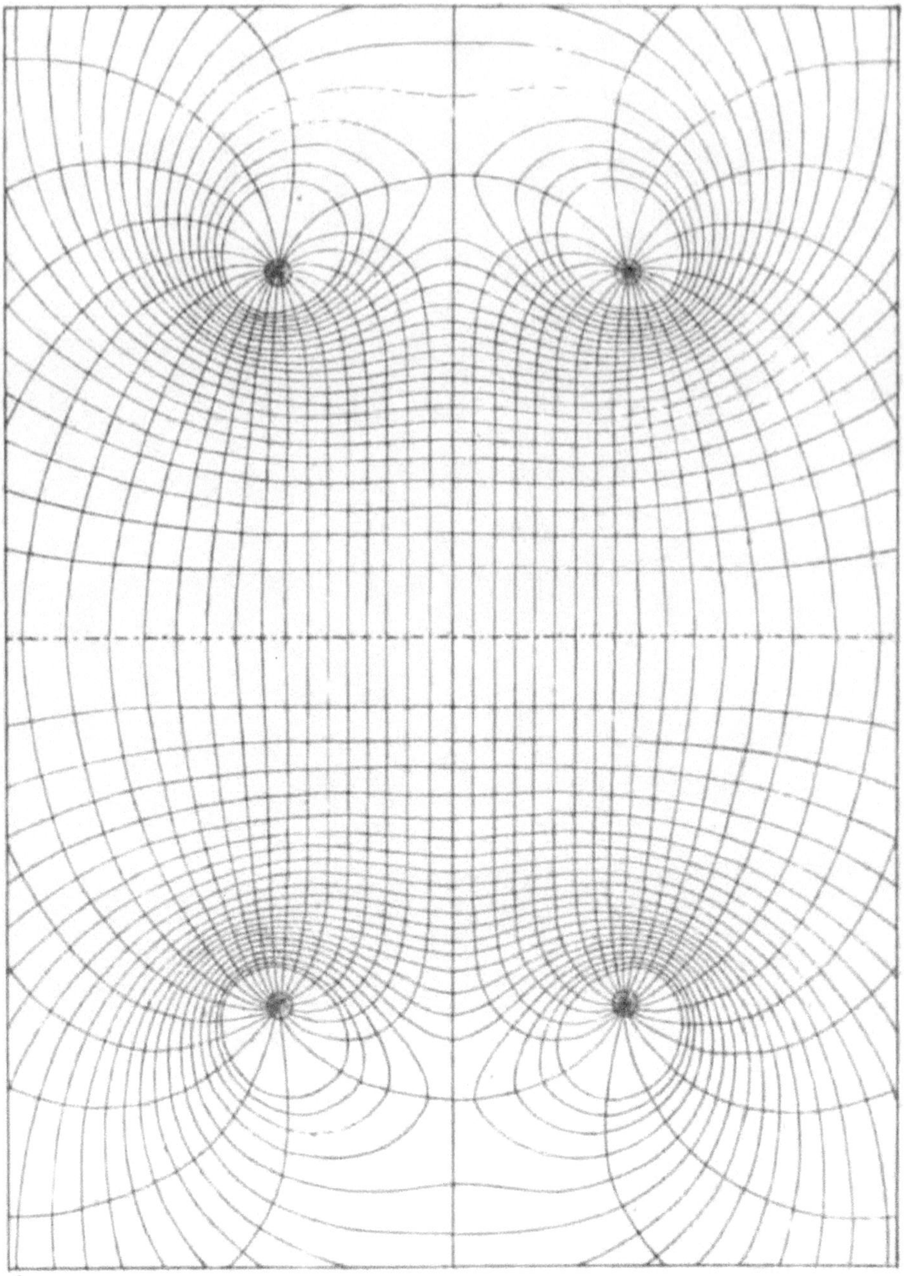

Two Circular Currents.

FIG XX.

Art. 225.

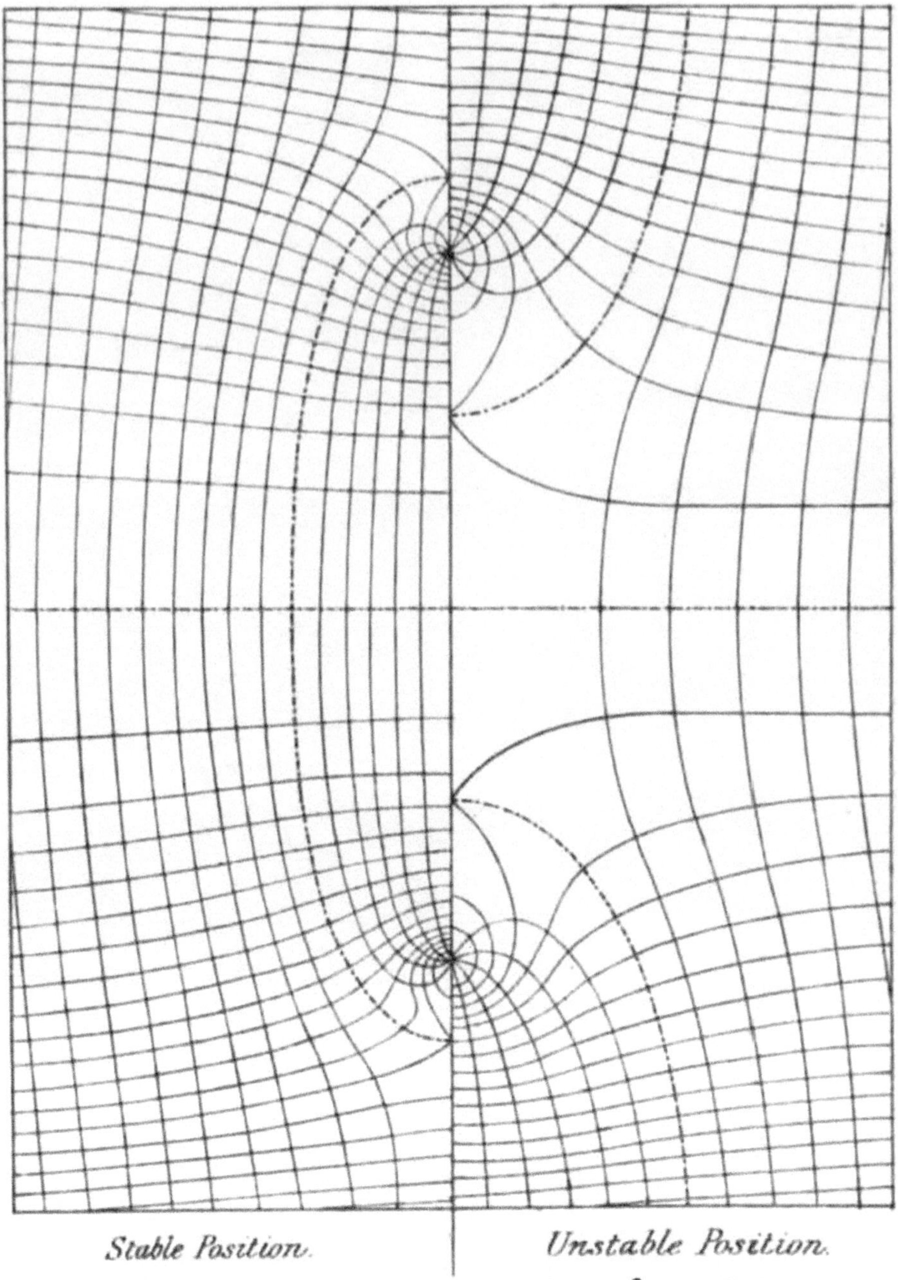

Circular Current in uniform field of Force.

THE SCIENTIFIC PAPERS OF JAMES CLERK MAXWELL
IN 2 VOLUMES

"Maxwell was a professor in King's College from 1860 to 1865, and this period of his life is distinguished by the production of his most important papers. The second memoir on Colours made its appearance in 1860."

- Royal Naval College, Greenwich, August, 1890.

A TREATISE ON ELECTRICITY AND MAGNETISM
IN 2 VOLUMES AND IN THREE EDITIONS

"He deserves to be remembered with affection wherever the name of James Clerk Maxwell is honoured or beloved."

- Lewis Campbell, August, 1882.

THE AUTHENTIC BIOGRAPHY OF JAMES CLERK MAXWELL
WRITTEN BY: LEWIS CAMPBELL AND WILLIAM GARNETT

"James Clerk Maxwell's biography has been written by 'Professors Lewis Campbell and William Garnett with so much skill and appreciation of their subject that nothing further remains to be told."

- Royal Naval College, Greenwich, August, 1890.

ISBN: 978-1773752341

To give an exact insight into the theory of Relativity to those readers who, from a general scientific and philosophical point of view.

NOTES

Made in United States
Cleveland, OH
05 November 2026

47670540R00269